Anisotropy of Metamaterials

Anisotropy of Metamaterials: Beyond Conventional Paradigms provides a comprehensive introduction to the mathematical modeling of metamaterials based on the macroscopic complex-valued permittivity tensor of dispersed random composites. Key topics include physical and mathematical theory, computer simulations, constructive homogenization, classification of dispersed random composites, and their applications in cancer recognition. Image processing and machine learning algorithms are used. The book also discusses the precision of various effective medium approximations, including Bruggeman and Maxwell-Garnett formulas. New analytical, approximate, and exact formulas and bounds for the macroscopic permittivity and piezoelectric constants of composites are derived. This book is a valuable tool for academics and professionals in photonics, presenting sustainable materials for sensing, health diagnostics, and cancer detection methodologies.

Key features:

- Offers key insights into the current trends and techniques in the study of the macroscopic properties of metamaterials, aiming at stimulating new avenues of research.
- Presents examples of image analysis, the primary tool for non-destructive metamaterials analysis.
- Discusses the applications of Machine Learning to image processing, illustrated using specific code in Python programming language.

Vladimir Mityushev is a professor at the Cracow University of Technology and a leader of the research group at www.materialica.plus. His expertise lies in the fields of mathematical modeling, computer simulations, porous media, permeability, diffusion, effective properties of composites with deterministic and random structures, representative volume elements, biomathematics, bioinformatics, and industrial mathematics.

Professor Tatjana Gric has been engaged in the investigation of waveguide devices (waveguide modulators, filters, etc.), namely in proposing their electrodynamical analysis, since 2009. Another major goal of her studies is plasmonics as the examination of the interaction between electromagnetic field and free electrons in a metal. During the past few years, she has been working on the investigation of nanostructured composites and their fascinating properties. She has authored and co-authored over 70 articles in refereed journals and conference proceedings. She also holds a 1LT patent.

Radosław Kycia is Head of the Computer Science Division at the Faculty of Computers Science and Telecommunications at the Tadeusz Kościuszko Cracow University of Technology. He specializes in multidisciplinary research involving computer science, physics, and mathematics. Currently, he is interested in machine learning and data analysis.

Natalia Rylko is an Associate Professor at the Cracow University of Technology. Her academic and scientific endeavors revolve around interdisciplinary research encompassing physics, material science, computer simulations, and machine learning. Her primary field lies in the study of the effective properties of composites and their wide-ranging applications within the field of material sciences.

Series in Optics and Optoelectronics

Handbook of Optoelectronics, Second Edition
Concepts, Devices, and Techniques (Volume One)
John P. Dakin, Robert Brown

Handbook of GaN Semiconductor Materials and Devices
Wengang (Wayne) Bi, Haochung (Henry) Kuo, Peicheng Ku, Bo Shen

Handbook of Optoelectronic Device Modeling and Simulation (Two-Volume Set)
Joachim Piprek

Handbook of Optoelectronics, Second Edition (Three-Volume Set)
John P. Dakin, Robert G. W. Brown

Optical MEMS, Nanophotonics, and Their Applications
Guangya Zhou, Chengkuo Lee

Thin-Film Optical Filters, Fifth Edition
H. Angus Macleod

Laser Spectroscopy and Laser Imaging
An Introduction
Helmut H. Telle, Ángel González Ureña

Fourier Optics in Image Processing
Neil Collings

Holography
Principles and Applications
Raymond K. Kostuk

An Introduction to Quantum Optics, Second Edition
Photon and Biphoton Physics
Yanhua Shih

Polarized Light and the Mueller Matrix Approach, Second Edition
José J. Gil, Razvigor Ossikovski

Introduction to Holography, Second Edition
Vincent Toal

Optical Coherence Tomography in Dentistry
Scientific Developments to Clinical Applications
Anderson S. L. Gomes, Denise M. Zezell, Cláudia C. B. O. Mota, John M. Girkin

Holography: Principles and Applications, Second Edition
Raymond K. Kostuk

CAD-Based Optical Design with Quadoa
Rafael G. González-Acuña

Anisotropy of Metamaterials: Beyond Conventional Paradigms
Vladimir Mityushev, Tatjana Gric, Radoslaw Kycia, Natalia Rylko

For more information about this series, please visit: *https://www.crcpress.com/Series-in-Optics-and-Optoelectronics/book-series/TFOPTICSOPT*

Anisotropy of Metamaterials
Beyond Conventional Paradigms

Vladimir Mityushev, Tatjana Gric,
Radoslaw Kycia, and Natalia Rylko

CRC Press
Taylor & Francis Group
Boca Raton London New York

CRC Press is an imprint of the
Taylor & Francis Group, an **informa** business

Designed cover image: © 2025 Vladimir Mityushev, Tatjana Gric, Radoslaw Kycia and Natalia Rylko

First edition published 2025
by CRC Press
2385 NW Executive Center Drive, Suite 320, Boca Raton FL 33431

and by CRC Press
4 Park Square, Milton Park, Abingdon, Oxon, OX14 4RN

CRC Press is an imprint of Taylor & Francis Group, LLC

ISBN: 978-1-032-61800-5
ISBN: 978-1-032-64661-9
ISBN: 978-1-032-64663-3

DOI: 10.1201/9781032646633

Typeset in TexGyreTermesX font
by KnowledgeWorks Global Ltd.

Publisher's note: This book has been prepared from camera-ready copy provided by the authors.

Contents

PART III Macroscopic Properties of 2D Piezoelectric Composites

PART IV Image Analysis and Machine Learning

Preface

This book is useful for specialists applying classic formulas such as those by Bruggeman, Maxwell-Garnett, and other effective medium approximations to estimate the macroscopic permittivity of composites. Literature often suggests that Bruggeman's formula provides a reliable approximation for certain classes of composites, with modifications suited for other types. Analyzing known analytical formulas and numerical simulations reveals a broad range of models. An appropriate model could be applied, or a new model could be developed when purely numerical methods are restricted by time and memory constraints. This established standard method of developing empirical models supported by experimental data is a widely adopted approach to studying heterogeneous media.

This book is devoted to an alternative approach based on rigorous investigations centered on assessing the precision of suggested formulas, including classic approximations. This task involves advanced mathematical analysis of the homogenization problem, in particular, the generalized alternating method of Schwarz. After the application of Schwarz's method, it becomes clear that the concentration of inclusions f and contrast parameters ϱ_k can be taken as key precision parameters. The precision of the Maxwell-Garnett approximation is found, revealing that many other popular formulas for dispersed media derived from effective medium approximations are asymptotically equivalent to the Maxwell-Garnett approximation. Seemingly distinct formulas differ mainly in their redundant tail terms in f and ϱ_k out of the considered precision. Therefore, they are versions of the Maxwell-Garnett approximation written in various asymptotically equivalent forms.

Schwarz's method yields the decomposition of the effective constants into a linear combination of pure geometric objects (structural sums) with coefficients depending on the permittivity of components. This book systematically develops the general method for two-dimensional, multi-phase deterministic, and random dispersed composites with complex permittivity depending on frequency. Exact and approximate analytical formulas explicitly demonstrate the influence of the location of inclusions and their shapes on the effective constants. The theory of analytical Representative Volume Element (*a*RVE) is developed based on structural sums, which is particularly valuable in metamaterials theory, where geometry plays a crucial role. Although advanced boundary value problem theories, including scalar Riemann-Hilbert and vector-matrix $\mathbb{R}$-linear problems for multiply connected domains, are discussed at a rigorous mathematical level, the resulting formulas are explicit and directly applicable for physicists in simulations. Analogous constructive formulas for elastic composites and viscous fluids can be found in [176, Chapter 9] and [67, 93]. An accompanying discussion of what is a *mathematical model* can also be found in [176].

The obtained theoretical formulas are applied within the metamaterial formalism approach, enhancing methods for cancer recognition outlined in [96]. Metamaterial at an optical frequency can be considered a kind of man-made architectural material possessing unusual properties. One of its remarkable features is that the equivalent permittivities in different directions differ from each other and can be varied by designing its structure elaborately. It is worth mentioning that this property, enhanced by the presence of the disorder, promises that metamaterial may serve surface plasmons propagating with more flexible performances than conventional metal or ordered metamaterial. For instance, in dealing with photonic crystals, some disorders can dramatically degrade the reflection and transmission properties based on constructive interferences.

Applying metamaterial formalism to cancer recognition involves sophisticated image analysis and statistical processing of medical images. The book contains two chapters related to Digital Image Processing methods and their applications to cancer cell distinctions and, in general, to granulometry. The first one is an elementary hands-on introduction to image processing methods to distinguish granules occurring in material science, biology, and medicine—cancer therapy. The required knowledge is a basic understanding of Python programming. The second chapter is more advanced and describes an example code that trains a simple Neural Networks model for detecting cancer cells in histopathological images. Then, such a model is applied to the parts of the larger image where there are suspected “granules”. Neural Network analyzes such subimage and labels it as a cancer/no-cancer region. This chapter can be interesting for newcomers to histopathological photo segmentation and using Neural Networks for image classification. Together, both parts can be a good introduction to Digital Image Processing methods in medical, biological, and material science applications.

This interdisciplinary work incorporates physical, mathematical, computational, and biological terminology that requires clarification. The term *cell* has distinct meanings in the theory of composites and biology. To prevent confusion, the term *periodicity cell Q* is used in homogenization contexts, while the biological term *cell* corresponds to the term *inclusion* in homogenization theory.

Complete code examples from the book can be downloaded from `https://github.com/rkycia/AnisotropyOfMetamaterialsBook`.

Vladimir Mityushev
Tatjana Gric
Radoslaw Kycia
Natalia Rylko

Kraków-Vilnius, September 2024

Authors

Professor **Vladimir Mityushev** at the Cracow University of Technology leads the research group Materialica+. His expertise encompasses mathematical modeling, computer simulations, and industrial mathematics. He has provided a complete solution to the Riemann-Hilbert problem for multiply connected domains, applying it to the analytical theory of representative volume elements for dispersed composites. His interdisciplinary research spans a wide array of fields, including the effective properties of composites with deterministic and random structures, elliptic partial differential equations, symbolic computations, boundary value problems, asymptotic methods, packing, deterministic and random graphs, biomathematics, bioinformatics, porous media, permeability, diffusion, elasticity, heat conduction, fracture mechanics, electroosmotic phenomena, and viscous flow in wavy channels.

Professor **Tatjana Gric** (female), since 2009, has been engaged in the investigation of waveguide devices (waveguide modulators, filters, etc.), namely on proposing their electrodynamical analysis of various complex structures (e.g., tissue-like structures). Another major goal of her studies is plasmonics the examination of the interaction between electromagnetic field and free electrons in a metal. The optically-active nanostructures have been simulated, and their fundamental photonic properties have been explored. During the past few years, she has been working on the investigation of nanostructured composites and their fascinating properties. She has authored and co-authored over 75 articles in refereed journals and conference proceedings.

Radoslaw Kycia, PhD, MBA (ORCID: 0000-0002-6390-4627) obtained his PhD in Theoretical Physics from Jagiellonian University. He specializes in multidisciplinary projects involving Computer Science, Mathematics, and Physics. He collaborated with CERN, Brookhaven National Laboratory, and Masaryk University. He is the Head of the Computer Science Department at the Faculty of Computer Science and Telecommunications of the Cracow University of Technology.

Natalia Rylko pursued her Doctorate in Technical Sciences, specializing in Mechanics and Thermomechanics of Fiber Composite Materials, at Poznan University of Technology. Currently, she serves as the Vice Dean of the Department of Computer Science and Telecommunications at the Cracow University of Technology, where she is actively involved in educational and scientific endeavors. Her research in Materials Engineering focuses on applying advanced mathematical techniques, machine learning, and analyzing large datasets, including intricate composite material structures depicted in images.

Part I

Physics of Metamaterials

1 Electromagnetics of Metamaterials

1.1 GOVERNING EQUATIONS DESCRIBING ELECTROMAGNETIC BEHAVIOR OF MATERIALS

1.1.1 MAXWELL'S EQUATIONS

Artificial electromagnetic materials, often called metamaterials, demonstrated their properties if visible for an electromagnetic field to which they provide unprecedented responses. One should rely on electromagnetic field theory to analyze and predict this behavior. A short introduction to electromagnetic wave theory and a reference to relevant textbooks are provided to demonstrate the present branch of physics. The electromagnetic phenomena can be described by Maxwell's equations, first formulated by James Clerk Maxwell in the 1860s. Their main purpose is to relate electric and magnetic field quantities [42, 112, 138]

$$\begin{aligned}
\nabla \times \mathbf{E}(\mathbf{x},t) &= -\frac{\partial}{\partial t}\mathbf{B}(\mathbf{x},t),\\
\nabla \times \mathbf{H}(\mathbf{x},t) &= \frac{\partial}{\partial t}\mathbf{D}(\mathbf{x},t) + \mathbf{J}(\mathbf{x},t),\\
\nabla \cdot \mathbf{D}(\mathbf{x},t) &= \rho(\mathbf{x},t),\\
\nabla \cdot \mathbf{B}(\mathbf{x},t) &= 0.
\end{aligned} \tag{1.1}$$

All quantities are functions of the spatial coordinate $\mathbf{x} = (x_1, x_2, x_3)$ and time t. The spatial and temporal differentiation operators are denoted by ∇ and $\frac{\partial}{\partial t}$. The quantities in these equations and their units in the SI system are the following: **E** is the electric field vector $[V/m]$, **H** the magnetic field vector $[A/m]$, **D** the electric flux density (displacement) vector $[As/m^2]$, **B** the magnetic flux density (displacement) vector $[Vs/m^2$ or tesla $T]$, **J** the current density vector $[A/m^2]$, and ρ the scalar charge density $[As/m^2]$. All the quantities are real-valued due to the fact that they all are physical and observable. The main question to answer is how the electromagnetic field quantities **E** and **H** in a material result in electric and magnetic flux densities, **D** and **B**. Consequently, Maxwell's equations must be supplemented by constitutive relations, which aim to get a deep insight into the interaction of electromagnetic fields, including light, with matter. They relate the flux densities to the fields, and both **D** and **B** may depend on the electric and the magnetic fields **E** and **H**, respectively

$$\mathbf{D} = \mathbf{D}(\mathbf{E},\mathbf{H}) \quad \text{and} \quad \mathbf{B} = \mathbf{B}(\mathbf{E},\mathbf{H}). \tag{1.2}$$

Herein, the functional dependency is dictated by the nature of the material in which the fields exist, and thus, it may be rather complicated, especially in the case of

DOI: 10.1201/9781032646633-1

certain metamaterial types. Due to the absence of any matter, the constitutive relations become trivial in a vacuum and are as follows:

$$\mathbf{D}(\mathbf{x},t) = \varepsilon_0 \mathbf{E}(\mathbf{x},t), \tag{1.3}$$

$$\mathbf{B}(\mathbf{x},t) = \mu_0 \mathbf{H}(\mathbf{x},t), \tag{1.4}$$

with the free-space permittivity ε_0 and permeability μ_0 which have the values

$$\varepsilon_0 = \frac{1}{c^2\mu_0} 8.854 \times 10^{-12} \frac{As}{Vm}, \tag{1.5}$$

$$\mu_0 = 4\pi \times 10^{-7} \frac{Vs}{Am}. \tag{1.6}$$

The parameter c appearing above presents the speed of light in a vacuum ($c = 299792458 \frac{m}{s}$). Any contribution to the fluxes that takes place due to the polarization of the matter is then usually included as electric polarization $\mathbf{P}(\mathbf{x},t)$ and magnetization $\mathbf{M}(\mathbf{x},t)$. Then the constitutive relations (1.3)–(1.4) become

$$\mathbf{D}(\mathbf{x},t) = \varepsilon_0 [\mathbf{E}(\mathbf{x},t) + \mathbf{P}(\mathbf{x},t)], \tag{1.7}$$

$$\mathbf{B}(\mathbf{x},t) = \mu_0 [\mathbf{H}(\mathbf{x},t) + \mathbf{M}(\mathbf{x},t)], \tag{1.8}$$

where $\mathbf{P}(\mathbf{x},t)$ and $\mathbf{M}(\mathbf{x},t)$ depend again explicitly on the electric and magnetic fields.

1.1.2 MAXWELL'S EQUATIONS IN THE FREQUENCY DOMAIN

The complex vector fields stand as the best option for treating electromagnetic field solutions for given frequencies. Let us deal with the case when all fields oscillate harmonically in time with an angular frequency denoted here as ω. The notation of $\mathrm{i} = \sqrt{-1}$ as the imaginary unit is applied in physical science; moreover, j is often used in electrical engineering. The time differentiation becomes an algebraic multiplication by $\mathrm{i}\omega$ assuming the time-dependence to be $\exp(\mathrm{i}\omega t)$ since all fields share this time-dependence.

Consider complex vector functions satisfying Maxwell's equations (1.1) of the special form

$$\mathbf{E}(\mathbf{x},t) = \hat{\mathbf{E}}(\mathbf{x})\mathrm{e}^{\mathrm{i}\omega t},\ \mathbf{B}(\mathbf{x},t) = \hat{\mathbf{B}}(\mathbf{x})\mathrm{e}^{\mathrm{i}\omega t},\ \mathbf{H}(\mathbf{x},t) = \hat{\mathbf{H}}(\mathbf{x})\mathrm{e}^{\mathrm{i}\omega t}, \ldots \tag{1.9}$$

Substitution of (1.9) into (1.1) yields

$$\begin{aligned}
\nabla \times \hat{\mathbf{E}}(\mathbf{x},\omega) &= -\mathrm{i}\omega \hat{\mathbf{B}}(\mathbf{x},\omega), \\
\nabla \times \hat{\mathbf{H}}(\mathbf{x},\omega) &= \mathrm{i}\omega \hat{\mathbf{D}}(\mathbf{x},\omega) + \hat{\mathbf{J}}(\mathbf{x},\omega), \\
\nabla \cdot \hat{\mathbf{D}}(\mathbf{x},\omega) &= \hat{\rho}(\mathbf{x},\omega), \\
\nabla \cdot \hat{\mathbf{B}}(\mathbf{x},\omega) &= 0.
\end{aligned} \tag{1.10}$$

Here, the dependence on ω is shown since the values $\hat{\mathbf{E}},\ldots$ satisfy equations (1.10) with the parameter ω. Therefore, they have to depend on ω.

Note that the price to be paid is that these fields (which are now independent of time but depend instead on the frequency) are complex-valued. One can calculate the real-valued time-dependent fields, which are physical observable quantities as follows

$$\mathbf{E}(\mathbf{x},t) = \mathrm{Re}\ [\hat{\mathbf{E}}(\mathbf{x},\omega)\mathrm{e}^{\mathrm{i}\omega t}], \quad \ldots. \tag{1.11}$$

Consequently, it is worthwhile mentioning that the time dependence for complex vectors can be $\exp(-\mathrm{i}\omega t)$. Here, the sign of the imaginary parts changes when moving from one notation to the other, causing difficulties in interpreting the dissipative character (the sign of the imaginary part of material parameters) of metamaterials.

1.1.3 TEMPORAL DISPERSION

The dielectric response in materials takes place because of the charge separation mechanisms having various origins and taking different amounts of time to develop. Moreover, any response from matter drastically depends on the frequency of the electromagnetic field. For instance, electronic polarization is more sensitive to higher frequencies than ionic polarization, and in the case of metamaterials made of macroscopic artificial molecules, the response may be very strong in the microwave range. The described phenomenon may be referred to as temporal dispersion. Consequently, the polarization at an instant t is written as a convolution of overall past events. It is worthwhile noting that polarization (usually connected to causality) will not affect future events. By assuming for the moment a linear (the flux densities depend only linearly on the fields) and local response of the medium (the flux at point $\mathbf{x}$ is only induced by the field at the same spatial position), and stationary (material properties do not change in time), the constitutive relation can be expressed as a convolution with a response function $\mathbf{R}_e(\mathbf{x},t)$ being a tensor

$$\mathbf{D}(\mathbf{x},t) = \varepsilon_0 \int_{-\infty}^{t} \mathbf{R}_e(\mathbf{x},t-\tau)\mathbf{E}(\mathbf{x},\tau)\ \mathrm{d}\tau. \tag{1.12}$$

Analogously,

$$\mathbf{B}(\mathbf{x},t) = \mu_0 \int_{-\infty}^{t} \mathbf{R}_m(\mathbf{x},t-\tau)\mathbf{H}(\mathbf{x},\tau)\ \mathrm{d}\tau. \tag{1.13}$$

Here, the time can be considered from $-\infty$ to $+\infty$. If the initial time $t = 0$ is taken with the corresponding initial conditions, we have to define $\mathbf{E}(\mathbf{x},t) = 0,\ldots$ for negative t.

It has to be stressed that the former response function fully describes any linear response of the medium. Lifting any assumptions as formulated above complicates the functional dependency, but substantial modifications of the functional relations will not take place. If, e.g., the assumption on the locality would not be valid anymore, one would need to take the convolution over the entire spatial domain. Moreover, when the assumption on the linear response is not valid, one would need to consider the higher-order response functions, which are multiplied by squares or cubes of the electric or magnetic field. Although written here only for electric polarization, dual expressions exist for the magnetization response of natural magnetics.

The description of the induced polarization in the time domain is apparently complicated and, from a practical perspective, not very handy. As the equation is a convolution in the time domain, the equation is, however, reduced to some simple algebraic multiplication in the frequency domain. Therefore, by Fourier transform of (1.12)–(1.13) yields the constitutive relations in the frequency domain

$$\hat{\mathbf{D}}(\mathbf{x},\omega) = \varepsilon_0 \chi_e(\mathbf{x},\omega)\hat{\mathbf{E}}(\mathbf{x},\omega), \tag{1.14}$$

$$\hat{\mathbf{B}}(\mathbf{x},\omega) = \mu_0 \chi_m(\mathbf{x},\omega)\hat{\mathbf{H}}(\mathbf{x},\omega), \tag{1.15}$$

where the electric and magnetic susceptibilities $\chi_{e,m}(\mathbf{x},\omega)$ are defined by means of the Fourier transform

$$\chi_{e,m}(\mathbf{x},\omega) = \int_{-\infty}^{\infty} \mathbf{R}_{e,m}(\mathbf{x},\tau)\exp(\mathrm{i}\omega\tau)\mathrm{d}\tau. \tag{1.16}$$

Analogously,

$$\hat{\mathbf{P}}(\mathbf{x},\omega) = \varepsilon_0 \chi_e(\mathbf{x},\omega)\hat{\mathbf{E}}(\mathbf{x},\omega), \tag{1.17}$$

$$\hat{\mathbf{M}}(\mathbf{x},\omega) = \chi_m(\mathbf{x},\omega)\nu\hat{\mathbf{H}}(\mathbf{x},\omega). \tag{1.18}$$

The presence of the "memorable" medium is caused by the susceptibility being a function of frequency $\chi_{e,m}(\omega)$; in other words, the polarization density is not only proportional to the instantaneous field amplitude but also dependable on its past values.

The temporal dispersion also brings about losses expressed by the imaginary parts. Both the real and imaginary parts of $\chi_{e,m}(\omega)$ are functions of the frequency. It is important to point out a rigorous connection between the frequency behavior of the real and imaginary parts given by the basic physical principle of causality. Based on it, the polarization response of a medium to an electric excitation cannot precede its cause. This principle was explicitly expressed above by taking the convolution in the time domain only over past events. The Kramers–Kronig relations follow from the principle of causality. The former have to be fulfilled for the real and imaginary parts of the susceptibility function [127]

$$\mathrm{Re}\,\chi(\omega) = -\frac{2}{\pi} PV \int_0^{\infty} \frac{\omega' \mathrm{Im}\,\chi(\omega')}{\omega'^2 - \omega^2} \mathrm{d}\omega', \tag{1.19}$$

$$\mathrm{Im}\,\chi(\omega) = -\frac{2\omega}{\pi} PV \int_0^{\infty} \frac{\mathrm{Re}\,\chi(\omega')}{\omega'^2 - \omega^2} \mathrm{d}\omega'. \tag{1.20}$$

Here *PV* stands for the principal value of the singular integral, which means that in the integration, the singular point $\omega' = \omega$ is symmetrically excluded [79, 193]. These relations, connected to the Hilbert transform, are useful in the analysis of the dispersive properties of materials.

Below, we write $\mathbf{E}(\mathbf{x}),\ldots$ instead of $\hat{\mathbf{E}}(\mathbf{x},\omega),\ldots$ for shortness.

1.1.4 CONSTITUTIVE RELATIONS

The functional dependency of the constitutive relations in the frequency domain may have arbitrary higher-order terms, whereas each of them has a particular meaning. By allowing electro-magnetic coupling and by also allowing for a static contribution to the polarization of the materials, one may obtain constitutive relations such as

$$\mathbf{P}(\mathbf{E},\mathbf{H}) = \mathbf{P}_s + \varepsilon_0 \boldsymbol{\chi}_e(\omega)\mathbf{E} + \boldsymbol{\xi}(\omega)\mathbf{H} + \boldsymbol{\beta}_e(\omega)\mathbf{H}\mathbf{H} + \boldsymbol{\gamma}_e(\omega)\mathbf{H}\mathbf{E} + \ldots, \tag{1.21}$$

$$\mathbf{M}(\mathbf{E},\mathbf{H}) = \mathbf{M}_s + \boldsymbol{\chi}_m(\omega)\mathbf{H} + \boldsymbol{\varsigma}(\omega)\mathbf{E} + \boldsymbol{\beta}_m(\omega)\mathbf{E}\mathbf{H} + \boldsymbol{\gamma}_m(\omega)\mathbf{E}\mathbf{E} + \ldots. \tag{1.22}$$

Here, $\mathbf{P}_s$ and $\mathbf{M}_s$ are electric polarization and magnetization existing in the absence of external (high frequency) fields. Such response may take place in, e.g., ferroic and multi-ferroic materials; $\boldsymbol{\chi}_e(\omega)$ and $\boldsymbol{\chi}_m(\omega)$ are electric and magnetic susceptibilities as introduced above. Tensors $\boldsymbol{\xi}$ and $\boldsymbol{\varsigma}$ express the induction of electric polarization by a magnetic field or magnetization by an electric field, which is called the linear magneto-electric effect. Higher-order terms are tensor products.

Today a lot of research is done aiming to understand the magneto-electric effect found in several oxide materials as well as in reciprocal bi-anisotropic materials and metamaterials. While higher-order terms appear in principle in these equations, for most relevant materials, it is fully sufficient to deal with only a few terms of the expansions (1.21)–(1.22). Then, only some frequency-dependent constants arise in the constitutive relations as proportionality coefficients which are usually taken as material constants. In order to find solutions to Maxwell's equations, one would need to deal with the constitutive relations, providing a link between the micro- or nanostructure of a fabricated metamaterial that may be composed of arbitrary complex shaped cells and an electromagnetic material with particular properties in which a group of well documented physical phenomena can be observed. Obtaining the largest possible degree of freedom over these material parameters provides wide avenues for researchers working in the field of metamaterials.

1.1.5 DESCRIPTION OF THE BEHAVIOR OF THE METAMATERIAL

The constitutive relations model for a particular design of a metamaterial provides a fertile ground for the analysis of the electromagnetic behavior of metamaterials. It is worthwhile noting that the former approach is far from trivial, and in many cases, metamaterials cannot be modeled by conventional material parameters [149]. Next, one can solve Maxwell's equations equipped with the appropriate constitutive relations for an infinite space filled with this metamaterial in the absence of sources and determine all the eigenwaves that can exist in the medium. By enforcing appropriate boundary conditions (in simple cases of weak or negligible spatial dispersion, continuity of the tangential electric and magnetic field components), the amplitudes of these eigenmodes are found for a finite-size sample of a given shape, and the field inside the metamaterial sample is expressed as a superposition of eigenmodes. It has to be stressed that one can express this superposition in the temporal domain if finite temporal pulses are used to illuminate the metamaterial or in the frequency domain if finite spatial beams are used to illuminate it. At the same time, the amplitudes of the eigenmodes used to decompose the reflected field are obtained, and so a solution to the entire problem is found.

1.2 CLASSES OF EFFECTIVELY CONTINUOUS MATERIALS

1.2.1 ISOTROPIC MEDIA

The material response expressed only by a scalar relating the field and displacement vectors is treated as the simplest one in terms of the constitutive relations (1.2) and is written as follows:

$$\mathbf{D} = \varepsilon \mathbf{E}, \tag{1.23}$$

$$\mathbf{B} = \mu \mathbf{H}, \tag{1.24}$$

where ε is the permittivity and μ is the permeability of the medium. Consequently, the medium is isotropic (the response is not affected by the direction of the field vector).

Let us consider a plane wave solution of Maxwell's equations in a homogeneous (the material parameters ε and μ are constants, and there are no sources ($J = 0$, $\rho = 0$)). It follows from equations (1.10) that

$$\nabla^2 \mathbf{E}(\mathbf{x}) + \omega^2 \mu\varepsilon \mathbf{E}(\mathbf{x}) = 0. \tag{1.25}$$

A Cartesian coordinate system can be chosen in the case of a plane wave; in doing so, the fields depend only on one coordinate, say x_3. The former leads to the following expression

$$\frac{d}{dx_3} \mathbf{E}(x_3) + \omega^2 \mu\varepsilon \mathbf{E}(x_3) = 0. \tag{1.26}$$

Introducing $k = \omega\sqrt{\mu\varepsilon}$ one can write the two solutions

$$\mathbf{E}(\mathbf{x}) = \mathbf{E}_1 \exp(-ikx_3), \quad \mathbf{E}(\mathbf{x}) = \mathbf{E}_2 \exp(+ikx_3). \tag{1.27}$$

Using (1.11), one can see that the first solution is a wave propagating in the positive x_3 direction, and the second one propagates in the negative x_3 direction. Note that in equations (1.27), the amplitude and polarization of the electric field are determined by the constant vectors $\mathbf{E}_1$ and $\mathbf{E}_2$. For example, a real vector $\mathbf{E}_1$ denotes a linearly polarized field, and if $\mathbf{E}_1$ is a complex vector, with real and imaginary parts being equal in length and perpendicular to each other, the polarization is circular. The electric field has to be free of divergence, as required by the third equation (1.10) for $\rho = 0$; consequently, vectors $\mathbf{E}_1$ and $\mathbf{E}_2$ are orthogonal to the axis x_3. Therefore, these wave solutions are termed transverse electromagnetic waves.

The material parameters ε and μ are real numbers if the medium is lossless. On the other hand, in the case of the two material parameters having different signs, k may be either real or purely imaginary. If k is real, the waves in (1.27) are dramatically affected by the phase change, and the wavelength λ_{ph} in the x_3-direction after which the phase attains the same value is given by the condition $k\lambda_{ph} = 2\pi$, or in terms of the material parameters,

$$\lambda_{ph} = \frac{2\pi}{\omega\sqrt{\mu\varepsilon}} = \frac{2\pi c}{\omega n}. \quad (1.28)$$

Here the refractive index n is defined by the latter equality. Hence, the phase wavelength becomes shorter for higher frequencies and also for materials with higher permittivity and permeability.

On the other hand, dielectric (magnetic) losses result in the permittivity (permeability) being complex. Then also, k becomes complex, and the exponential decay of amplitude of the propagating wave (1.27) is caused by the k imaginary part. The medium is lossless if no dissipation occurs.

1.2.2 MATERIALS DESCRIBED ONLY BY PERMITTIVITY

First, we will consider materials that can be described by the permittivity ε only. Dealing with such a material, it is meant that the material has a permeability equal to that of a vacuum and thus that the relation between **B** and **H** is $\mathbf{B} = \mu_0\mathbf{H}$. Dielectrics, conductors, and piezoelectric materials are the most common examples of such natural materials.

1.2.2.1 Dielectrics

The incident electromagnetic field moves the charges, and in doing so, a polarization of the medium is induced in the case of dielectrics. The induced polarization is a complex function depending on the electric field but can be expanded into a Taylor series.

Assuming the electric field to be small, only a few expansion terms need to be employed. From the perspective of linear electromagnetics and optics, one can truncate the Taylor series for dielectric materials, and in the first approximation, we get

$$\mathbf{D} = \varepsilon_0\mathbf{E} + \mathbf{P}. \quad (1.29)$$

The extent to which the incident EM field is capable of separating the positive and negative charges in the material are expressed by the polarization **P**. The material type makes a significant on **P** having a specific relation to **E**. A bound electron in a dielectric would bounce forwards and backward with the oscillating incident field as its driving force, and this would be expressed as a harmonic oscillation ($\lambda = c\frac{2\pi}{\omega}$). The polarization would then also oscillate harmonically, being the product of unit-volume displaced charge and distance. For each frequency, the response will differ, and this is expressed as follows:

$$\mathbf{P}(\omega) = \varepsilon_0\chi(\omega)\mathbf{E}(\omega). \quad (1.30)$$

It is worthwhile noting that the frequency domain is considered in the frame of the current chapter, doing so, one gets a chance to get a deep insight into the wave phenomena. Combining (1.29) and (1.30) we find

$$\varepsilon(\omega) = \varepsilon_0(1 + \chi(\omega)). \quad (1.31)$$

and we have derived an expression for the permittivity in a simple dielectric material. The parameter χ is called the electric susceptibility. If the dielectric is not isotropic, then χ is a tensor in (1.30). This case holds for many metamaterials.

1.2.2.2 Conductors

Influenced by an incident electromagnetic wave described by an oscillating current density $\mathbf{J}$ being the product of the charge density per unit volume and the velocity of charge movement, the free electrons in a metal will oscillate harmonically. When the effect of the electromagnetic wave on the metamaterials is linear, $\mathbf{J}$ and $\mathbf{E}$ are proportional

$$\mathbf{J}(\omega) = \sigma(\omega)\mathbf{E}(\omega). \tag{1.32}$$

The proportionality scalar $\sigma(\omega)$ is called conductivity.

To derive an expression for the permittivity, we remember that $\mathbf{J}$ in (1.29) is proportional to the charge displacement $\mathbf{x}$ and $\mathbf{J}$ is proportional to the charge speed $\mathbf{v}$ (that is, $\mathbf{P} = Ne\mathbf{x}$ and $\mathbf{J} = Ne\mathbf{v}$, with N being the particle density and e being the electric charge). In the complex notation for time-harmonic fields $\mathbf{v} = \mathrm{i}\omega\mathbf{x}$. We can express $\mathbf{P}$ as a function of $\mathbf{J}$ and insert the result in (1.32) to get

$$\varepsilon(\omega) = \varepsilon_0 + \frac{\sigma(\omega)}{\mathrm{i}\omega}. \tag{1.33}$$

In the real world, a certain material might have polarization due to both free and bound charges, and we describe that as follows:

$$\hat{\varepsilon}(\omega) = \varepsilon(\omega) + \frac{\sigma(\omega)}{\mathrm{i}\omega}. \tag{1.34}$$

Here $\varepsilon(\omega)$ describes the polarization response due to bound charges being also a complex number. The permittivity in a conductor possesses a large imaginary part (the second term, proportional to the conductivity σ), and this means that the losses are large. But this also means that the penetration depth of the electromagnetic waves into a material sample is small. The fields attenuate rapidly within the conductor, and the fields are confined to a thin surface layer; this allows the definition of surface current and surface impedance, which are notions often used by electronic engineers. The quantity $\hat{\varepsilon}(\omega)$ is often called the generalized permittivity.

1.2.2.3 Piezoelectric Materials

A phenomenon in materials where electric fields can be created by mechanical compression in some directions is called piezoelectricity. The former can be explained by a relative displacement of positive and negative charges resulting in an induced dipole. On thermodynamic grounds, it is expected that for such materials also, the opposite effect takes place, which is indeed the case: the application of a potential difference in the material creates a mechanical distortion. This effect is often characterized as the direct piezoelectric effect. The mechanical relation linking the stress and strain is cross-coupled to the electrical effect, and this appears in the (electrical) constitutive equations. The constitutive relations in the case of piezoelectric materials are more complicated than for ordinary anisotropic materials due to the mechanical stresses and strains themselves being second-order tensors, and hence their relations with the electric field and flux density have to be described by cross-coupling third-order tensors.

1.2.3 MAGNETIC MATERIALS

For magnetic materials the permeability is the dominating parameter expressing the relation between **B** and **H**

$$\mathbf{B} = \mu_0(\mathbf{H}+\mathbf{M}) = \mu\mathbf{H}, \tag{1.35}$$

where **M** is the magnetic moment density (magnetization) and μ is the permeability. (Almost in all magnetic materials, the medium is also electrically polarizable, and the name magneto-dielectric is used.)

Materials described by permittivity and permeability are called magneto-dielectric.

Consider magneto-dielectric materials with negative permittivity and permeability. The sign of the real parts of permittivity and permeability is not limited by any physical law and can be negative. The real part of the permittivity is negative, for example, in metals, the former phenomenon takes place due to the presence of the free electrons moving in the applied electric field and creating a polarization that is out of phase with the applied field. The plasma frequency denotes the resonant frequency of free-electron oscillations and depends on the electron density and effective electron mass. For metals, this frequency is in the ultraviolet range, leading to very large negative values of the real part of the permittivity at the microwave frequencies (range of GHz) and below. However, at these frequencies, the imaginary part of the permittivity dominates due to the very high conductivity of metals. In natural materials, negative permeability occurs in materials with resonant phenomena. The real part of permeability may take negative values near a resonance of sufficient strength, for instance, in ferrites near the ferromagnetic resonance frequency. Manipulation of magnetic charges is not an option in the case of artificial materials, but a negative permeability response can be realized using electrically small resonant metal inclusions of special shapes, e.g., split rings. In the visible part of the spectrum, this is the only possibility of realizing negative permeability (as well as any relative permeability values different from 1).

1.2.4 ANISOTROPIC AND BI-ANISOTROPIC MATERIALS

Isotropic media do not possess any special direction in their structure (a high degree of spatial symmetry), leading to the electromagnetic response (and any other response, for that matter) being unaffected by how the material is rotated with respect to the field. For isotropic media, the material parameter tensors are multiplies of the unit tensor, and scalar multiplications suffice.

Metamaterials are often anisotropic due to broken geometric symmetry. The impact on the constitutive parameters is that a tensor description is necessary. Relations

(1.23)–(1.24) cannot be used for materials that are no longer considered isotropic. The next step is to deal with permittivity and permeability as tensors. Such materials are called anisotropic materials.

Materials are called bi-anisotropic if a linear relation exists between all four vectors of the electromagnetic fields. The constitutive relations for bi-anisotropic materials read

$$\mathbf{D} = \boldsymbol{\varepsilon}\mathbf{E} + \boldsymbol{\xi}\mathbf{H}, \tag{1.36}$$

$$\mathbf{B} = \boldsymbol{\zeta}\mathbf{E} + \boldsymbol{\mu}\mathbf{H}. \tag{1.37}$$

In these relations, the four material parameter tensors necessary to characterize the material are the permittivity $\boldsymbol{\varepsilon}$, the permeability $\boldsymbol{\mu}$, and the magneto-electric coupling is expressed by two magneto-electric tensors $\boldsymbol{\xi}$ and $\boldsymbol{\zeta}$. Note that the most general bi-anisotropic material requires 36 complex parameters for a full constitutive electromagnetic description.

We want to linger for a second on magneto-electric coupling. It can be observed in materials where a magnetic field creates electric polarisation in the medium and vice versa. An electric excitation brings forth a magnetic response (bi-anisotropic materials introduced above). Materials are called bi-anisotropic if they lack a preferred direction but show magneto-electro coupling. In this case, the amount of free parameters in (1.36)–(1.37) decreases to four.

Under these conditions, one can divide materials into subclasses when the criteria are isotropy-anisotropy and whether the medium displays magneto-electric coupling. Classes of linear electromagnetic materials and the number of free material parameters in their full characterization [49]. The bi-anisotropic theory holds for most metamaterials, but many parameters might be needed to describe the material.

1.2.5 SPIN-WAVE/MAGNONIC METAMATERIALS

Periodic magnetic structures exhibit band gaps if "probed" with a wave of the order or smaller than the period, but they show effectively continuous properties if "probed" with a large-scale wave (microwaves or long-wavelength spin waves). The same theory, where "effectively continuous" behavior is the long-wavelength limit of the band gap theory, is used to describe both phenomena theoretically.

The effective parameters are those that are required to describe the collective spin waves as if the array were a continuous material. These parameters determine the dispersion of spin waves in continuous samples, e.g., the effective exchange constant, effective anisotropy, effective magnetization, and effective gyro-magnetic ratio [124]. Traditionally, the research of spin waves in magnetic nanostructures has always been linked to measuring spin wave spectra or associated high-frequency magnetic susceptibilities. Nevertheless, for instance, media containing nano-dots may possess fascinating, effective material properties. Indeed, effective electromagnetic properties of a medium containing ferromagnetic inclusions were theoretically investigated, e.g., in [28]. However, the response of the magnetic inclusions was entirely because of the uniform ferromagnetic resonance and, therefore, bound to lie in the low GHz range. The useful frequency range would significantly expand if non-uniform resonances (so-called spin-wave resonances) were exploited. To take full advantage of the latter, strategies improving the coupling of essentially uniform electromagnetic fields to non-uniform spin waves will have to be adopted from the field of spin dynamics, and this is, at the moment, a topic of advanced research.

1.2.6 DECOMPOSITION OF FIELDS INTO BLOCH WAVES

Waves that, once excited, can propagate through an infinite material without any further external excitation having any amplitude are called eigenwaves. In order to excite a given eigenwave, an incoming plane wave must have the same wavenumber along the interface between metamaterial and surrounding space as the eigenwave. In order to pave the way for the described situation, the structures need to be simultaneously infinite and perfectly lossless. When dealing with structures with non-zero losses, eigenmode excitation will only occur if the incident wave has a non-zero energy component with the same wavelength. The former is expressed as requiring the incident field to have a non-zero projection on the eigenmode. The eigenwaves can be plane, spherical, and cylindrical waves in homogeneous isotropic media. In the case of materials with periodically positioned inhomogeneities, the eigenwaves are of a different type, and these waves are called Bloch waves or Bloch modes. The eigenmode properties reflect the periodicity of the structures, and they are periodic functions of the spatial coordinates, with constant phase shifts between fields in consecutive unit cells.

The eigenmodes can be described via the dependence of the phase shifts per unit cell along the coordinate directions on the frequency. For 3D-periodic structures, this function can be visualized as a hyper-surface in the four-dimensional space made of the three wavenumbers and the frequency. For a given eigenmode, let us denote by ψ_x, ψ_y and ψ_z the phase shifts between consecutive cells along x, y and z axes, respectively, and by a, b and c the corresponding periods. Then, the eigenmode is described by the wavevector $\mathbf{k} = (k_x, k_y, k_z)$, where the wavenumbers $k_x = \psi_x/a$, $k_y = \psi_x/b$ and $k_z = \psi_x/c$. The eigensurface corresponds to the function $\omega = \omega(\mathbf{k})$, called the dispersion relation. The phase velocity of an eigenmode is, per definition, in the same direction as the wave vector, and its value equals $\omega/|\mathbf{k}|$. Hence, it can be written as:

$$\mathbf{v}_{phase} = \mathbf{k}\frac{\omega}{|\mathbf{k}|^2}. \tag{1.38}$$

The eigensurface corresponds to 2D surfaces in a 3D domain such as $\omega(k_x, k_y)$ in the case of a doubly infinite slab of a metamaterial. Materials behave as bandgap structures when, for some frequencies, there is no propagating solution, which is expressed by the fact that there are no corresponding real solutions for that particular phase shift.

1.3 RESONANCES IN PERMITTIVITY AND PERMEABILITY

The basic dispersion models will be introduced and the resonance notion be explained in the following sections.

1.3.1 LORENTZ MODEL FOR DIELECTRICS

It is worthwhile mentioning, that the charges are displaced by an incident electric field in dielectric materials. The amplitude is presented by the restoring forces on the electrons and is drastically influenced by the time-variation of the field, and this dipole (formed by the displaced electrons and the remaining ion) will oscillate. The induced dipole moment might be very large and drastically affected by the frequency in case of frequencies around a specific frequency ω_e. Strong scattering and absorption phenomena will be caused by this exceptional interaction (called 'resonance') of an incident electromagnetic field with the structure. The behavior of the oscillating dipole can be described based on a simple driven harmonic oscillator, and the solution gives the relation between the polarization $\mathbf{P}$ and electric field $\mathbf{E}$ (constitutive relation) expressed by the susceptibility χ_e. A complex Lorentzian function governs the solution to the differential equation describing forced resonance, and the relation is as follows

$$\mathbf{P}(\omega) = \frac{\varepsilon_0 F}{\omega_e^2 - \omega^2 + \mathrm{i}\gamma\omega}\mathbf{E}(\omega) = \varepsilon_0 \chi(\omega)\mathbf{E}(\omega), \tag{1.39}$$

in which F is a phenomenological strength of the resonance, and γ is the damping factor. The amplitude of this function depends on the detuning of the frequency from resonance and the damping. As in all resonance processes in nature and technology, the driven harmonic oscillator can follow the excitation at frequencies smaller than the resonance frequency, and in these cases, it is said to be in-phase (here we speak about the case of negligible losses: $\gamma \approx 0$ in (1.39). At frequencies larger than the resonance frequency, the response in the material is 180° out of phase relative to the excitation. The lower the damping, the sharper the resonance, and in that case, the spectral domain in which this relative phase difference between the illuminating wave field and the dipolar excitation arises will be narrow. The resonance peak is also higher, and the energy transfer from the illuminating wave field to the dipole will be strongest directly at the resonance in the case of small damping. The fact that for frequencies slightly smaller than the resonance frequency, the susceptibility is large and positive, whereas it is negative at frequencies larger than the resonance is suggested by the change in 180° of the phase around the resonance frequency. The relation (1.31) is used to obtain the relative permittivity as the defining quantity.

1.3.2 DRUDE MODEL FOR METALS

The Drude model stands for a special case of the Lorentz model. The Drude model is often used to describe the optical properties of noble metals, comprising most optical metamaterials. The Drude model is also employed to model the properties of wire media for waves with no field dependence along the wires [27,57]. The Drude behavior arises from (1.39) by letting the electrons be free and setting the resonance frequency to zero, $\omega_0 = 0$. Introduce the parameter $\nu = 0.2\omega_{phase}$. Then

$$\varepsilon(\omega) = \varepsilon_0 - \varepsilon_0 \frac{\omega_{phase}^2}{\omega^2 - \mathrm{i}\omega\nu}, \tag{1.40}$$

where $\varepsilon = \varepsilon_0$ at infinity. At low frequencies, the imaginary part dominates the response. Indeed, the typical conductivity behavior for metals at low frequencies $\varepsilon(\omega) \approx \frac{i\sigma}{\omega}$ is included in the model (1.40) in the limit $\omega \longrightarrow 0$, where the conductivity $\sigma = \frac{\varepsilon_0 \omega_{phase}^2}{\nu}$.

1.4 QUASISTATIC LIMIT

Consider a domain $Q = G \cup G_1 \cup G_2 \cup \ldots$ divided onto mutually disjoint simply connected subdomains G_k $(k = 1,2,\ldots)$ and multiply connected domain G in the space. In the main part of the book, we consider plane 2D domains.

We restrict ourselves by Maxwell's equations (1.10) with $\mathbf{J}(\mathbf{x}) = \mathbf{0}$ and $\rho(\mathbf{x}) = 0$

$$\nabla \times \mathbf{E}(\mathbf{x}) = -\mathrm{i}\omega\mathbf{B}(\mathbf{x}),\ \nabla \times \mathbf{H}(\mathbf{x}) = \mathrm{i}\omega\mathbf{D}(\mathbf{x}),\ \nabla \cdot \mathbf{D}(\mathbf{x}) = 0,\ \nabla \cdot \mathbf{B}(\mathbf{x}) = 0. \tag{1.41}$$

Moreover, we will consider these equations of electromagnetic wave propagation in the quasistatic limit when the wavelengths are much larger than the characteristic size of the microstructure. More precise explanations can be found in [155, Chapter 11]. In this case, the terms proportional to ω are neglected and (1.41) is reduced to equations

$$\nabla \times \mathbf{E}(\mathbf{x}) = \mathbf{0}, \quad \nabla \times \mathbf{H}(\mathbf{x}) = \mathbf{0}, \quad \nabla \cdot \mathbf{D}(\mathbf{x}) = 0, \quad \nabla \cdot \mathbf{B}(\mathbf{x}) = 0. \tag{1.42}$$

The constitutive relations have the form

$$\mathbf{D}(\mathbf{x}) = \boldsymbol{\varepsilon}(\mathbf{x})\mathbf{E}(\mathbf{x}), \quad \mathbf{B}(\mathbf{x}) = \boldsymbol{\mu}(\mathbf{x})\mathbf{H}(\mathbf{x}). \tag{1.43}$$

The equations of electromagnetic wave propagation in the quasistatic limit (1.42) and (1.43) are reduced to the usual equations of conductivity but with complex fields and complex tensors $\boldsymbol{\varepsilon}$ and $\boldsymbol{\mu}$.

The further part of the book deals with a two-dimensional (2D) dispersed composite in the framework of quasistatic limit. In this case, the pairs fields **E**, **D** and **H**, **B** can be considered independently. Select the equations concerning the first pair

$$\nabla \times \mathbf{E}(\mathbf{x}) = \mathbf{0}, \quad \nabla \cdot \mathbf{D}(\mathbf{x}) = 0, \quad \mathbf{D}(\mathbf{x}) = \boldsymbol{\varepsilon}(\mathbf{x})\mathbf{E}(\mathbf{x}). \tag{1.44}$$

In order to precisely describe equations for a 2D dispersed composite, we consider a domain Q divided onto the non-overlapping inclusions G_k $(k = 1,2,\ldots,N)$ and the host domain G on a plane. The plane corresponds to considered pictures; for instance, in Figure 2.1. The plane coordinates are fixed as $\mathbf{x} = (x_1,x_2)$. Usually, the inclusions G_k are supposed to be simply connected domains; the host G is a multiply connected domain. Let the tensor $\boldsymbol{\varepsilon}(\mathbf{x})$ be constant and diagonal in every component G, G_k $(k = 1,2,\ldots,N)$, i.e. $\boldsymbol{\varepsilon}(\mathbf{x}) = \varepsilon_k I$ for $\mathbf{x} \in G_k$ $(k = 1,2,\ldots,N)$ and $\boldsymbol{\varepsilon}(\mathbf{x}) = \varepsilon_{host} I$, where ε_{host} and ε_k are complex constants, I is the identity matrix.

Equations (1.44) hold in the components of Q. It is convenient to rewrite them in the form

$$\nabla \times \mathbf{E}_k(\mathbf{x}) = \mathbf{0},\ \nabla \cdot \mathbf{D}_k(\mathbf{x}) = 0,\ \mathbf{D}_k(\mathbf{x}) = \varepsilon_k \mathbf{E}_k(\mathbf{x}),\ \mathbf{x} \in G_k\ (k = 1,2,\ldots,N), \tag{1.45}$$

$$\nabla \times \mathbf{E}(\mathbf{x}) = \mathbf{0}, \quad \nabla \cdot \mathbf{D}(\mathbf{x}) = 0, \quad \mathbf{D}(\mathbf{x}) = \varepsilon_{host}\mathbf{E}(\mathbf{x}), \quad \mathbf{x} \in G. \tag{1.46}$$

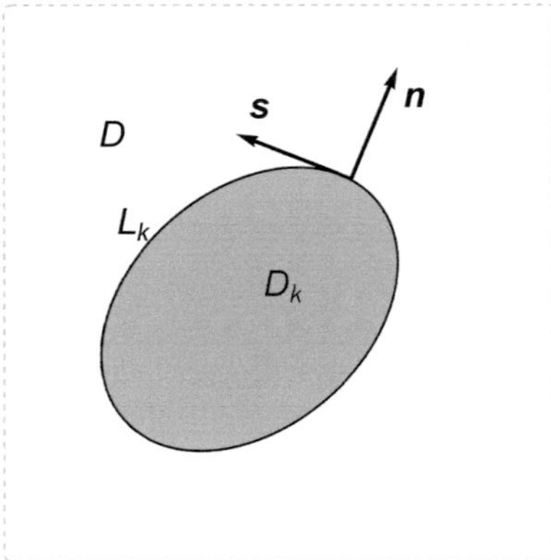

Figure 1.1 The curve L_k divides the inclusion D_k and the host medium D. The outward normal and tangent unit vectors, **n** and **s**, respectively, are defined on L_k.

Consider the complex fields $\mathbf{E}_k(\mathbf{x})$ and $\mathbf{D}_k(\mathbf{x})$ in the fixed inclusion G_k. The first equation (1.45) implies that there exists a scalar function $u_k(\mathbf{x})$ such that

$$\mathbf{E}_k(\mathbf{x}) = \nabla u_k(\mathbf{x}), \quad \mathbf{x} \in G_k. \tag{1.47}$$

The complex function $u_k(\mathbf{x})$ is called the potential; it is defined up to an additive arbitrary constant in the simply connected domain G_k. Substitute the third equation (1.45) into the second equation and obtain $\nabla \cdot \mathbf{E}_k(\mathbf{x}) = 0$ since ε_k is a constant. Substitution of (1.47) into the later equation yields Laplace's equation

$$\nabla^2 u_k(\mathbf{x}) = 0, \quad \mathbf{x} \in G_k. \tag{1.48}$$

The third equation (1.45) yields $\mathbf{D}_k(\mathbf{x}) = \varepsilon_k \nabla u(\mathbf{x})$ in G_k.

Analogous formulas take place in the domain G

$$\nabla^2 u(\mathbf{x}) = 0, \quad \mathbf{x} \in G, \tag{1.49}$$

where

$$\mathbf{E}(\mathbf{x}) = \nabla u(\mathbf{x}), \quad \mathbf{D}(\mathbf{x}) = \varepsilon_{host} \nabla u(\mathbf{x}), \quad \mathbf{x} \in G. \tag{1.50}$$

It should be noted that the potential $u(\mathbf{x})$ in the multiply connected domain G is not uniquely defined up to an additive constant [185, Section 2.4].

Let $L_k = \partial G_k$ denote the smooth positively oriented boundary of the inclusion and $\mathbf{n} = (n_1, n_2)$ the unit outward normal vector to L_k. Let $\mathbf{s} = (-n_2, n_1)$ denote the unit tangent vector to L_k. These vectors are displayed in Figure 1.1. If the free surface charge density at the boundary vanishes, the contact between the inclusion and the host is perfect. This implies that the tangential components of the electric field and the normal components of the electric displacement are continuous on L_k

$$\mathbf{E}_{k\,\mathbf{s}}(\mathbf{x}) = \mathbf{E}_{\mathbf{s}}(\mathbf{x}), \quad \mathbf{D}_{k\,\mathbf{n}}(\mathbf{x}) = \mathbf{D}_{\mathbf{n}}(\mathbf{x}), \quad \mathbf{x} \in L_k. \tag{1.51}$$

Following [93, Chapter 2], one can rewrite the transmission condition in terms of the potentials

$$u(t) = u_k(t), \quad \frac{\partial u}{\partial \mathbf{n}}(t) = \varepsilon_k \frac{\partial u_k}{\partial \mathbf{n}}(t), \quad t \in L_k. \tag{1.52}$$

Part II

Mathematical Model of Complex Permittivity in Composites

2 Method of Schwarz

2.1 ON SCHWARZ'S METHOD IN THE CONSTRUCTIVE THEORY OF COMPOSITES

The theory of composites includes a broad spectrum of applied mathematics, in particular, asymptotic homogenization theory and both analytical and numerical computation of effective material properties. Homogenization theory leads to the statement of the periodic boundary value problem. It provides a rationale for employing local fields derived from the boundary value problem to determine the effective constants. The periodic statement yields the proper definition of effective constants through the integrals over the periodicity cell from the local fields [22, 113]. A composite with one inclusion per unit periodicity cell Q is called a regular periodic composite. A composite with $N > 1$ inclusions per unit periodicity cell is called a periodic composite. In the homogenization theory of random composites [94, 113, 223] based on the theory of measure, the periodicity is replaced by the spatial stationary process when the statistical properties of the medium are invariant under translations. In this case, a unit periodicity cell Q exists, and it represents the considered random composite [93, Chapter 3]. This is the postulate of the homogenization theory of random composites. Therefore, the proper way to determine effective constants of a random composite is to consider a periodized part of the structure [214, 215] with increasing N, keeping the distribution of location up to the stabilization of the computed constants. In this part, we show that the violation of periodization can lead to methodologically misleading results.

The homogenization theory is associated with the concepts of existence and uniqueness of effective constants. The estimation of the effective constants is the next problem of homogenization. In this part, we analyze approximate and exact analytical formulas. We call this area the constructive theory of composites, emphasizing the derivation of formulas for effective constants. Leave aside now other ways to achieve the same goal, namely, the theory of bounds [51, 155], pure numerical simulations, and the theory of Representative Volume Element (RVE) discussed in Chapter 4.

Some mathematical theories and algorithms were first proposed due to physical intuition. The way from a physical idea to a rigorous mathematical method may be extended and contradictory. In this part, we discuss the reasonable approximations for effective permittivity (conductivity, permeability, elastic constants, etc.). Often, new formulas derived using intuition are reduced to the classical lower-order approximations with redundant higher-order tails; some formulas are correct, and some are wrong. The absence of precision analysis and numerical investigations of conditionally convergent series explains the main theoretical discrepancy.

DOI: 10.1201/9781032646633-2

In the present chapter, we determine the complex effective permittivity tensor for 2D (two-dimensional) composites

$$\boldsymbol{\varepsilon}_{\perp} = \begin{pmatrix} \varepsilon_{11} & \varepsilon_{12} \\ \varepsilon_{21} & \varepsilon_{22} \end{pmatrix}. \tag{2.1}$$

In the case of 3D fibrous composites, the effective permittivity tensor is decomposed onto the transversal part perpendicular to fibers $\boldsymbol{\varepsilon}_{\perp}$ and the longitudinal permittivity $\boldsymbol{\varepsilon}_{||}$ parallel to fibers.

One of the seemingly natural and straightforward ways to compute $\boldsymbol{\varepsilon}_{\perp}$ is based on the determination of the local fields around a finite collection of M inclusions in the space with the further suggestion that M tends to infinity. It is not equivalent to the limit $N \to \infty$ in the extending periodicity cell since the obtained series converges conditionally. Therefore, the order of its summation can lead to any result. This is the crucial reason for diverse models to determine $\boldsymbol{\varepsilon}_{\perp}$ and announce a new model (formula) obtained as a partial sum of absolutely divergent series. First, this problem of the conditionally convergent series was resolved by Rayleigh [206] in 1892. Rayleigh did not present any explanation and just indicated the proper method of summation (Eisenstein's approach proposed in 1848 [231]). Some historical notes can be found in [172].

We come back to the permittivity (2.1) in Section 2.3 after introduction to Schwarz's method. In 1869-1870, Hermann Schwarz proposed the alternating method for solving the Dirichlet problem for the union of two overlapping domains. S.G. Mikhlin [150] extended the classical alternating method to boundary value problems for a multiply connected domain with holes/inclusions. The method is based on the decomposition of the considered domain with complex geometry onto simply connected domains and subsequent estimations of the field in a simply connected domain induced by fields in other domains. Though the main idea by Schwarz presents in Mikhlin's approach, the case of non-overlapping inclusions essentially differs from the original method for overlapping domains and decomposition methods [90, 104, 218]. The generalized alternating method of Schwarz can be presented as an infinite sequence of all the mutual interactions between the inclusions in the considered composite. In this book, we use the term Schwarz's method for shortness but only mean non-overlapping inclusions considered by Mikhlin.

In 1949, Mikhlin [150] proved the convergence of Schwarz's method for a doubly connected domain. Schwarz's method was independently applied to two circular inclusions in many works without referencing the original method. For example, it is called the iterating analytic self-maps of disks in [48]. The authors of [107] called Schwarz's method and its implementation by "heterogenization technique." The complete closed-form[1] solution to the problem for two circular holes was derived in [213].

Schwarz's method was slightly modified in [184, 185] so that its convergence was established for any multiply connected domain for Laplace's and other equations. The

[1]Unfortunately, the terms "closed-form" and "exact" solutions have been incongruously applied in some works. So, we are forced to stress that [213] contains a closed-form solution in the proper sense [17].

sequential steps of Schwarz's method can be presented in symbolic form. This leads to analytical approximate formulas for the local fields and for the effective constants of composites. The interactions between circular (spherical) inclusions were written by Kelvin's reflections on circles, and the effective constants were obtained in terms of Eisenstein functions [67, 93, 185].

2.2 GENERAL SCHWARZ'S SCHEME

Integral equations in a Banach space corresponding to Schwarz's method were derived in [150] and modified in [184, 185]. Below, we present these equations in the general operator form on the potentials u_k in G_k, omitting some mathematical details described in the next sections. Let u_0 denote the given external potential. For simplicity, consider a two-phase dispersed composite with non-overlapping inclusions G_k of permittivity ε_1 embedded in the host of permittivity ε. In this Chapter, up to Section 2.4, we will delve into Schwarz's method, presenting its basic idea, taking for simplicity real constants ε_1 and ε. Introduce the dimensionless contrast parameter

$$\varrho = \frac{\varepsilon_1 - \varepsilon}{\varepsilon_1 + \varepsilon}, \tag{2.2}$$

which $|\varrho| \leq 1$ for ε_1 and ε changing from zero to $+\infty$.

The potential u_k is related to others u_m by the linear operator equation [180]

$$u_k = \varrho F_k u_k + \varrho \sum_{m \neq k} F_m u_m + u_0, \quad \text{in } G_k, \quad k = 1, 2, \ldots, N, \tag{2.3}$$

where $F_m u_m$ denotes the field in the domain G_k induced by the inclusion G_m. The term $F_k u_k$ produces the self-induced field. Equation (2.3) will be written explicitly in Section 2.5. The physical contrast parameter ϱ is a multiplier of the bounded operators F_m. The operators F_m are determined by the microstructure of the considered composite, which implicitly depends on the concentration of inclusions f. After undergoing a constructive homogenization procedure, the explicit dependence on the variable f can be represented.

Various explicit and implicit iterative schemes [90], including the splitting of the operators F_m, can be applied to equation (2.3). For instance, the self-induced field can be included in determining the field u_k. This scheme can be expressed in terms of the inverse operator

$$u_k = \varrho (I - \varrho F_k)^{-1} \left[\sum_{m \neq k} F_m u_m + u_0 \right] \quad \text{in } G_k, \quad k = 1, 2, \ldots, N, \tag{2.4}$$

where I stands for the identity operator. The construction of the inverse operator is reduced to a solution to the problem for one inclusion G_k. In the present chapter, we develop an asymptotic method in order to find the inverse operator in a constructive form.

The method of successive approximations for equation (2.3) leads to the contrast expansion

$$u_k = u_0 + \varrho \sum_{k_1} F_{k_1} u_0 + \varrho^2 \sum_{k_1,k_2} F_{k_1} F_{k_2} u_0 + \varrho^3 \sum_{k_1,k_2} F_{k_1} F_{k_2} F_{k_3} u_0 + \dots \quad \text{in } G_k \cup L_k, \tag{2.5}$$

where $k, k_s = 1,2,\dots,N$ for $s = 1,2,\dots$.

After determining the local field, the effective constants can be found following the homogenization theory [22]. In the case of dispersed composites, the effective constants can be calculated by the averaged local field in inclusions [93, Chapter 3]. The local field can be found using various numerical and analytical methods, for instance, by FEM [66, 126], by integral equations [98, 150], by asymptotic and perturbation methods [19, 21, 64]. The effective properties can be estimated by the Hashin-Shtrikman bounds and their generalizations [51, 155]. Contrast and cluster expansions [67, 93, 155, 226] are applied to derive the macroscopic properties of composites.

We concentrate our attention on the asymptotic analysis of Schwarz's method for dispersed composites. This study yields a constructive method to derive analytical formulas for the effective constants with the exactly calculated precision orders [169, 180]. The 2D permittivity problem is equivalent to conductivity and an anti-plane elastic problem. The elasticity problem requires two contrast parameters [173] that complicate computations. The same scheme outlined by equations (2.3)–(2.4) can be applied to 3D composites, including thermoelastic, piezoelectric, and other physical phenomena [16, 67, 93, 211].

Schwarz's method can be outlined as follows. First, we order the steps of the method by the sequence $\{\{0\},\{k_1\},\{k_1,k_2\},\dots,\{k_1,k_2,\dots,k_s\},\dots\}$, where k_s runs over the numbers of inclusions $1,2,\dots,N$ and $s = 0,1,\dots$. The number s is called the level of step. The zero-level potential $U_k^{(0)}$ in G_k is equal to the external potential $u_0 = x_1$ for which $\mathbf{E}_0 = (-1,0)$ is calculated as the gradient $\mathbf{E}_0 = -\nabla u_0$. Let us fix an inclusion G_{k_1} and introduce the potential U_{k_1} induced by $U_m^{(0)}$ $m = 1,2\dots,N$. The induced terms $F_m U_m^{(0)}$ correspond to the first iteration in equations (2.3). This first step has the level $s = 1$ and includes N elements U_{k_1} $(k_1 = 1,2\dots,N)$. The second step of level $s = 2$ includes N^2 elements U_{k_1,k_2} $(k_1,k_2 = 1,2\dots,N)$. The element U_{k_1,k_2} in the inclusion G_{k_1} goes back to the impact of the inclusion G_{k_2} onto G_{k_1}, i.e., first, the impact of the zero potential $U_{k_1}^{(0)}$ in G_{k_1} to the potential U_{k_2} in G_{k_2} and next the impact of U_{k_2} to the field in G_{k_1}. It is worth noting that the elements U_{k_1,k_2} and U_{k_2,k_1} do not coincide since they are defined in different inclusions. For example, the genesis of $U_{1,3,2,3,4}$ is demonstrated in Figure 2.2 for a square periodicity cell displayed in Figure 2.1.

The potential in G_k is represented in the form of series equivalent to (2.5) for $|\varrho| < 1$ up to an additive constant c_k

$$u_k(\mathbf{x}) = u_0(\mathbf{x}) + \sum_{s=1}^{\infty} \varrho^s \sum_{k_1,k_2,\dots,k_s} [U_{k_1,k_2,\dots,k_s}(\mathbf{x}) - U_{k_1,k_2,\dots,k_s}(\mathbf{w})] + c_k, \quad \mathbf{x} \in G_k \cup L_k \quad (k = 1,2,\dots,N), \tag{2.6}$$

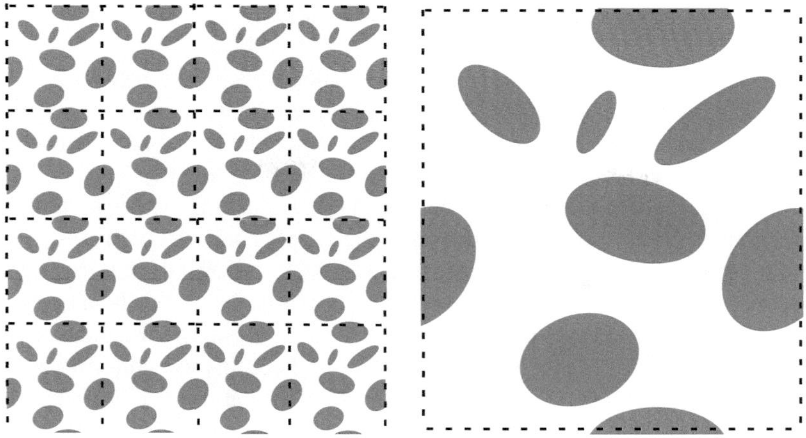

Figure 2.1 Doubly periodic composite and its representative square cell Q (RVE).

where $\mathbf{w}$ is a fixed point outside of $\cup_k^N (G_k \cup L_k)$. The additional constant term $U_{k_1,k_2,\ldots,k_s}(\mathbf{w})$ is introduced in order to obtain the uniformly convergent series for a high dielectric constant of inclusions ($\varrho = 1$) and for insulating inclusions ($\varrho = -1$). The order of summation by levels is important for $|\varrho| = 1$ because the series (2.6) may be conditionally convergent [93, 185].

In the case of circular inclusions, the above scheme coincides with the image method based on the Kelvin transform. The elements of series (2.6) are constructed by the classic Schottky group of inversions on circles L_k [167]. The series (2.6) becomes the θ_2-Poincaré series which determines an authomorphic function [159].

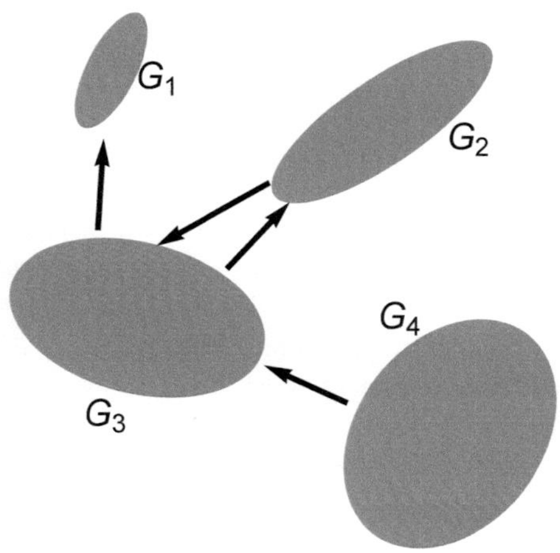

Figure 2.2 The element $U_{1,3,2,3,4}$ in G_1 emerges due to the sequence of impacts of the zero field in G_4 to G_3, the resulting field in G_3 to G_2, the resulting field in G_2 to G_3 and finally the resulting field in G_3 to G_1.

Computationally effective methods for the Poincaré series were developed in [178] and other works.

The constructive applications of contrast expansion (2.5) in the theory of composites are usually accompanied by a comment on sufficiently small $|\varrho|$. Such an assumption for dispersed composites is redundant since the series (2.5) converges absolutely in the unit disk $|\varrho| < 1$ of the complex plane [93, 180, 184].

Schwarz's method is reduced to a system of integral equations whose kernels are Green's functions of separate inclusions occupying simply connected domains [180]. The similar Lippmann-Schwinger type equation [187] is stated over the total boundary of inclusions. Both methods ultimately lead to the same series (2.5). The methodology outlined in [187] is based on absolute convergence, i.e., on the Neumann series estimated by the norm of the integral operator. It turns out that the uniform convergence of Schwarz's method gives more general results than absolute convergence. It requires subtle construction of the conditionally convergent series (2.6) dependent on the order of summation.

We now define the class of *dispersed composite* used in the present book. Simple illustrations are insufficient since the main parameters used in asymptotic formulas must be introduced precisely. First, every domain G_k or its convex hull is assumed to be a bounded convex set. Fix a domain G_k and describe its properties. Its boundary, closed curve L_k, can be locally parametrized by a $C^{(1,\alpha)}$ function except at a finite set of points where one-sided derivatives along L_k exist but can be different. Let a_k denote the gravitational center of the domain G_k. It will be convenient to consider a_k as a complex number

$$a_k = \frac{1}{|G_k|} \int_{G_k} z \, \mathrm{d}x_1 \mathrm{d}x_2 \quad (z = x_1 + \mathrm{i}x_2). \tag{2.7}$$

Introduce the generalized radius of G_k as $r_k = \sup_{z \in L_k} |z - a_k|$. Hereafter, the distance $|z - w|$ is considered in the plane torus topology. It is assumed that the closed domains $(G_k \cup L_k)$ for $k = 1, 2, \ldots, N$ are mutually disjoint and

$$r_k + r_m < |a_k - a_m|, \quad \text{for } k \neq m \quad (k, m = 1, 2, \ldots, N). \tag{2.8}$$

We call such a set $\{G_1, G_2, \ldots, G_N\}$ by inclusions in a dispersed composite. The exterior domain G to the inclusions is called the host or matrix of the dispersed composite.

Introduce the maximum generalized radius $r = \max_{m=1,2,\ldots,N} r_m$. It will be convenient to estimate the effective constants using the concentration of inclusions

$$f = \sum_{k=1}^{N} |G_k|, \tag{2.9}$$

where $|G_k|$ stands for the area of the domain G_k in the periodicity cell of the normalized unit area $|Q| = 1$. It follows from inequality $|G_k| \leq \pi r_k^2$ that

$$|G_k| = O(r^2), \quad \text{as } r \to 0 \quad \Leftrightarrow \quad N|G_k| = O(f), \quad \text{as } f \to 0. \tag{2.10}$$

Here, N is considered as a constant; the value $|G_k|$ is of order $O\left(\pi(\frac{r_k}{r})^2 r^2\right)$. Therefore, only two geometrical equivalent infinitesimal parameters, r^2 and f, can be considered in asymptotic formulas.

An important consequence of Schwarz's method is the decomposition theorem, which can be loosely formulated as follows

Theorem 1 (Decomposition theorem) *The effective properties tensor* $\boldsymbol{\varepsilon}_\perp$ *can be represented as a linear combination of the pure geometrical parameters of inclusions with the coefficients depending on the local physical constants.*

Consider for simplicity the formulas (2.6) for the local field in the inclusion G_k. The tensor $\boldsymbol{\varepsilon}_\perp$ can be calculated by averaging the local field in inclusions per a representative volume element

$$\boldsymbol{\varepsilon}_\perp = \sum_{s=1}^{\infty} \varrho^s \sum_{k,k_1,k_2,\dots,k_s} \mathbf{c}_{k,k_1,k_2,\dots,k_s}, \tag{2.11}$$

where the tensors $\mathbf{c}_{k,k_1,k_2,\dots,k_s}$ describe the mutual interactions between inclusions and depend only on their concentration, location, and shapes; ϱ^s are the coefficients expressed only on the physical constants. Equation (2.11) for multi-phase composites becomes

$$\boldsymbol{\varepsilon}_\perp = \sum_{s=1}^{\infty} \sum_{k,k_1,k_2,\dots,k_s} \varrho_k \varrho_{k_1} \cdots \varrho_{k_s} \mathbf{c}_{k,k_1,k_2,\dots,k_s}, \tag{2.12}$$

where ϱ_k stands for the contrast parameter of the materials occupied G_k and G.

Explicit formulas for the tensor $\boldsymbol{\varepsilon}_\perp$ can be found in Chapter 4. The decomposition (2.12) emerges due to the contrast expansion of $\boldsymbol{\varepsilon}_\perp$. Its first constructive form was derived in [171, Section 4.2]. Theorem 1 is the theoretical base of the *a*RVE theory outlined in Section 4.1.7.

2.3 MATHEMATICAL MODELING AND DETERMINATION OF EFFECTIVE PROPERTIES

The most popular analytical formulas for the effective permittivity of dispersed composites are the Maxwell-Garnett [140] and Brugemann [45] approximations, see equations (2.13) on page 27 and (3.31) on page 70. We now postpone formulas associated with the famous neutral Hashin-Shtrikman assemblage [50, 155] for coated composites and discuss analytical formulas for two-phase composites based on the self-consistent method (SCM). A typical modern review of various analytical formulas for the effective properties of composites, including elastic media, begins with a discussion of these formulas and their numerous extensions. Lots of items are laid out on the counter and offered for usage. Some contradict others; some violate the Hashin-Shtrikman bounds; some disturb the symmetry of tensors. However, everything becomes allowed if the magic word "model" is used. There is no theoretical justification when creating a new model, giving way to empirical reasons. Experiments might confirm the model.

We will not discuss a rich assortment of models and narrow our focus to the specific two-phase 2D macroscopically isotropic composites in order to derive precision of various modifications of SCM such as effective medium approximation, mean-field, Mori-Tanaka methods, etc. It will be shown that some popular modifications of SCM are covered by the first-order approximation of Schwarz's method. This investigation explains plenty of illusory different formulas which are reduced to the same Maxwell-Garnett type estimation for dilute composites. It will be demonstrated that some higher-order extensions of SCM violate homogenization principles and lead to methodologically misleading results.

SCM has multiple meanings in different fields of science and engineering. It is related to the many-body problem in physics [227], assembly in statistical mechanics [60], molecular dynamics simulations [11, 141], biological fields and others [87, 99, 232]. We pay particular attention to self-consistent method in the theory of composites by Maxwell [139, p.365] developed by Hill [106], Beran [30], Kröner [123] and many others, and its relations to Schwarz's method outlined in the next sections of this chapter.

J.C. Maxwell proposed SCM in 1873 [139]. S.G. Mikhlin developed the alternating method of Schwarz in 1949 [150]. Mikhlin's result can be considered a mathematical embodiment of Maxwell's approach. It is summarized in Section 2.6.6 in the framework of Schwarz's method. We are interested in the Maxwell-Garnett approximation derived in 1904 [140] and suitable for the complex dielectric permittivity. It is worth noting that the same formula for real conductivity (permittivity) was known as the Clausius-Mossotti approximation [186] published in 1850. It coincides with Maxwell's approximation [139]. This formula arose in the works by Lorenz (1869) and Lorentz (1879) in another context[2]. Perhaps the universality of mathematical modeling was not clear in the XIX century. The historical notes [128, 155] and references therein clarify the series of the above formulas.

The main result of the present chapter consists of applying Schwarz's method to composites and deriving new higher-order approximations for the tensor $\boldsymbol{\varepsilon}_{\perp}$. In particular, we demonstrate that Maxwell's SCM and its modern modifications coincide with the first-order approximation of Schwarz's method.

An extensive theoretical review of SCMs for a dispersed composite can be found in [198, 199] where the term "grain composite" was used. After these works, many engineers implemented SCMs and obtained numerical estimations of the effective constants for various composites. However, the principal theoretical question of the limitations of SCMs was still open. The argumentation of SCM validity was based on the condition of dilute concentration in [106, 123, 139, 198, 199]. It was noted that applying SCM to non-dilute composites, i.e., to the second-order terms in f, may yield physically unacceptable predictions [56, 72].

The obtained result for a self-consistent approach in a 2D permittivity statement has to be expected in general consideration, in particular for 2D-3D conductive and

[2]James Clerk Maxwell and J.C. Maxwell Garnett are different persons, L. Lorenz and H.A. Lorentz, too.

elastic composites, the macroscopic viscosity of suspensions, and so forth [67, 93], due to the universality of mathematical modeling. Hence, the precision analysis for the 2D permittivity of the present book can be extended to general composites and porous media. Analytical formulas for the effective properties were derived in [67, 93, 180] in the form of power series in the concentration of inclusions f and contrast parameter ϱ. The series was truncated to polynomials in $f^{1/2}$ and ϱ with the exactly found coefficients symbolically depending on the location and shape of inclusions. Moreover, the order precision $O(|\varrho|^m f^{p/2})$ of the derived formulas was justified. Here, m denotes the number of iterations in Schwarz's method, and p is related to the concentration approximation used at every iteration.

The literature presents numerous formulas for effective properties, often viewed as perturbations or extensions of lower-order formulas in f and ϱ to higher-order terms, albeit frequently without justification. Some of these formulas, although formally distinct, describe composites under similar deterministic or probabilistic frameworks. For instance, an apparent contradiction between "two different models" concerning the effective permittivity of elliptical inclusions was clarified in [181]. Some researchers overlook inconsistencies in analytical formulas and instead rely on numerical comparisons of specific datasets. We revisit this numerical comparison in Section 3.2.3, highlighting that such comparisons often boil down to assessing whether a linear dependence can approximate a parabolic one in f.

This raises a fundamental question: why do some formulas applied to the same composite differ? Typically, the issue stems from confusing terminology, where the term "model" is incorrectly used in place of "formula." It is crucial not to conflate a model with the resulting formula in exact sciences. If a mathematically defined problem yields two different answers, it does not imply the consideration of two distinct models. The correct research approach defines the model as a priori and conducts the mathematical study as a posteriori, not the reverse. This principle equally applies to probabilistic problems involving random composites, where the distribution of inclusions must be explicitly defined within the problem statement. Subsequently, the mathematical expectation of the effective constants, uniquely determined for a well-described class of random composites, must be calculated, as demonstrated in [195].

In numerous studies, algorithms simulate random composites incorporating random variables whose probability distributions are often obscured within machine codes. Furthermore, simulation outcomes can vary depending on the employed protocols [122, 226]. Such differences may lead to varying results for the effective constants, which correspond naturally to different spatial distributions of inclusions.

To the end of the above discussion, we have the exact coefficients in our formulas [67, 93] in ϱ and f. Hence, we can compare formulas from the previous publications with the asymptotic estimations in order to make conclusions concerning the validity of the formulas called "models." Such a comparison of the Maxwell-Garnett formula (Clausius-Mossotti approximation) for equal disks yields the asymptotic formula

$$\varepsilon_e = \frac{1+\varrho f}{1-\varrho f} + O(f^3), \tag{2.13}$$

where the tensor $\boldsymbol{\varepsilon}_\perp$ takes the form $\varepsilon_e I$, the scalar ε_e denotes the macroscopic permittivity, I the identity matrix. The exact precision up to $O(f^3)$ was established in [179].

The Hashin-Shtrikman bounds for 2D two-phase macroscopically isotropic composites for real normalized permittivity of components ε_1 and $\varepsilon = 1$ when $\varepsilon_1 > 1$ can be written in the form [103]

$$\varepsilon_{HS}^- \leq \varepsilon_e \leq \varepsilon_{HS}^+, \tag{2.14}$$

where

$$\varepsilon_{HS}^- = 1 + \frac{2f(\varepsilon_1 - 1)}{(1-f)(\varepsilon_1 - 1) + 2}, \quad \varepsilon_{HS}^+ = \varepsilon_1 + \frac{2(1-f)(1-\varepsilon_1)\varepsilon_1}{f(1-\varepsilon_1) + 2\varepsilon_1}. \tag{2.15}$$

The bounds (2.15) in terms of ϱ and f becomes

$$\varepsilon_{HS}^- = \frac{1+\varrho f}{1-\varrho f}, \quad \varepsilon_{HS}^+ = \frac{1+\varrho}{1-\varrho}\,\frac{1-\varrho(1-f)}{1+\varrho(1-f)}. \tag{2.16}$$

It is worth noting that ε_{HS}^- coincides with the Maxwell-Garnett approximation (2.13).

In this Chapter, we show that Schwarz's method can be implemented in the form of different iterative schemes known in the theory of composites as the contrast and cluster expansions [93,155,226]. The contrast expansion for 2D permittivity problems leads to a power series in ϱ. The cluster expansion means a series in $f^{1/2}$. Frequently, the authors of SCMs do not care about the precision in ϱ and $f^{1/2}$, and derive, in the best case, the same formula asymptotically equivalent to the Maxwell-Garnett formula. The analysis of precision [67,93,168,173,179] demonstrates a discrepancy between the rigorously derived formulas for the effective permittivity of disks/spheres and some formulas obtained by SCM.

It was proved in [168, 173] that an extension of Maxwell's approach to a finite collection of inclusions (cluster) may give, at most, the effective properties of dilute clusters. A simple feature of an expansion of the effective permittivity in f was formulated in [179]. Any general formula for 2D two-phase macroscopically anisotropic dispersed composites which does not include the location of disks holds up to $O(f^2)$; for macroscopically isotropic composites up to $O(f^3)$. This criterion of lower-order formula could be extended to multi-phase composites except at the Hashin-Shtrikman type assemblages [50, 53] with neutral inclusions. For instance, a set of valuable formulas for multi-phase macroscopically anisotropic composites [143] is valid up to $O(f^2)$.

2.4 TRANSMISSION CONDITIONS AND $\mathbb{R}$-LINEAR PROBLEM FOR 2D COMPOSITES

We now proceed to implement Schwarz's method for boundary value problems of dispersed 2D periodic composites described in Section 1.4. Consider a multi-phase composite with non-overlapping inclusions embedded in the 2D matrix. The composite is supposed to be represented by a periodicity cell Q introduced as follows. The plane coordinates $\mathbf{x} = (x_1, x_2)$ will be identified with the complex variable $z = x_1 + \mathrm{i}x_2$.

Let ω_1 and ω_2 be the fundamental periods on the complex plane $\mathbb{C}$. Let Re and Im stand for the real and imaginary parts of values. Without loss of generality, the complex numbers ω_1 and ω_2 can be chosen in such a way that $\omega_1 > 0$, Im $\omega_2 > 0$ and

$$\omega_1 \text{Im}\, \omega_2 = 1. \tag{2.17}$$

Equation (2.17) means that the area of Q is normalized to unity. Let the fundamental parallelogram Q be defined by the vertices $\pm\frac{\omega_1}{2}$ and $\pm\frac{\omega_2}{2}$

$$Q \equiv Q_{(0,0)} = \left\{ z = t_1\omega_1 + t_2\omega_2 \in \mathbb{C} : -\frac{1}{2} < t_1, t_2 < \frac{1}{2} \right\}. \tag{2.18}$$

The points $m_1\omega_1 + m_2\omega_2$ $(m_1, m_2 \in \mathbb{Z})$ generate a doubly periodic lattice Q where $\mathbb{Z}$ stands for the set of integer numbers. The lattice Q generates the periodic cells $Q_{(m_1,m_2)} = Q_{(0,0)} + m_1\omega_1 + m_2\omega_2$. For instance, the square array is generated by two translation vectors expressed by complex numbers

$$\omega_1 = 1, \quad \omega_2 = \text{i}. \tag{2.19}$$

Consider N non-overlapping simply connected domains G_k in the unit periodicity cell Q with piece-wise Lyapunov's boundaries L_k and the multiply connected domain $G = Q \setminus \cup_{k=1}^{N} (G_k \cup L_k)$, the complement of all the closures of G_k to Q (see Figure 2.1). Each simple closed curve L_k leaves G_k to the left. Every curve L_k is smooth except at a finite set of points called vertices. Following [193], we say that a function belongs to Muskhelishvili's class H if it is Hölder continuous on all smooth closed arcs of L_k.

Consider a doubly periodic multi-phase composite when the host $G + m_1\omega_1 + m_2\omega_2$ and the inclusions $G_k + m_1\omega_1 + m_2\omega_2$ are occupied by dielectric materials. Let the permittivity of the host is normalized to unity, and the permittivity of kth inclusion be a complex number $\varepsilon_k = \varepsilon'_k + \text{i}\varepsilon''_k$, where ε'_k = Re ε_k and ε''_k = Im ε_k. One can consider ε_k as the ratio of the permittivity of the kth inclusion to the permittivity of the matrix, where the dimension permittivities can be complex. One can assume that the constants ε_k take the values from a set $\mathcal{M}$ which contains less than N elements. Let $\#\mathcal{M} = M$, i.e., the composite is $(M+1)-$phases and $\varepsilon_k = \varepsilon^{(j)}$, if $j = 1, 2, \ldots, M$. In this case, formulas similar to the formulas (4.2.26)–(4.2.27) from [93] can be derived for a multi-phase composite by the method developed in [162]. These formulas can be considered as extensions of formulas [217] to higher-order contrast parameters terms.

The external flux $\mathbf{E}_0$ can be defined by the given constant jump of the potential per periodicity cell along, for instance, the horizontal x_1-axis. Then $\mathbf{E}_0$ can be taken as the vector $(-1, 0)$, and the jump of potential per cell in the x_1-direction holds 1 and in the x_2-direction 0.

Let $u = u' + iu''$ and $u_k = u'_k + iu''_k$ denote the potentials in G and G_k, respectively, where for instance u' = Re u and u'' = Im u in G. The complex functions u and u_k satisfy Laplace's equation in the corresponding domains and are continuously differentiable in their closures except at the vertices $W_k \subset L_k$ where they belong to Muskhelishvili's class H [193].

The perfect contact between the components is expressed by equations (1.52)

$$u(t) = u_k(t), \quad \frac{\partial u}{\partial \mathbf{n}}(t) = \varepsilon_k \frac{\partial u_k}{\partial \mathbf{n}}(t), \quad t \in L_k \quad (k = 1,2,\ldots,N), \tag{2.20}$$

where the normal derivative $\frac{\partial}{\partial \mathbf{n}}$ on L_k is used. Two complex relations (2.20) can be written in the extended real form

$$\begin{aligned} u'(t) = u_k'(t), \quad & \frac{\partial u'}{\partial \mathbf{n}}(t) = \varepsilon_k' \frac{\partial u_k'}{\partial \mathbf{n}}(t) - \varepsilon_k'' \frac{\partial u_k''}{\partial \mathbf{n}}(t), \\ u''(t) = u_k''(t), \quad & \frac{\partial u''}{\partial \mathbf{n}}(t) = \varepsilon_k'' \frac{\partial u_k'}{\partial \mathbf{n}}(t) + \varepsilon_k' \frac{\partial u_k''}{\partial \mathbf{n}}(t), \\ & t \in L_k \quad (k = 1,2,\ldots,N). \end{aligned} \tag{2.21}$$

Following the homogenization theory [22], we must consider a composite in the plane torus topology. Hence, such a structure can be considered as a doubly periodic representative unit cell Q; see the MMM principle by Hashin [102] and its constructive application to random composites in Chapter 3 of [93].

The external field is modeled by two conditions

$$u(z+\boldsymbol{\omega}_1) - u(z) = \xi_1, \quad u(z+\boldsymbol{\omega}_2) - u(z) = \xi_2, \tag{2.22}$$

where ξ_1 and ξ_2 are two complex vectors that determine the jump of the external field in the directions $\boldsymbol{\omega}_1$ and $\boldsymbol{\omega}_2$, respectively. It follows from the theory of homogenization [22] that the external potential $u_{ext}(z)$ has to be taken as a linear function in x_1 and x_2. A linear function in x_1 and x_2 can be represented as an $\mathbb{R}$-linear function in the complex variable $z = x_1 + \mathrm{i}x_2$

$$u_{ext}(z) = \mathbf{a}z + \mathbf{b}\bar{z}, \tag{2.23}$$

where $\mathbf{a}$ and $\mathbf{b}$ are constant vectors.

The theory of homogenization implies that the problem (2.21)–(2.22) has to be solved for two linearly independent vectors ξ_1 and ξ_2 in order to completely determine the effective permittivity tensor. One can consider the normalized external fields parallel to the coordinate axes determined by the external potential (2.23) with $\mathbf{a} = \mathbf{b} = (\frac{1}{2},0)^\top$ and $\mathbf{a} = -\mathbf{b} = (\frac{1}{2\mathrm{i}},0)^\top$, respectively, where $^\top$ denotes the transposition. These fields yield two pairs

$$\xi_1 = (\boldsymbol{\omega}_1,0)^\top,\ \xi_2 = (\mathrm{Re}\,\boldsymbol{\omega}_2,0)^\top \quad \text{and} \quad \xi_1 = (0,0)^\top,\ \xi_2 = (0,\mathrm{Im}\,\boldsymbol{\omega}_2)^\top. \tag{2.24}$$

The corresponding calculations can be found in [93, Section 2.2] for the real fields.

In the present chapter and most others, we consider the square array with (2.19) for definiteness. In this case, the function $u(t)$ satisfies the normalized jump conditions per unit periodic square cell Q

$$u(z+1) - u(z) = (1,0)^\top, \quad u(z+i) - u(z) = (0,0)^\top. \tag{2.25}$$

The conditions (2.25) mean that the external complex flux $\mathbf{E}_0 = (-1,0)$ is applied. Here, the components of $\mathbf{E}_0$ are complex numbers. More precisely, for instance, the real part of first relation (2.25) are formally written in the same form

$$\operatorname{Re} u(z+1) - \operatorname{Re} u(z) = (1,0)^\top, \quad \operatorname{Im} u(z+1) - \operatorname{Im} u(z) = (0,0)^\top. \tag{2.26}$$

Analogous relations occur for the second problem when $\mathbf{E}_0 = (0,-1)$.

We now reduce the problem (2.21) to a vector-matrix $\mathbb{R}$-linear problem. The second column of (2.21) suggests to introduce the non-degenerate real matrix

$$\boldsymbol{\alpha}_k = \begin{pmatrix} \varepsilon_k' & -\varepsilon_k'' \\ \varepsilon_k'' & \varepsilon_k' \end{pmatrix}. \tag{2.27}$$

A similar matrix was introduced in [52] to develop variational principles for the complex effective tensor.

Introduce the vector complex potentials

$$\boldsymbol{\varphi}(z) = \begin{pmatrix} u'(z) + iv'(z) \\ u''(z) + iv''(z) \end{pmatrix}, \quad z \in G \tag{2.28}$$

and

$$\boldsymbol{\varphi}_k(z) = \frac{1}{2}(I + \boldsymbol{\alpha}_k) \begin{pmatrix} u_k'(z) + iv_k'(z) \\ u_k''(z) + iv_k''(z) \end{pmatrix}, \quad z \in G_k. \tag{2.29}$$

where $v'(z)$, $v''(z)$ and $v_k'(z)$, $v_k''(z)$ denote the imaginary parts of the components of the analytic (meromorphic) vector-functions $\boldsymbol{\varphi}(z)$ and $\boldsymbol{\varphi}_k(z)$, respectively, I denotes the identity matrix.

The harmonic and analytic functions are related by the vector relations The complex vector flux is introduced as the derivatives $\boldsymbol{\psi}(z) = \frac{d}{dz}\boldsymbol{\varphi}(z)$ and $\boldsymbol{\psi}_k(z) = \frac{d}{dz}\boldsymbol{\varphi}_k(z)$. It can be expressed through the partial derivatives of u', u'' and u_k', u_k'' calculated by (2.28) and (2.29), respectively,

$$\boldsymbol{\psi}(z) = \begin{pmatrix} \frac{\partial u'}{\partial x_1} - \mathrm{i}\frac{\partial u'}{\partial x_2} \\ \\ \frac{\partial u''}{\partial x_1} - \mathrm{i}\frac{\partial u''}{\partial x_2} \end{pmatrix}, \quad z \in G, \tag{2.30}$$

and

$$\boldsymbol{\psi}_k(z) = \frac{1}{2}(I + \boldsymbol{\alpha}_k) \begin{pmatrix} \frac{\partial u_k'}{\partial x_1} - \mathrm{i}\frac{\partial u_k'}{\partial x_2} \\ \\ \frac{\partial u_k''}{\partial x_1} - \mathrm{i}\frac{\partial u_k''}{\partial x_2} \end{pmatrix}, \quad z \in G_k. \tag{2.31}$$

Using the formula

$$2(I + \boldsymbol{\alpha}_k)^{-1} = \frac{2}{|1+\varepsilon_k|^2}\left(I + \boldsymbol{\alpha}_k^T\right) = \frac{2}{|1+\varepsilon_k|^2} \begin{pmatrix} 1+\varepsilon_k' & \varepsilon_k'' \\ -\varepsilon_k'' & 1+\varepsilon_k' \end{pmatrix}, \tag{2.32}$$

introduce the contrast matrix parameter corresponding to the particular case (2.2)

$$\boldsymbol{\beta}_k = -(I - \boldsymbol{\alpha}_k)(I + \boldsymbol{\alpha}_k)^{-1} = \frac{1}{|1+\varepsilon_k|^2} \begin{pmatrix} |\varepsilon_k|^2 - 1 & -2\varepsilon_k'' \\ 2\varepsilon_k'' & |\varepsilon_k|^2 - 1 \end{pmatrix}. \tag{2.33}$$

Then, the conditions (2.21) can be written in the form of vector-matrix $\mathbb{R}$-linear problem

$$\boldsymbol{\varphi}(t) = \boldsymbol{\varphi}_k(t) - \boldsymbol{\beta}_k \overline{\boldsymbol{\varphi}_k(t)}, \quad t \in L_k \quad (k = 1, 2, \ldots, N). \tag{2.34}$$

It follows from (2.25) that the vector-function $\boldsymbol{\varphi}(z)$ satisfies the quasi-periodicity conditions

$$\boldsymbol{\varphi}(z+1) = \boldsymbol{\varphi}(z) + \begin{pmatrix} 1 + \mathrm{i}d_{11} \\ \mathrm{i}d_{21} \end{pmatrix}, \ \boldsymbol{\varphi}(z+i) = \boldsymbol{\varphi}(z) + \begin{pmatrix} \mathrm{i}d_{12} \\ \mathrm{i}d_{22} \end{pmatrix}, \tag{2.35}$$

where d_{ij} are undetermined real constants that could be found during the solution to the boundary value problem (2.34).

The eigenvalues of the matrix $\boldsymbol{\beta}_k$ are

$$\varrho_k = \frac{\varepsilon_k - 1}{\varepsilon_k + 1}, \quad \overline{\varrho_k} = \frac{\overline{\varepsilon_k} - 1}{\overline{\varepsilon_k} + 1}. \tag{2.36}$$

The matrix $\boldsymbol{\beta}_k$ admits the decomposition

$$\boldsymbol{\beta}_k = \mathbf{T} \Lambda_k \mathbf{T}^{-1}, \tag{2.37}$$

where

$$\mathbf{T} = \begin{pmatrix} \mathrm{i} & -\mathrm{i} \\ 1 & 1 \end{pmatrix}, \quad \mathbf{T}^{-1} = \frac{1}{2} \begin{pmatrix} -\mathrm{i} & 1 \\ \mathrm{i} & 1 \end{pmatrix}, \quad \Lambda_k = \begin{pmatrix} \varrho_k & 0 \\ 0 & \overline{\varrho_k} \end{pmatrix}. \tag{2.38}$$

Equation (2.37) implies that

$$\boldsymbol{\beta}_k \boldsymbol{\beta}_m = \mathbf{T} \begin{pmatrix} \varrho_k \varrho_m & 0 \\ 0 & \overline{\varrho_k}\,\overline{\varrho_m} \end{pmatrix} \mathbf{T}^{-1}. \tag{2.39}$$

In the case of real ε_k, the matrix (2.33) becomes diagonal $\boldsymbol{\beta}_k = \varrho_k I$, and the vector-matrix $\mathbb{R}$-linear problem (2.34) can be decomposed onto two the same scalar $\mathbb{R}$-linear problems systematically studied in [93]. The contrast expansion method equivalent to Schwarz's method can be applied to the vector-matrix problem to determine the local fields. It is suggested that the convergence will hold for $|\varrho_k| \leq 1$.

In the subsequent sections of this chapter, we assume that permittivity takes real values. Consequently, the vector complex potentials mentioned earlier share two identical coordinates. This enables us to reduce vector-matrix operations to scalar transformations, as elaborated in [93]. We will proceed with our analysis under the assumption of real permittivity, deferring the exploration of the complex case to Chapter 2.7.

2.5 SCHWARZ'S METHOD FOR DISPERSED PERIODIC COMPOSITES WITH REAL-VALUED PERMITTIVITY

It is noted at the end of the previous section that Schwarz's method can be applied to the vector-matrix $\mathbb{R}$-linear problem (2.34). As a result, the local fields can be found by power series in the matrices $\boldsymbol{\beta}_k$, perhaps truncated, but in symbolic form. Such an expression can be helpful in the computation of the effective permittivity tensor $\boldsymbol{\varepsilon}_\perp$. However, as we progress through the subsequent Chapter 2.7, we will discover that solving the scalar $\mathbb{R}$-linear problem is adequate after the application of the general theorem of analytic continuation to the complex domain of ε_k in formulas for the tensor $\boldsymbol{\varepsilon}_\perp$ with real ε_k. Therefore, we can adopt the following strategy. First, we obtain the symbolic form of $\boldsymbol{\varepsilon}_\perp$ by considering real ε_k and solving the scalar $\mathbb{R}$-linear problem discussed in this chapter. This step alone yields the desired result. Next, we extend the obtained analytical exact and approximate formulas by formally considering complex values of ε_k. This approach simplifies matrix transformations and reduces them to more manageable scalar manipulations.

In addition, we will offer a lucid illustration of the methodological limitations of EMA in the scalar case. This demonstration will be comprehensively presented towards the chapter's conclusion, following our rigorous investigation of Schwarz's method in Sections 2.5–2.6.5.

So, hereafter, non-overlapping inclusions G_k of real permittivity ε_k ($k = 1,2,\dots,N$) are embedded in a periodic square cell Q of unit area. The following equations establish a relationship between the scalar complex and real potentials.

$$\operatorname{Re}\varphi(z) = u(z),\ z \in G,\ \operatorname{Re}\varphi_k(z) = \frac{\varepsilon_k + 1}{2} u_k(z),\ z \in G_k\ (k = 1,2,\dots,N). \tag{2.40}$$

The complex potentials $\varphi(z)$ and $\varphi_k(z)$ are analytic in G and G_k, respectively, and continuously differentiable in the closures of the considered domains except at the vertices of L_k where the limit values of derivatives $\varphi'(z)$ and $\varphi_k'(z)$ belong to Muskhelishvili's class H. The function $\varphi(z)$ satisfies the quasi-periodicity conditions

$$\varphi(z+1) = \varphi(z) + 1,\ \varphi(z+i) = \varphi(z) + id, \tag{2.41}$$

d is an undetermined real constant that should be found while solving the problem [93, Chapter 3, Section 4.3]. The perfect contact between the components (transmission condition) is written as the scalar $\mathbb{R}$-linear problem [40, 185]

$$\varphi(t) = \varphi_k(t) - \varrho_k \overline{\varphi_k(t)}, \quad t \in L_k \quad (k = 1,2,\dots,N). \tag{2.42}$$

Introduce a space $\mathcal{H}(G^+)$ consisting of functions analytic in $G^+ = \cup_{k=1}^n G_k$ and Hölder continuous in the closure of G^+ endowed the norm [40]

$$||\omega|| = \sup_{t\in L} |\omega(t)| + \sup_{t_{1,2}\in L} \frac{|\omega(t_1)| - \omega(t_2)|}{|t_1 - t_2|^\alpha}, \tag{2.43}$$

where $0 < \alpha \leq 1$ and $L = \cup_{k=1}^N L_k$. The space $\mathcal{H}(G_k)$ of functions analytic in a fixed G_k and Hölder continuous in $G_k \cup L_k$ can also be considered. The problem (2.41)–(2.42) is considered in the space $\mathcal{H}(G^+)$ [185].

Let $\widehat{\mathbb{C}} = \mathbb{C} \cup \{\infty\}$ denote the extended complex plane. Let $h(t)$ be a function Hölder continuous on L. Consider the Cauchy-type integral on the complex plane [193]

$$\Phi(z) = \frac{1}{2\pi i} \int_L \frac{h(t)}{t - z} \mathrm{d}t, \quad z \in \widehat{\mathbb{C}} \backslash L. \tag{2.44}$$

Its generalization to a class of doubly periodic functions (on the plane torus) has the form [98]

$$\Phi(z) = \frac{1}{2\pi i}\int_L h(t)E_1(t-z)\,\mathrm{d}t,\ z \in G^+ \cup D. \tag{2.45}$$

It is defined by the Eisenstein function $E_1(z)$ expressed through the Weierstrass function $\zeta(z)$ described in Appendix A. The function (2.45) is double periodic up to constant jumps per the periodicity cell

$$\Phi(z+\boldsymbol{\omega_j}) - \Phi(z) = -\frac{\delta_j}{\pi i}\int_L h(t)\,\mathrm{d}t \quad (j = 1,2), \tag{2.46}$$

where δ_j are given by (A.20) in Appendix A.

Let the domain G^- be the complement of the closure of all the inclusion domains to the extended complex plane, i.e., $G^- = \widehat{\mathbb{C}}\backslash(G^+ \cup L)$. If $h(z)$ belongs to $\mathcal{H}(G^+)$, according to Chauchy's formula, we have [79, 193]

$$\frac{1}{2\pi i}\int_L \frac{h(t)}{t-z}\mathrm{d}t = \begin{cases} h(z), & z \in G^+, \\ 0, & z \in G^-. \end{cases} \tag{2.47}$$

If $h(z)$ belongs to $\mathcal{H}(G^-)$, we have [79, 193]

$$\frac{1}{2\pi i}\int_L \frac{h(t)}{t-z}\mathrm{d}t = \begin{cases} h(\infty), & z \in G^+, \\ -h(z)+h(\infty), & z \in G^-. \end{cases} \tag{2.48}$$

The same formulas (2.47)–(2.48) hold for the Cauchy's integral (2.45) with the other kernel on the torus topology with $G^- = G$.

The limit boundary values $\Phi^+(t) = \lim_{G_k \ni z \to t}\Phi(z)$ and $\Phi^-(t) = \lim_{G \ni z \to t}\Phi(z)$ of the Cauchy-type integral (2.44) satisfy Sochocki's formulas [79, 193]

$$\Phi^+(t) = \frac{1}{2}h(t) + \frac{1}{2\pi i}\int_L \frac{h(\tau)}{\tau - t}\mathrm{d}\tau,\ \Phi^-(t) = -\frac{1}{2}h(t) + \frac{1}{2\pi i}\int_L \frac{h(\tau)}{\tau - t}\mathrm{d}\tau,\ t \in L_k. \tag{2.49}$$

The singular integral $\frac{1}{2\pi i}\int_L \frac{h(\tau)}{\tau - t}\mathrm{d}\tau$ from (2.49) is defined as the principal value integral [79, 193]. The same formulas hold for (2.45) with the other kernel

$$\begin{aligned} \Phi^+(t) &= \tfrac{1}{2}h(t) + \tfrac{1}{2\pi i}\int_L h(\tau)E_1(\tau - t)\mathrm{d}\tau, \\ \Phi^-(t) &= -\tfrac{1}{2}h(t) + \tfrac{1}{2\pi i}\int_L h(\tau)E_1(\tau - t)\mathrm{d}\tau,\ t \in L_k. \end{aligned} \tag{2.50}$$

Application of the Cauchy-type integral (2.45) to (2.42) yields the following system of integral equations [93]

$$\varphi_k(z) = \sum_{m=1}^{N} \frac{1}{2\pi i}\int_{L_m} \varrho_m \overline{\varphi_m(t)} E_1(t-z)\,\mathrm{d}t + z + c_k,\ z \in G_k\ (k = 1,2,\ldots,N), \tag{2.51}$$

where c_k are undetermined constants. Equation (2.51) can be considered as an explicit form of the general equation similar to (2.3)

$$u_k = \sum_{m=1}^{N} \varrho_m F_m u_m + u_0, \quad z \in G_k\ (k = 1,2,\ldots,N). \tag{2.52}$$

This equation expresses the charge (energetic) balance between the field u_k in the inclusion G_k from one side and the external field u_0 and the fields $F_m u_m$ induced by the inclusions G_m from the other side. Schwarz's method can be considered as an iterative scheme applied to equations (2.51) or to (2.52).

Koiter developed the theory of singular integral equations for doubly periodic problems [118], see also [54] and historical notes in [98, p. 55–58]. The theory of singular integral equations for doubly periodic problems was further developed by Filshtinsky [98,172] who used the Weierstrass function $\zeta(z)$ instead of the Eisenstein function $E_1(z)$. It was established in [40,93] that the system (2.51) for $|\varrho_m| \leq 1$ with fixed c_k has a unique solution in $\mathcal{H}(G^+)$. This solution can be found by the method of successive approximations converging in $\mathcal{H}(G^+)$, i.e., uniformly in $\cup_{k=1}^{N}(G_k \cup L_k)$.

When the functions $\varphi_k(z)$ are determined, the complex potential in the domain G is calculated by the formula [93]

$$\varphi(z) = \sum_{m=1}^{N} \frac{\varrho_k}{2\pi i} \int_{L_m} \overline{\varphi_m(t)} E_1(t-z)\, \mathrm{d}t + z + c_k,\ z \in G. \tag{2.53}$$

Using Sochocki's formulas (2.50) one can subtract the limit values of (2.53) from (2.51) on L_k in order to check the relation $\varphi_k(t) - \varphi(t) = \varrho\, \overline{\varphi_k(t)}$, $t \in L_k$, equivalent to (2.42).

The normalized effective permittivity tensor in the considered case of real ε_k can be calculated by the formula (3.2.42) of the book [93]. Two components of the tensor are given in the form convenient in the further consideration of complex ε_k

$$\begin{aligned} \varepsilon_{11} &= 1 + 2\mathrm{Re}\ \textstyle\sum_{k=1}^{N} \varrho_k \int_{G_k} \frac{d\varphi_k}{dz}(\xi_1 + i\xi_2)\, \mathrm{d}\xi_1 \mathrm{d}\xi_2, \\ \varepsilon_{12} &= -2\mathrm{Im}\ \textstyle\sum_{k=1}^{N} \varrho_k \int_{G_k} \frac{d\varphi_k}{dz}(\xi_1 + i\xi_2)\, \mathrm{d}\xi_1 \mathrm{d}\xi_2. \end{aligned} \tag{2.54}$$

2.6 IMPLEMENTATION OF SCHWARZ'S METHOD

2.6.1 EXPLICIT AND IMPLICIT ITERATIVE SCHEMES

The method of successive approximations can be applied to equations (2.51) by means of two different iterative schemes. The explicit iteration scheme has the form

$$\varphi_k^{(0)}(z) = z, \tag{2.55}$$

$$\varphi_k^{(p)}(z) = \textstyle\sum_{m=1}^{N} \frac{\varrho_m}{2\pi i} \int_{L_m} \overline{\varphi_m^{(p-1)}(t)} E_1(t-z)\, \mathrm{d}t + z,\ z \in G_k,\ (p = 1, 2, \ldots). \tag{2.56}$$

The implicit iteration scheme to equations (2.51) with the same zero approximation (2.55) has the form

$$\begin{aligned} &\varphi_k^{(p)}(z) - \tfrac{\varrho_k}{2\pi i} \textstyle\int_{L_k} \overline{\varphi_k^{(p)}(t)} E_1(t-z)\mathrm{d}t = \\ &\textstyle\sum_{m \neq k} \frac{\varrho_m}{2\pi i} \int_{L_m} \overline{\varphi_m^{(p-1)}(t)} E_1(t-z)\mathrm{d}t + z, \quad z \in G_k,\ (p = 1, 2, \ldots). \end{aligned} \tag{2.57}$$

The implicit scheme corresponds to equation (2.3).

In order to reduce computations, equations (2.55)–(2.57) are written without additive constants c_k. First, it is related to the observation that we need the derivative $\varphi_k'(z)$ to compute the effective constants (2.54). Second, an additive constant C in the approximation $\varphi_k^{(p-1)}(z)$ yields an additive constant in the next approximation $\varphi_k^{(p)}(z)$. This fact is established by the residue theorem

$$\frac{1}{2\pi i}\int_{L_m} \overline{C}E_1(t-z)\,\mathrm{d}t = \delta_{mk}\overline{C}, \quad z \in G_k, \tag{2.58}$$

where δ_{mk} stands for the Kronecker delta. Application of (2.55)–(2.57) leads to a power series in ϱ_k. This is the reason why the method is also called contrast expansion [93, 155, 226].

Let h_0 be given and h unknown functions from the space $\mathcal{H}(G_k)$ for a fixed k. The following integral equation has to be solved in every iteration step of (2.57)

$$h(z) - \frac{\varrho_k}{2\pi i}\int_{L_k} \overline{h(t)}E_1(t-z)\mathrm{d}t = h_0(z), \quad z \in G_k. \tag{2.59}$$

Introduce the compact integral operators in the space $\mathcal{H}(G_k)$

$$(P_k h)(z) = \frac{1}{2\pi i}\int_{L_k} \overline{h(t)}E_0(t-z)\mathrm{d}t \quad z \in G_k, \tag{2.60}$$

where the function $E_0(z)$ analytic in the periodicity cell Q is determined by its Taylor series

$$E_0(z) = E_1(z) - \frac{1}{z} = \sum_{k=1}^{\infty} S_{2k} z^{2k-1}. \tag{2.61}$$

Here, equation (A.30) from Appendix A is used.

Write equation (2.59) in the operator form

$$h - \varrho_k(\mathcal{S}h + P_k h) = h_0, \tag{2.62}$$

where

$$\mathcal{S}h(z) = \begin{cases} \frac{1}{2\pi i}\int_{L_k} \frac{\overline{h(t)}}{t-z}\mathrm{d}t, & z \in G_k, \\ \frac{1}{2}\overline{h(z)} + \frac{1}{2\pi i}\int_{L_k} \frac{\overline{h(t)}}{t-z}\mathrm{d}t, & z \in L_k. \end{cases} \tag{2.63}$$

Here, the boundary values of $\mathcal{S}h(z)$ on L_k are written in accordance with Sochocki's formulas (2.49)

$$\lim_{\substack{\zeta \to z \\ \zeta \in G_k}} \frac{1}{2\pi i}\int_{L_k} \frac{\overline{h(t)}}{t-\zeta}\mathrm{d}t = \frac{1}{2}\overline{h(z)} + \frac{1}{2\pi i}\int_{L_k} \frac{\overline{h(t)}}{t-z}\mathrm{d}t, \; z \in L_k. \tag{2.64}$$

Consider the integral equation similar to (2.62)

$$h - \varrho_k \mathcal{S}h = h_0. \tag{2.65}$$

The singular operator $\mathcal{S}$ is bounded in the space $\mathcal{H}(L_k)$ of Hölder continuous functions on the curve L_k [229]. The space $\mathcal{H}(G_k)$ of functions analytically continued from L_k into G_k can be considered as a closed subspace of $\mathcal{H}(L_k)$. The following two theorems justify the approximation schemes developed in the next sections

Theorem 2 ([185]) *The operators $\mathcal{S}$ and $\mathcal{S}+P_k$ are bounded in $\mathcal{H}(G_k)$.*

Theorem 3 ([40]) *Equation (2.65) has the unique solution, which can be written by the inverse operator correctly defined and bounded in $\mathcal{H}(G_k)$ for $|\varrho_k| < 1$*

$$h = (I - \varrho_k\mathcal{S})^{-1}h_0 = h_0 + \varrho_k\mathcal{S}h_0 + \varrho_k^2\mathcal{S}^2h_0 + \ldots. \tag{2.66}$$

An analogous assertion takes place for equation (2.59) [93].

The both methods (2.56) and (2.57) converge absolutely for any fixed $|\varrho_k| < 1$ [183, 184]. In the case $|\varrho_k| = 1$, the uniform convergence can be established by means of the regularization (2.6).

Equations (2.56) can be written in the form

$$\varphi_k^{(p)}(z) = \sum_{m=1}^{N} \frac{\varrho_m}{2\pi i}\int_{L_m} \overline{\varphi_m^{(p-1)}(t)}\frac{\mathrm{d}t}{t-z} + \sum_{m=1}^{N} \varrho_m\left(P_m\overline{\varphi_m^{(p-1)}}\right)(z) + z, \quad z \in G_k,\ (p = 1, 2, \ldots). \tag{2.67}$$

Equations (2.57) can be modified by splitting the integral term with pth iteration onto two terms with pth and $(p-1)$th iterations as follows

$$\begin{aligned} \varphi_k^{(p)}(z) - \frac{\varrho_k}{2\pi i}\int_{L_k} \overline{\varphi_k^{(p)}(t)}\frac{\mathrm{d}t}{t-z} &= \varrho_k\left(P_k\overline{\varphi_k^{(p-1)}}\right)(z) \\ &+ \sum_{m\neq k} \frac{\varrho_m}{2\pi i}\int_{L_m} \overline{\varphi_m^{(p-1)}(t)}E_1(t-z)\mathrm{d}t + z, \quad z \in G_k,\ (p = 1, 2, \ldots). \end{aligned} \tag{2.68}$$

The iterative scheme (2.68) can be treated as a mixed scheme when the self-induced charge of kth inclusions is partly included in the pth approximation and partly in the previous $(p-1)$th approximation.

Thus, Schwarz's method can be implemented by means of explicit and implicit iterative schemes with different modifications. The scheme (2.56) corresponds to the contrast expansion, (2.57) to the expansion in the concentration of inclusions called cluster expansion in [226]. These two principal schemes with modifications like (2.68) have a lot of other names usually related to self-consistent methods. One can meet manipulations with the distorted balance of precision as in the difference scheme and Mori-Tanaka method.

For example, consider the approximation of (2.68) neglecting the terms with P_k for a two-phase composites ($\varrho_k = \varrho$)

$$\begin{aligned} \varphi_k^{(p)}(z) - \varrho\frac{1}{2\pi i}\int_{L_k} \overline{\varphi_k^{(p)}(t)}\frac{\mathrm{d}t}{t-z} &\approx \\ \varrho\sum_{m\neq k} \frac{1}{2\pi i}\int_{L_m} \overline{\varphi_m^{(p-1)}(t)}E_1(t-z)\mathrm{d}t + z, &\quad z \in G_k,\ (p = 1, 2, \ldots). \end{aligned} \tag{2.69}$$

Such an approximation can be accompanied by engineering arguments about the balance of flux produced by one inclusion and the flux by all the others. This is true up to $O(\varrho^2)$. Let the problem is solved in the first iteration. It can be solved analytically or numerically, it doesn't matter. But the problem is solved up to $O(\varrho^2)$. Now, let the second iteration is applied. The obtained second-order solution contains the term ϱ^2 analytically or numerically. The higher-order terms of f accompanying ϱ^2 can arise. And so forth. After a few iterations, one arrives at a formula or numerical data with a redundant tail of high-order terms in ϱ and f.

Equations (2.69) can be "improved" by the approximation of integral operator $\varrho\left(P_k\overline{\varphi_k^{(p-1)}}\right)(z) \approx \varrho C_k$ for some constants C_k. Such a type of approximation was proposed for elastic composites in [197]. It was based on the integral approximation, hence, was of order $O(f)$. This approximation was used in [135] for a square array of circular inclusions. Though the applied method gave the proper result of order $O(f)$, formulas for the effective constants were written with the tails up to $O(f^4)$.

Other schemes based on various mathematical manipulations, such as Brugemann's differential scheme, can be considered as a separate method. However, if one examines the result within the obtained precision, it becomes clear that the result coincides with the dilute Clausius-Mossotti (Maxwell) approximation written in another asymptotic form. This question will be discussed below.

2.6.2 CONTRAST EXPANSION (EXPLICIT SCHEME)

In the present section, we pay attention to the explicit iteration scheme (2.55)–(2.56). Denote $\varrho = \max_{k=1,2,\ldots,N}|\varrho_k|$. One can see that every iteration (2.56) increases the contrast parameters precision by the multiplier of order $O(|\varrho|)$. The parameter ϱ is used with the modulus in order to avoid confusion in other places when $\varrho = \varrho_k$ for a two-phase composite. The integration operator increases the precision by $O(f)$, since

$$F \mapsto |G_k|\left(\frac{1}{|G_k|}\int_{G_k} F\,\mathrm{d}x_1\mathrm{d}x_2\right), \tag{2.70}$$

where the operator in the parentheses is bounded in $\mathcal{H}(G_k)$. The multiplier $|G_k|$ is of order f because of (2.9). Therefore,

$$\frac{d\varphi_k}{dz}(z) = \frac{d\varphi_k^{(p)}}{dz}(z) + O(|\varrho|^{p+1} f^p), \quad z \in G_k \cup L_k. \tag{2.71}$$

Integrating this relation we estimate the integral

$$\int_{G_k} \frac{d\varphi_k}{dz}(z)\,\mathrm{d}x_1\mathrm{d}x_2 = \int_{G_k} \frac{d\varphi_k^{(p)}}{dz}(z)\,\mathrm{d}x_1\mathrm{d}x_2 + O(|\varrho f|^{p+1}). \tag{2.72}$$

This integral will be applied to determine the effective constants by (2.53). The multiplier ϱ_m in (2.56) guarantees the required precision in ϱ. We will use the expansion of the kernel $E_1(t-z)$ in terms of order $f^{1/2}$ that requires a more careful study of precision in concentration. Hereafter in this section, the clear behavior of precision in ϱ at every step is omitted for shortness.

We now proceed to develop a symbolic algorithm for the effective constants (2.54) for a given p. It follows from (2.56) that the functions $\varphi_k^{(q)}(z)$ for $q = p-1$ have to be determined up to $O(f^{q+1/2})$. The kernel $E_1(t-z)$ in the integral from (2.56) for $\varphi_k^{(q)}(z)$ has to be estimated with the same precision.

Introduce the integral frequently met in the theory of analytic functions [229]

$$J_k(z) = \frac{1}{2\pi i}\int_{L_k} \frac{\overline{t}}{t-z}\mathrm{d}t, \quad z \in G_k. \tag{2.73}$$

This function is analytic in G_k and Hölder continuous in $G_k \cup L_k$ [193, 229]. It is worth noting that the limit values of $J_k(z)$ as $z \to L_k$ after the application of Sochocki's formula (2.49) can be written by means of the singular integral

$$J_k(\tau) = \frac{\overline{\tau}}{2} + \frac{1}{2\pi i}\int_{L_k} \frac{\overline{t}}{t-\tau}\mathrm{d}t, \quad \tau \in L_k. \tag{2.74}$$

The iterations (2.67) include the integral

$$I_{mk}^{(q)}(z) = \frac{1}{2\pi i}\int_{L_m} \overline{\varphi_m^{(q)}(t)}\, E_1(t-z)\,\mathrm{d}t. \tag{2.75}$$

It follows from $\varphi_m^{(0)}(t) = t$ and (2.61) that the singular part of the integral $I_{kk}^{(0)}(z)$ coincides with $J_k(z)$. The integral (2.75) will be estimated below in two cases.

i) First, it is assumed that $m \neq k$. Consider the simple expression

$$t - z = (t-a_m) + (a_m - a_k) - (z - a_k), \tag{2.76}$$

where $|a_m - a_k|$ is the dominating term in comparison with $|t-a_m|$ and $|z-a_k|$ due to (2.8). The value $|t-a_m+a_k-z|$ is of order $O(r) = O(f^{1/2})$. By virtue of (2.76) and (A.30) from Appendix A the Taylor approximation of $E_1(t-z)$ holds near $a_m - a_k$

$$E_1(t-z) = \sum_{n=0}^{2q+1}(-1)^n E_{n+1}(a_m-a_k)(t-a_m+a_k-z)^n + O(f^{q+1}). \tag{2.77}$$

Using (2.77) we estimate the integral (2.75)

$$I_{mk}^{(q)}(z) = \sum_{n=0}^{2q+1}(-1)^n E_{n+1}(a_m-a_k)\frac{1}{2\pi i}\int_{L_m}\overline{\varphi_m^{(q)}(t)}(t-a_m+a_k-z)^n\mathrm{d}t + O(f^{q+1}). \tag{2.78}$$

ii) Consider now the case $m = k$. Using (2.61) we approximate $E_1(t-z)$ in G_k by expression

$$E_1(t-z) = \frac{1}{t-z} - \sum_{n=1}^{q+1} S_{2n}(t-z)^{2n-1} + O(f^{q+1}). \tag{2.79}$$

Substitution of (2.79) into (2.75) yields

$$I_{kk}^{(q)}(z) = \frac{1}{2\pi i}\int_{L_k} \overline{\varphi_k^{(q)}(t)}\frac{\mathrm{d}t}{t-z} - \sum_{n=1}^{q+1}\frac{S_{2n}}{2\pi i}\int_{L_k}\overline{\varphi_k^{(q)}(t)}(t-z)^{2n-1}\mathrm{d}t + O(f^{q+1}). \quad (2.80)$$

The asymptotic formulas (2.78) and (2.80) yield the estimation for

$$\varphi_k^{(q)}(z) = z + \sum_{m=1}^{N}\varrho_m I_{mk}^{(q-1)}(z). \quad (2.81)$$

2.6.3 SECOND ITERATION IN CONTRAST EXPANSION

The general method described above can be implemented in symbolic form and simplified within a fixed precision. We illustrate the general scheme for $p = 2$. Even such a low-order approximation yields a new analytical approximation for the effective permittivity tensor. We will use the formulas (2.54) in the following approximation

$$\begin{aligned}\varepsilon_{11} &= 1 + 2\mathrm{Re}\ \textstyle\sum_{k=1}^{N}\varrho_k\int_{G_k}\frac{d}{dz}\varphi_k^{(2)}(\xi_1+i\xi_2)\ \mathrm{d}\xi_1\mathrm{d}\xi_2 + O(|\varrho|^3 f^{7/2}),\\ \varepsilon_{12} &= -2\mathrm{Im}\ \textstyle\sum_{k=1}^{N}\varrho_k\int_{G_k}\frac{d}{dz}\varphi_k^{(2)}(\xi_1+i\xi_2)\ \mathrm{d}\xi_1\mathrm{d}\xi_2 + O(|\varrho|^3 f^{7/2}).\end{aligned} \quad (2.82)$$

It will be seen below that it is sufficient to take the approximation $\frac{d}{dz}\varphi_k^{(2)}$ up to $O(|\varrho|^2 f^{5/2})$ in order to reach the specified in (2.82) precision.

First, investigate the precision in ϱ. The idea can be easily represented for a two-phase composite when the constant $\varrho = \varrho_k$ for all $k = 1, 2, \ldots, N$. The iterative scheme (2.56) yields $\varphi_k^{(0)} = \alpha_0$ and $\varphi_k^{(1)} = \alpha_1\varrho + \beta_1$, where the terms $\alpha_0, \alpha_1, \beta_1$ are constants in ϱ. The second iteration can be schematically written as $\varphi_k^{(2)} = \varrho(\alpha_1\varrho + \beta_1) + \alpha_0$. After substitution into (2.82), one can see that one may neglect the coefficient α_1. Therefore, it is sufficient to determine $\varphi_k^{(1)}$ up to $O(1)$, as $\varrho \to 0$,

$$\varphi_k^{(1)}(z) = \sum_{m=1}^{N}\frac{\varrho_m}{2\pi i}\int_{L_m}\bar{t}\,E_1(t-z)\ \mathrm{d}t + z = z + O(1), \quad z \in G_k. \quad (2.83)$$

For completeness, we have to note that the following singular integral is a bounded function in $G_k \cup L_k$

$$\frac{1}{2\pi i}\int_{L_k}\bar{t}\,E_1(t-z)\ \mathrm{d}t = J_k(z) - \frac{1}{2\pi i}\int_{L_k}\bar{t}\,E_0(t-z)\ \mathrm{d}t. \quad (2.84)$$

Here, equations (2.61) and (2.73) are used. The second approximation has the form

$$\varphi_k^{(2)}(z) = \sum_{m=1}^{N}\frac{\varrho_m}{2\pi i}\int_{L_m}\bar{t}\,E_1(t-z)\ \mathrm{d}t + z + O(|\varrho|^2), \quad z \in G_k. \quad (2.85)$$

Below, we omit the precision in ϱ for shortness and keep track of the precision in f. Differentiate equation (2.85)

$$\frac{d}{dz}\varphi_k^{(2)}(z) = 1 + \frac{1}{\pi}\sum_{m=1}^{N} \varrho_m \frac{d}{dz} g_m(z), \tag{2.86}$$

where the analytic in G_m function is introduced

$$g_m(z) = \frac{1}{2i}\int_{L_m} \bar{t} E_1(t-z)\mathrm{d}t, \quad z \in G_k. \tag{2.87}$$

The complex Green formulas hold for a function $w(z,\overline{z})$ continuously differentiable in a smooth closed domain G [229]

$$\int_G \frac{\partial w}{\partial \overline{z}}\,\mathrm{d}\xi_1 \mathrm{d}\xi_2 = \frac{1}{2i}\int_{\partial G} w\,\mathrm{d}t, \quad \int_G \frac{\partial w}{\partial z}\,\mathrm{d}\xi_1 \mathrm{d}\xi_2 = -\frac{1}{2i}\int_{\partial G} w\,\mathrm{d}\bar{t}. \tag{2.88}$$

Introduce the averaged contrast parameter over inclusions

$$\langle \varrho \rangle = \sum_{k=1}^{N} \varrho_k |G_k|. \tag{2.89}$$

Using (2.86) and (2.82) we obtain

$$\begin{aligned} \varepsilon_{11} &= 1 + 2\langle\varrho\rangle + \mathrm{Re}\,\tfrac{2}{\pi}\textstyle\sum_{k=1}^N \sum_{m=1}^N \varrho_k \varrho_m h_{km} + O(|\varrho|^3 f^{7/2}), \\ \varepsilon_{12} &= -\mathrm{Im}\,\tfrac{2}{\pi}\textstyle\sum_{k=1}^N \sum_{m=1}^N \varrho_k \varrho_m h_{km} + O(|\varrho|^3 f^{7/2}). \end{aligned} \tag{2.90}$$

where

$$h_{km} = \int_{G_k} \frac{dg_m}{dz}(z)\,\mathrm{d}\xi_1 \mathrm{d}\xi_2 = -\frac{1}{2i}\int_{L_k} \bar{t}\, g_m(t)\,\mathrm{d}t \tag{2.91}$$

by the second Green formula (2.88).

i) Consider the integral (2.87) when $m \neq k$. Introduce for shortness the designations $E_{p,mk} = E_p(a_m - a_k)$ and $T = t - a_m$, $Z = z - a_k$. Using (A.26) from Appendix A we represent the function $E_1(t-z)$ by the Taylor formula

$$\begin{aligned} &E_1(t-z) = E_{1,mk} - E_{2,mk}(T-Z) + E_{3,mk}(T-Z)^2 - E_{4,mk}(T-Z)^3 \\ &+E_{5,mk}(T-Z)^4 + O(f^{5/2}). \end{aligned} \tag{2.92}$$

Substitute (2.92) into (2.87)

$$g_m(z) = \frac{1}{2i}\int_{L_m} \bar{t}\,[f_1(z) + f_2(z,t)]\,\mathrm{d}t + O(f^{5/2}), \tag{2.93}$$

where the following designations are introduced

$$\begin{aligned} &f_1(z) = E_{1,mk} + E_{2,mk}Z + E_{3,mk}Z^2 + E_{4,mk}Z^3 + E_{5,mk}Z^4, \\ &f_2(z,t) = -(E_{2,mk} + 2ZE_{3,mk} + 3Z^2E_{4,mk} + 4E_{5,mk}Z^3)T \\ &+(E_{3,mk} + 3E_{4,mk}Z + 6E_{5,mk}Z^2)T^2 - (E_{4,mk} + 4E_{5,mk}Z)T^3 + E_{5,mk}T^4. \end{aligned} \tag{2.94}$$

Remark 1 *The result (2.94) and others below are obtained due to symbolic computations with the package Mathematica®. The corresponding symbolic computations can increase the precision [67, Chapter 2].*

Using the first Green formula (2.88) introduce the static complex moments of the domain G_m

$$\begin{gathered} s_{qm} = \frac{1}{2i}\int_{L_m} \bar{t}(t-a_m)^q \mathrm{d}t = \int_{G_m} (t-a_m)^q \mathrm{d}x_1 \mathrm{d}x_2 = \\ \frac{1}{2i}\int_{L_m} (\overline{t-a_m})(t-a_m)^q \mathrm{d}t \quad (q=0,1,\ldots) \end{gathered} \tag{2.95}$$

and calculate

$$s_{0m} = \frac{1}{2i}\int_{L_m} \bar{t}\,\mathrm{d}t = \int_{G_m} \mathrm{d}x_1 \mathrm{d}x_2 = |G_m|. \tag{2.96}$$

The dimensionless complex static moments can be introduced as follows

$$s_{qm}^{(0)} = s_{qm} s_{0m}^{-\frac{q}{2}-1}. \tag{2.97}$$

In particular, (2.97) yields $s_{0m}^{(0)} = 1$. It follows from (2.95) and (2.97) that

$$s_{qm} = O(f^{\frac{q}{2}+1}), \quad s_{qm}^{(0)} = O(1). \tag{2.98}$$

Continue to use the designations

$$E_{p,mk} = E_p(a_m - a_k), \quad p = 2,3,\ldots; \quad m,k = 1,2,\ldots,N, \tag{2.99}$$

including the case $m = k$. Here, it is assumed for shortness that $E_p(a_m - a_k) = S_p$ when a_m coincides with a_k. The lattice sums S_p are introduced in Appendix A. Then, equation (2.93) can be written in the form

$$\begin{aligned} &g_m(z) = f_1(z)s_{0m} - (E_{2,mk} + 2ZE_{3,mk} + 3Z^2E_{4,mk} + 4Z^3E_{5,mk})s_{1m} \\ &+(E_{3,mk} + 3E_{4,mk}Z + 6E_{5,mk}Z^2)s_{2m} \\ &-(E_{4,mk} + 4E_{5,mk}Z)s_{3m} + E_{5,mk}s_{4m} + O(f^{7/2}), \end{aligned} \tag{2.100}$$

where the estimation (2.98) is used and $Z = z - a_k$. Calculate

$$\begin{aligned} &\frac{dg_m}{dz}(z) = (E_{2,mk} + 2E_{3,mk}Z + 3E_{4,mk}Z^2 + 4E_{5,mk}Z^3)s_{0m} \\ &-2(E_{3,mk} + 3ZE_{4,mk} + 6Z^2E_{5,mk})s_{1m} + 3(E_{4,mk} + 4E_{5,mk}Z)s_{2m} \\ &-4E_{5,mk}s_{3m} + O(f^3). \end{aligned} \tag{2.101}$$

The differentiation operator decreases the precision on $f^{1/2}$, since $|z - a_k| = O(f^{1/2})$. Using the first Green formula (2.88), calculate the integral (2.91) with $\frac{dg_m}{dz}(z)$ given by the approximation (2.101)

$$\begin{aligned} &h_{km} = (E_{2,mk}s_{0k} + 2E_{3,mk}s_{1k} + 3E_{4,mk}s_{2k} + 4E_{5,mk}s_{3k})s_{0m} \\ &-2(E_{3,mk}s_{0k} + 3E_{4,mk}s_{1k} + 6E_{5,mk}s_{2k})s_{1m} \\ &+3(E_{4,mk}s_{0k} + 4E_{5,mk}s_{1k})s_{2m} - 4E_{5,mk}s_{0k}s_{3m} + O(f^4). \end{aligned} \tag{2.102}$$

Write (2.102) in the form

$$\begin{aligned} h_{km} &= E_{2,mk} s_{0k} s_{0m} + 2E_{3,mk}(s_{1k}s_{0m} - s_{1m}s_{0k}) \\ &+3E_{4,mk}(s_{2k}s_{0m} + s_{2m}s_{0k} - 2s_{1m}s_{1k}) \\ &+4E_{5,mk}[s_{3k}s_{0m} - s_{0k}s_{3m} + 3(s_{1k}s_{2m} - s_{2k}s_{1m})] + O(f^4). \end{aligned} \tag{2.103}$$

ii) Consider the integral (2.87) when $m = k$. It is equal to the integral (2.84)

$$g_k(z) = J_k(z) - \frac{1}{2i}\int_{L_k} \bar{t}\, E_0(t-z)\, \mathrm{d}t. \tag{2.104}$$

Using (A.30) from Appendix A, we find for the considered periodic square unit cell

$$E_0'(t-z) = \pi + 3S_4(t-z)^2 + 7S_8(t-z)^6 + O(f^{7/2}). \tag{2.105}$$

Substitute (2.105) into (2.104) and use the first formula (2.88)

$$g_k'(z) = J_k'(z) + \pi s_{0k} + 3S_4 \int_{G_k} (t-z)^2\, \mathrm{d}\xi_1 \mathrm{d}\xi_2 + O(f^{5/2}). \tag{2.106}$$

The term $(t-z)^6$ will be neglected due to the considered precision.

Calculate the integral from (2.106)

$$\int_{G_k} [t - a_k - (z - a_k)]^2\, \mathrm{d}\xi_1 \mathrm{d}\xi_2 = s_{2k} - 2s_{1k}(z - a_k) + s_{0k}(z - a_k)^2. \tag{2.107}$$

Then, h_{kk} defined by (2.91) can be calculated by the asymptotic formula

$$h_{kk} = \int_{G_k} J_k'(z)\, \mathrm{d}\xi_1 \mathrm{d}\xi_2 + \pi s_{0k}^2 + 6S_4(s_{2k}s_{0k} - s_{1k}^2) + O(f^4). \tag{2.108}$$

Therefore, the values h_{km} can be found by (2.102) and (2.108).

Introduce the value

$$\mathcal{J} = \sum_{k=1}^{N} \varrho_k^2 \int_{G_k} J_k'(z)\, \mathrm{d}\xi_1 \mathrm{d}\xi_2 = -\frac{1}{2\mathrm{i}} \sum_{k=1}^{N} \varrho_k^2 \int_{L_k} J_k(t)\, \mathrm{d}\bar{t}. \tag{2.109}$$

Applying Sochocki's formulas (2.49) to $J_k(z)$ defined by (2.73) and substituting the result into (2.109) we obtain

$$\mathcal{J} = \frac{1}{4\pi} \sum_{k=1}^{N} \varrho_k^2 \int_{L_k} \mathrm{d}\bar{t} \int_{L_k} \frac{\bar{\tau}}{\tau - t}\, \mathrm{d}\tau. \tag{2.110}$$

We now proceed to calculate the double sum $\frac{1}{\pi}\sum_{k=1}^{N}\sum_{m=1}^{N} \varrho_k \varrho_m h_{km}$ using the asymptotic formulas (2.103) and (2.108). It is convenient to split this sum into the following ones. Introduce the double sums

$$\mathcal{L} = \frac{1}{\pi} \sum_{k,m=1}^{N} \varrho_k \varrho_m |G_k||G_m| E_{2,mk} \tag{2.111}$$

and

$$\mathcal{V}(1) = \frac{2}{\pi} \sum_{k,m=1}^{N} \varrho_k \varrho_m |G_k||G_m| E_{3,mk} (s_{1k}^{(0)} |G_k|^{\frac{1}{2}} - s_{1m}^{(0)} |G_m|^{\frac{1}{2}}) \tag{2.112}$$

using the designation (2.99) and the normalized static moments (2.97). Analogously introduce

$$\begin{gathered}\mathcal{V}(2) = \tfrac{3}{\pi} \textstyle\sum_{k,m=1}^{N} \varrho_k \varrho_m |G_k||G_m| E_{4,mk} \times \\ \left(|G_k| s_{2k}^{(0)} + |G_m| s_{2m}^{(0)} - 2\left(|G_k||G_m|\right)^{\frac{1}{2}} s_{1m}^{(0)} s_{1k}^{(0)} \right)\end{gathered} \tag{2.113}$$

and

$$\begin{gathered}\mathcal{V}(3) = \tfrac{4}{\pi} \textstyle\sum_{k,m=1}^{N} \varrho_k \varrho_m |G_k||G_m| E_{5,mk} \times \\ \left[|G_k|^{\frac{3}{2}} s_{3k}^{(0)} - |G_m|^{\frac{3}{2}} s_{3m}^{(0)} + 3\left(|G_k|^{\frac{1}{2}} |G_m| s_{1k}^{(0)} s_{2m}^{(0)} - |G_m|^{\frac{1}{2}} |G_k| s_{2k}^{(0)} s_{1m}^{(0)} \right) \right].\end{gathered} \tag{2.114}$$

Put

$$\mathcal{V} = \mathcal{V}(1) + \mathcal{V}(2) + \mathcal{V}(3). \tag{2.115}$$

Then, equation (2.90) can be written in the form

$$\begin{aligned} \varepsilon_{11} &= 1 + 2\langle\varrho\rangle + 2\mathrm{Re}\ (\mathcal{J} + \mathcal{L} + \mathcal{V}) + O(|\varrho|^3 f^4), \\ \varepsilon_{12} &= -2\mathrm{Im}\ (\mathcal{J} + \mathcal{L} + \mathcal{V}) + O(|\varrho|^3 f^4). \end{aligned} \tag{2.116}$$

We again write formulas with the explicitly shown precision in ϱ. The asymptotic formula (2.116) is a result of the second iteration in the contrast expansion. The value $|G_k|$ can be considered as the volume fraction of the component having the contrast parameter ϱ_k.

Recall that the periodicity cell Q has the unit area, which means normalizing the centers a_k and using the scaled to the unit periodicity cell for the Eisenstein functions. Hence, the terms $\mathcal{L}$, $\mathcal{V}_1$, $\mathcal{V}_2$ and $\mathcal{V}_3$ are of orders $O(f^2)$, $O(f^{5/2})$, $O(f^3)$ and $O(f^{7/2})$, respectively. The term $\mathcal{J}$ from (2.109) has the order $O(f)$ except at some shapes when it is equal to zero. Therefore, one can introduce the dimensionless cumulative factors

$$\mathcal{J}_0 = f^{-1}\mathcal{J}, \quad \mathcal{L}_0 = f^{-2}\mathcal{L}, \quad \mathcal{V}_{0j} = f^{-\frac{j+4}{2}}\mathcal{V}_j. \tag{2.117}$$

Then, (2.116) can be written up to $O(|\varrho|^3 f^4)$ in the form

$$\begin{aligned} \varepsilon_{11} &= 1 + 2\langle\varrho\rangle + 2\mathrm{Re}\ \left[f\mathcal{J}_0 + f^2\mathcal{L}_0 + f^{5/2}\left(\mathcal{V}_{01} + f^{1/2}\mathcal{V}_{02} + f\mathcal{V}_{03}\right)\right], \\ \varepsilon_{12} &= -2\mathrm{Im}\ \left[f\mathcal{J}_0 + f^2\mathcal{L}_0 + f^{5/2}\left(\mathcal{V}_{01} + f^{1/2}\mathcal{V}_{02} + f\mathcal{V}_{03}\right)\right]. \end{aligned} \tag{2.118}$$

In the case of circular inclusions $J_k(z) = 0$, see [93], and all the static moments s_{qm} vanishes for $q \geq 1$. Hence, the terms $\mathcal{J}$ and $\mathcal{V}$ vanish.

In order to calculate the components ε_{22} and $\varepsilon_{21} = \varepsilon_{12}$ one can rotate the structure about the angle $\frac{\pi}{2}$ and apply the same method. Let the inclusion G_k be transformed into G_k^* after this rotation. Then, (2.116) yields

$$\begin{aligned} \varepsilon_{22} &= 1 + 2\langle\varrho\rangle + 2\text{Re }(\mathcal{J}^* + \mathcal{L}^* + \mathcal{V}^*) + O(|\varrho|^3 f^4), \\ \varepsilon_{21} &= \varepsilon_{12} = 2\text{Im }(\mathcal{J}^* + \mathcal{L}^* + \mathcal{V}^*) + O(|\varrho|^3 f^4), \end{aligned} \tag{2.119}$$

where the values with the asterisk correspond to the values from (2.116). The expressions for ε_{12} in (2.118) and ε_{21} in (2.119) have the opposite signs since after the rotation, the coordinate system x_1Ox_2 changes its orientation.

Let $\mathcal{J}$ be given by (2.110). Then,

$$\mathcal{J}^* = \frac{1}{4\pi}\sum_{k=1}^{N}\varrho_k^2 \int_{L_k^*} d\overline{t^*} \int_{L_k^*} \frac{\overline{\tau^*}}{\tau^* - t^*}\, d\tau^* = -\mathcal{J}, \tag{2.120}$$

The relation (2.120) is established by the change of variables $t^* = \mathrm{i}t$ and $\tau^* = \mathrm{i}\tau$ in the integral (2.120).

The next term is calculated by application of (A.37) from Appendix A

$$\mathcal{L}^* = \frac{1}{\pi}\sum_{m,k=1}^{N} \varrho_k\varrho_m|G_k||G_m|E_2^*(i(a_k - a_k)) = 2\langle\varrho\rangle^2 - \mathcal{L}. \tag{2.121}$$

One can check that the static moments (2.95) are transformed after rotation in the following way

$$s_{qm}^* = \mathrm{i}^q s_{qm} \quad (q = 1, 2, \ldots). \tag{2.122}$$

The same relation (2.122) holds for the normalized static moments $s_{qm}^{(0)*}$. Introduce the value

$$\mathcal{V}_1^* = \frac{2}{\pi}\sum_{k,m=1}^{N} \varrho_k\varrho_m|G_k||G_m|E_3(a_k^* - a_k^*)(s_{1k}^{(0)*}|G_k|^{\frac{1}{2}} - s_{1m}^{(0)*}|G_m|^{\frac{1}{2}}). \tag{2.123}$$

One can see that $\mathcal{V}_1^* = -\mathcal{V}_1$ due to (A.38) from Appendix A and (2.122). The same arguments yield $\mathcal{V}_2^* = -\mathcal{V}_2$ and $\mathcal{V}_3^* = -\mathcal{V}_3$, hence, $\mathcal{V}^* = -\mathcal{V}$.

Let I denote the unit tensor. The effective tensor can be calculated by (2.118) and (2.119) up to $O(|\varrho|^3 f^4)$

$$\begin{aligned} \boldsymbol{\varepsilon}_\perp &= (1 + 2\langle\varrho\rangle)I + 2f\begin{pmatrix} \text{Re }\mathcal{J}_0 & -\text{Im }\mathcal{J}_0 \\ -\text{Im }\mathcal{J}_0 & -\text{Re }\mathcal{J}_0 \end{pmatrix} \\ &+ 2f^2\begin{pmatrix} \text{Re }\mathcal{L}_0 & -\text{Im }\mathcal{L}_0 \\ -\text{Im }\mathcal{L}_0 & 2 - \text{Re }\mathcal{L}_0 \end{pmatrix} + 2f^{5/2}\begin{pmatrix} \text{Re }\mathcal{V}_{01} & -\text{Im }\mathcal{V}_{01} \\ -\text{Im }\mathcal{V}_{01} & -\text{Re }\mathcal{V}_{01} \end{pmatrix} \\ &+ 2f^3\begin{pmatrix} \text{Re }\mathcal{V}_{02} & -\text{Im }\mathcal{V}_{02} \\ -\text{Im }\mathcal{V}_{02} & -\text{Re }\mathcal{V}_{02} \end{pmatrix} + 2f^{7/2}\begin{pmatrix} \text{Re }\mathcal{V}_{03} & -\text{Im }\mathcal{V}_{03} \\ -\text{Im }\mathcal{V}_{03} & -\text{Re }\mathcal{V}_{03} \end{pmatrix}. \end{aligned} \tag{2.124}$$

The value $\mathcal{J}$ defined by (2.110) can be considered as the sum of cumulative shape factors of inclusions G_k. The value $\mathcal{L}$ defined by (2.112) represents the mutual locations of inclusions. Therefore, $\mathcal{L}$ can be considered as the cumulative location factor. The term $\mathcal{V}$ defined by (2.112)–(2.115) can be considered as the cumulative mixed location-shape factor. It consists of the terms of orders $O(f^{5/2})$, $O(f^3)$ and $O(f^{7/2})$ which may vanish.

The formula (2.124) can be easily implemented. For real permittivity, one can find numerical examples in [180] not accessible by FEM because of computational restrictions.

2.6.4 TWO-PHASE COMPOSITE

Consider a two-phase composite with the same contrast parameter $\varrho = \varrho_k$. Then, $\langle \varrho \rangle$ becomes ϱf. Following [180] introduce the parameters $\mathcal{J}'$, $\mathcal{L}'$, $\mathcal{V}'(j)$ and $\mathcal{V}'$ equal to $\mathcal{J}$, $\mathcal{L}$,$\mathcal{V}(j)$ and $\mathcal{V}$, respectively, after the substitution $\varrho_k = 1$ into (2.110)–(2.115). For instance,

$$\mathcal{L}' = \frac{1}{\pi} \sum_{k,m=1}^{N} |G_k||G_m| E_{2,mk}. \tag{2.125}$$

In the considered case, (2.124) becomes

$$\begin{aligned}
\boldsymbol{\varepsilon}_\perp &= (1+2\varrho f)I + 2f\varrho^2 \begin{pmatrix} \mathrm{Re}\, \mathcal{J}_0' & -\mathrm{Im}\, \mathcal{J}_0' \\ -\mathrm{Im}\, \mathcal{J}_0' & -\mathrm{Re}\, \mathcal{J}_0' \end{pmatrix} \\
&+ 2\varrho^2 f^2 \begin{pmatrix} \mathrm{Re}\, \mathcal{L}_0' & -\mathrm{Im}\, \mathcal{L}_0' \\ -\mathrm{Im}\, \mathcal{L}_0' & 2-\mathrm{Re}\, \mathcal{L}_0' \end{pmatrix} + 2\varrho^2 f^{5/2} \begin{pmatrix} \mathrm{Re}\, \mathcal{V}_{01}' & -\mathrm{Im}\, \mathcal{V}_{01}' \\ -\mathrm{Im}\, \mathcal{V}_{01}' & -\mathrm{Re}\, \mathcal{V}_{01}' \end{pmatrix} \\
&+ 2\varrho^2 f^3 \begin{pmatrix} \mathrm{Re}\, \mathcal{V}_{02}' & -\mathrm{Im}\, \mathcal{V}_{02}' \\ -\mathrm{Im}\, \mathcal{V}_{02}' & -\mathrm{Re}\, \mathcal{V}_{02}' \end{pmatrix} + 2\varrho^2 f^{7/2} \begin{pmatrix} \mathrm{Re}\, \mathcal{V}_{03}' & -\mathrm{Im}\, \mathcal{V}_{03}' \\ -\mathrm{Im}\, \mathcal{V}_{03}' & -\mathrm{Re}\, \mathcal{V}_{03}' \end{pmatrix}.
\end{aligned} \tag{2.126}$$

The component ε_{22} in the formula (91) from [180] contains an erroneous sign in $\mathcal{J}$ and $\mathcal{V}$ corrected here in (2.126).

The first two terms of (2.126) deserve special attention. The following universal formulas hold for any 2D two-phase dispersed composite

$$\begin{aligned}
\boldsymbol{\varepsilon}_\perp &= (1+2\varrho f)I + O(\varrho^2), \\
\boldsymbol{\varepsilon}_\perp &= (1+2\varrho f)I + 2f\varrho^2 \begin{pmatrix} \mathrm{Re}\, \mathcal{J}_0' & -\mathrm{Im}\, \mathcal{J}_0' \\ -\mathrm{Im}\, \mathcal{J}_0' & -\mathrm{Re}\, \mathcal{J}_0' \end{pmatrix} + O(f^2).
\end{aligned} \tag{2.127}$$

The first formula (2.127) can be independently justified by the Wiener bounds [233] written for two-phase composites

$$\left[(f\varepsilon_1^{-1} + (1-f)\varepsilon^{-1}) \right]^{-1} \leq \varepsilon_e \leq f\varepsilon_1 + (1-f)\varepsilon. \tag{2.128}$$

Here, the phases have the permittivity ε_1 and ε, the concentrations f and $1-f$, respectively. We now consider the normalized positive permittivity when $\varepsilon = 1$.

Substitution of $\varepsilon_1 = \frac{1+\varrho}{1-\varrho}$ and $\varepsilon = 1$ into (2.128) yields the lower and upper bounds coinciding up to $O(\varrho^2)$.

The term from the second formula (2.127) can be determined by the sum

$$\mathcal{J}_0' = \sum_{k=1}^{N} j_k, \tag{2.129}$$

where j_k means the dimensionless shape factor of kth inclusion

$$j_k = \frac{1}{4\pi f} \int_{L_k} \mathrm{d}\bar{t} \int_{L_k} \frac{\overline{\tau}}{\tau - t}\, \mathrm{d}\tau. \tag{2.130}$$

One can see that j_k is not changed under a translation $t \mapsto t + a$ since

$$j_k(a) - j_k(0) = \frac{1}{4\pi f} \int_{L_k} \mathrm{d}\bar{t} \int_{L_k} \frac{\overline{a}}{\tau - t}\, \mathrm{d}\tau = \frac{1}{4\pi f} \int_{L_k} \mathrm{d}\bar{t}\, \overline{a}\, \pi \mathrm{i} = 0. \tag{2.131}$$

The integral (2.130) is not changed under a scaling $t \mapsto ct$ since $f \mapsto c^2 f$.

We now proceed to consider a two-phase macroscopically isotropic dispersed composite. The tensor $\boldsymbol{\varepsilon}_\perp = \varepsilon_e I$ is expressed through the scalar value ε_e called the effective permittivity. Due to the properties of analytic functions, such a relation can take place for almost all ϱ and f, if the following conditions are fulfilled

$$\mathcal{J}_0' = \mathcal{V}_0' = 0, \quad \mathcal{L}_0' = 1. \tag{2.132}$$

In this case, (2.126) yields the Claudius–Mossotti approximation with the exactly written precision in concentration and contrast parameter

$$\varepsilon_e = 1 + 2\varrho f + 2\varrho^2 f^2 + O(|\varrho|^3 f^4) = \frac{1+\varrho f}{1-\varrho f} + O(|\varrho|^3 f^4). \tag{2.133}$$

We now end up analyzing the results obtained from the explicit scheme for arbitrary shapes of inclusions with real permittivity. These results will be developed in Chapter 4 for complex permittivity and circular shape.

2.6.5 FIRST ITERATION OF THE IMPLICIT SCHEME FOR A TWO-PHASE COMPOSITE

In the present section, we analyze the precision for the first iteration of implicit scheme (2.57)

$$\varphi_k^{(1)}(z) - \tfrac{\varrho}{2\pi i} \int_{L_k} \overline{\varphi_k^{(1)}(t)} E_1(t-z)\mathrm{d}t = \textstyle\sum_{m \neq k} \tfrac{\varrho}{2\pi i} \int_{L_m} \bar{t} E_1(t-z)\mathrm{d}t + z. \quad z \in G_k. \tag{2.134}$$

For simplicity, a two-phase composite is considered when $\varrho = \varrho_k$ for all $(k = 1, 2, \ldots, N)$. We are interested in the effective constant calculated by (2.54) up to $O(f^3)$ for an arbitrary ϱ. We now demonstrate that it is sufficient to determine $\varphi_k^{(1)}(z)$ up to $O(f^2)$.

Theorem 4 *Any approximation* $\varphi_k^{(p)}(z)$ *of implicit scheme* (2.57) *coincides with* $\varphi_k^{(1)}(z)$ *up to* $O(f^2)$.

Proof. Equation (2.134) can be simplified up to $O(f^2)$ to equation

$$\varphi_k^{(1)}(z) - \frac{\varrho}{2\pi i}\int_{L_k} \overline{\varphi_k^{(1)}(t)} E_1(t-z)\mathrm{d}t = z, \quad z \in G_k \tag{2.135}$$

due to the boundness of the integral operator established in Theorem 2. Here, we use the asymptotic formula $E_1(t-z) = E_1(a_m - a_k) + O(f^{1/2})$, as $f \to 0$, for dispersed composites. It is worth noting that the integral operator over L_k increases the precision by the multiplier f. The constant term $E_1(a_m - a_k)$ produces a constant that does not impact the final result, hence it can be discarded. Moreover, in order to find $\varphi_k^{(1)}(z)$ up to $O(f^2)$ it is sufficient instead of (2.135) to solve equation

$$\varphi_k^{(1)}(z) - \frac{\varrho}{2\pi i}\int_{L_k} \frac{\overline{\varphi_k^{(1)}(t)}}{t-z}\mathrm{d}t = z, \quad z \in G_k. \tag{2.136}$$

It is justified by means of the expansion

$$E_1(t-z) = \frac{1}{t-z} - \pi(t-z) - S_4(t-z)^3 + O(f^2). \tag{2.137}$$

The same arguments hold for an arbitrary iteration order p. More precisely, equation (2.57) can be written up to $O(f^2)$ in the form

$$\varphi_k^{(p)}(z) - \frac{\varrho}{2\pi i}\int_{L_k} \overline{\varphi_k^{(p)}(t)} E_1(t-z)\mathrm{d}t = z, \quad z \in G_k. \tag{2.138}$$

This equation coincides with (2.135) which has the unique solution for $|\varrho| < 1$ in the space $\mathcal{H}(G_k)$. It is also proved that equations (2.135) and (2.136) have solutions coinciding up to $O(f^2)$.

This proves the theorem.

Introduce the operator bounded in the space $\mathcal{H}(G_k)$ [229]

$$(\mathcal{S}_k h)(z) = \frac{1}{2\pi i}\int_{L_k} \frac{\overline{h(t)}}{t-z}\mathrm{d}t, \quad h \in \mathcal{H}(G_k). \tag{2.139}$$

Equation (2.136) has the unique solution in the space $\mathcal{H}(G_k)$ for $|\varrho| < 1$ [40]. It can be represented by the absolutely convergent in ϱ series [93, Chapter 2]

$$\varphi_k^{(1)}(z) = z + \sum_{\ell=1}^{\infty} \varrho^\ell \mathcal{S}_k^\ell z \equiv z + \varrho\phi_k^{(1)}(z), \quad z \in G_k \cup L_k, \tag{2.140}$$

where $\mathcal{S}_k^\ell z$ denotes the consequent application ℓ times of the operator $\mathcal{S}$ with the initial function $h(z) = z$.

Two components of the normalized effective permittivity tensor $\boldsymbol{\varepsilon}_\perp$ are calculated by the formula (2.54). Calculating the derivative $\frac{d}{dz}\varphi_k^{(1)}(z)$ and substituting the result into (2.54) we obtain the formulas valid up to $O(f^2)$

$$\begin{aligned}\varepsilon_{11} &= 1+2\varrho f+2\mathrm{Re}\ \textstyle\sum_{\ell=1}^{\infty}\varrho^{\ell+1}\sum_{k=1}^{N}\int_{G_k}\frac{dS_k^\ell z}{dz}(\xi_1+i\xi_2)\ d\xi_1 d\xi_2,\\ \varepsilon_{12} &= -2\mathrm{Im}\ \textstyle\sum_{\ell=1}^{\infty}\varrho^{\ell+1}\sum_{k=1}^{N}\int_{G_k}\frac{dS_k^\ell z}{dz}(\xi_1+i\xi_2)\ d\xi_1 d\xi_2.\end{aligned} \tag{2.141}$$

The solution of equation (2.136) is represented in terms of the series (2.140) in ϱ. An analogous representation holds for the solution of equation (2.135). Such a series yields the asymptotic formula (2.141) for effective constants up to $O(f^2)$. This confirm the precision up to $O(f^2)$ for $\varphi_k^{(1)}(z)$ chosen a priori at the beginning.

The integral equations (2.135) and (2.136) can be considered as a method to determine the complex potentials $\varphi_k^{(1)}(z)$ in inclusions. The complex potential $\varphi(z)$ in the host can be calculated by (2.42). Analysis of this method yields the series (2.141) in ϱ, i.e., shows the structure of approximation and its precision. The formula (2.141) cannot be considered as a closed form expression until the infinite set of compositions of integral operators $S_k^\ell z$ is not constructively found. One can apply another numerical method, FEM, to determine the local field in a composite with one inclusion G_k. But any method yields at most the term $1+2\varrho f$ plus the term of order $O(f)$ symbolically or numerically equivalent to the double sum $\sum_{\ell=1}^{\infty}\sum_{k=1}^{N}$ from (2.141) and does not concern the higher-order term $O(f^2)$.

2.6.6 MAXWELL'S SELF-CONSISTENT APPROACH

In the present Section, we outline Maxwell's self-consistent approach [139], and its relation to Schwarz's method is explained. Maxwell's approach is based on the one-inclusion problem in the whole plane. For simplicity, it is assumed that the homogenized composite is macroscopically isotropic.

Consider one inclusion G_1 of permittivity ε_1 embedded in the host material of permittivity ε. It is assumed that the permittivity is real and the boundary $L_1 = \partial G_1$ is a piece-wise simple Lyapunov curve. The perfect contact between the components is expressed by equations (2.20)

$$u(t) = u_1(t), \quad \frac{\partial u}{\partial \mathbf{n}}(t) = \varepsilon_1 \frac{\partial u_1}{\partial \mathbf{n}}(t), \quad t \in L_1, \tag{2.142}$$

where the functions $u_1(z)$ and $u(z)$ satisfy Laplace's equation in the domains G_1 and $G = \widehat{\mathbb{C}}\backslash(G_1 \cup L_1)$. Moreover, the functions are continuously differentiable in the closures of considered domains, and the function $u(z)$ has a singularity at infinity

$$u(z) \sim x_1, \quad \text{as } z \to \infty. \tag{2.143}$$

This singularity corresponds to the external potential (2.23) written in the complex form

$$u_{ext}(z) = \mathrm{Re}\ z. \tag{2.144}$$

In the considered case of a two-phase composite, we have one dimensionless scalar contrast parameter

$$\varrho = \frac{\varepsilon_1 - 1}{\varepsilon_1 + 1}. \tag{2.145}$$

The considered problem is reduced to the scalar $\mathbb{R}$-linear problem similar to (2.34)

$$\varphi(t) = \varphi_1(t) - \varrho\overline{\varphi_1(t)}, \quad t \in L_1, \tag{2.146}$$

where the functions $\varphi(z)$ and $\varphi_1(z)$ are analytic in the domains G and G_1, respectively. Moreover, these functions are continuously differentiable in the closures of considered domains, and the function $\varphi(z)$ has a pole at infinity

$$\varphi(z) \sim z, \quad \text{as } z \to \infty. \tag{2.147}$$

Apply the Cauchy-type operator (2.44) for L_1 to the boundary condition (2.146). Using (2.48) we obtain the integral equation equivalent to (2.135) in the case $\theta_1 = 0$

$$\varphi_1(z) - \frac{\varrho}{2\pi i}\int_{L_1} \frac{\overline{\varphi_1(t)}}{t-z}\mathrm{d}t = z, \quad z \in G_1. \tag{2.148}$$

Application of the Cauchy-type operator (2.146) for $z \in G$ yields

$$\varphi(z) = \frac{\varrho}{2\pi i}\int_{L_1} \frac{\overline{\varphi_1(t)}}{t-z}\mathrm{d}t + z, \quad z \in G. \tag{2.149}$$

Let $\varphi_1(z)$ be the solution to equation (2.148). Then, the complex potential $\varphi(z)$ is calculated by (2.149).

One can see that equations (2.135) and (2.148) are the same. Therefore, Maxwell's self-consistent approach leads to the same equation, obtained by applying the first-order iteration in the implicit scheme of Schwarz's method discussed in Section 2.6.5. It was established that the solution to equation (2.135) yields the formulas (2.141) valid up to $O(f^2)$. This justifies the validity of Maxwell's approach only in the first-order approximation in f. The claim often made in various works, stating that the first-order approximation could be extended to higher orders in f without accounting for higher-order terms $\varphi_1^{(p)}(z)$ $(p \geq 2)$, lacks any foundation.

Given its limitations, we continue to describe Maxwell's approach along the lines of [139]. Expand the complex potential $\varphi(z)$ near infinity in the Laurent series and take the coefficient on $-\frac{1}{z}$. The result value M_1 is called the dipole moment of the inclusion G_1 [13, 188, 209]. It follows from (2.149) that

$$M_1 = \frac{\varrho}{2\pi \mathrm{i}}\int_{L_1} \overline{\varphi_1(t)}\mathrm{d}t. \tag{2.150}$$

Considering the single inclusion G_k in the complex plane by the above method, we introduce the dipole moments of G_k and estimate it using (2.140)

$$M_k = \frac{\varrho}{2\pi \mathrm{i}}\int_{L_k} \overline{\varphi_k(t)}\mathrm{d}t = \varrho\frac{|G_k|}{\pi} + m_k', \quad k = 1, 2, \ldots, n, \tag{2.151}$$

where $m'_k = \frac{\varrho^2}{2\pi i} \int_{L_k} \overline{\varphi_k(t)} dt$ is of order $O(\varrho^2 f)$. It is assumed that $n = n(R_0)$ inclusions lie in the disk $|z| < R_0$.

Let the disk $|z| < R_0$ be occupied by a material of the real permittivity ε_e and embedded in a medium with normalized unit permittivity. This problem is a particular case of the above problem (2.142)–(2.143) and can be written in the form

$$U(t) = U_1(t), \quad \frac{\partial U}{\partial \mathbf{n}}(t) = \varepsilon_e \frac{\partial U_1}{\partial \mathbf{n}}(t), \quad |t| = R_0, \tag{2.152}$$

where the functions $U_1(z)$ and $U(z)$ satisfy Laplace's equation in the domains $|z| < R_0$ and $|z| > R_0$. The considered functions are continuously differentiable in the closures of considered domains, and the function $U(z)$ has the same singularity at infinity as $u(z)$

$$U(z) \sim x_1, \quad \text{as } z \to \infty. \tag{2.153}$$

The problem (2.152)–(2.153) is reduced to the $\mathbb{R}$-linear problem similar to (2.34)

$$\Phi(t) = \Phi_1(t) - \varrho_e \overline{\Phi_1(t)}, \quad |t| = R_0, \tag{2.154}$$

where

$$\varrho_e = \frac{\varepsilon_e - 1}{\varepsilon_e + 1}. \tag{2.155}$$

The functions $\Phi(z)$ and $\Phi_1(z)$ are analytic in the domains $|z| < R_0$ and $|z| > R_0$, respectively. Moreover, these functions are continuously differentiable in the closures of considered domains, and the function $\Phi(z)$ has a pole at infinity

$$\Phi(z) \sim z, \quad \text{as } z \to \infty. \tag{2.156}$$

The problem (2.154)–(2.156) has the following solution up to an additive arbitrary constant

$$\Phi(z) = z - \frac{\varrho_e R_0^2}{z}, \quad \Phi_1(z) = z. \tag{2.157}$$

Suppose that n inclusions G_k $(k = 1, 2, \ldots, n)$ are located inside the large disk $|z| < R_0$. Let the effective permittivity of the homogenized large disk $|z| < R_0$ be equal to the scalar ε_e. The dipole moment of $\Phi(z)$ equals $\varrho_e R_0^2$. Following Maxwell, we equate the sum of the dipole moments of G_k to the total dipole moment of the homogenized material of the large disk

$$\varrho_e R_0^2 = \sum_{k=1}^{n} M_k \equiv \varrho \sum_{k=1}^{n} \frac{|G_k|}{\pi} + nm', \tag{2.158}$$

where $m' = \frac{1}{n} \sum_{k=1}^{n} m'_k$. Taking into account (2.155) consider (2.158) as an equation on ε_e. Its solution yields the effective permittivity of the disk $|z| < R_0$

$$\varepsilon_e(n) = \frac{1 + R_0^{-2} \sum_{k=1}^{n} M_k}{1 - R_0^{-2} \sum_{k=1}^{n} M_k} = \frac{1 + \frac{\varrho}{\pi R_0^2} \sum_{k=1}^{n} |G_k| + nm' R_0^{-2}}{1 - \frac{\varrho}{\pi R_0^2} \sum_{k=1}^{n} |G_k| + nm' R_0^{-2}}. \tag{2.159}$$

It is assumed that $\varepsilon_e(n)$ tends to the effective permittivity of composite ε_e, as $n \to \infty \Leftrightarrow R_0 \to \infty$. In particular, it is assumed that the concentration is properly defined

$$f = \lim_{n\to\infty} \sum_{k=1}^{n} \frac{n|G_k|}{\pi R_0^2}. \tag{2.160}$$

Bear in mind that the homogenized medium is initially assumed macroscopically isotropic.

Remark 2 *The existence of the concentration, the limit* (2.160), *does not guarantee that a composite is homogenized, i.e., there exists the limit* $\varepsilon_e = \lim_{n\to\infty} \varepsilon_e(n)$ *even in the second-order approximation* $O(f^2)$ *[161]. This implies that any universal formula for* ε_e *holds only up to* $O(f^2)$ *except at neutral inclusions. The devil is in the details, in the term of order* f^2*. This mathematical curiosity arises in self-consistency. The concept becomes clear through the analysis of two half-planes* $x_1 > 0$ *and* $x_1 < 0$ *comprised of square and hexagonal arrays of disks having the same concentration. These arrays maintain the same concentration but have different effective permittivity. We come back to this discussion in Chapter 3 near Figure 3.2.*

Consider the case when $|G_1| = |G_2| = \ldots = |G_N|$ and $m_1' = m_2' = \ldots = m_N'$. Then, (2.159) yields

$$\varepsilon_e = \frac{1 + \varrho f + m' f \pi |G_1|^{-1}}{1 - \varrho f - m' f \pi |G_1|^{-1}} + O(f^2). \tag{2.161}$$

One can check that the term $m' f \pi |G_1|^{-1}$ is of order $O(\varrho^2 f)$.

Example 1 *Let equal disks of radius r be embedded in the host materials. Using the relation* $t = \frac{r^2}{\bar{t}}$ *on the circle* $|t| = r$ *rewrite equation* (2.148)

$$\varphi_1(z) - \frac{\varrho}{2\pi i} \int_{|t|=r} \frac{\overline{\varphi_1\left(\frac{r^2}{\bar{t}}\right)}}{t - z} dt = z, \quad |z| < r. \tag{2.162}$$

By the symmetry principle the function $\overline{\varphi_1\left(\frac{r^2}{\bar{t}}\right)}$ *is analytically continued into the domain* $|z| > r$*. Applying* (2.48) *we calculate the integral from* (2.162) *and obtain*

$$\varphi_1(z) = \varrho\overline{\varphi_1(0)} + z, \quad |z| < r. \tag{2.163}$$

Determine $\varphi(z)$ *by* (2.146) *up to an additive constant*

$$\varphi(z) = z - \frac{\varrho r^2}{z}, \quad |z| > r. \tag{2.164}$$

Therefore, the dipole for the disk $M_1 = \varrho r^2$ *and* $m_1' = 0$ *by* (2.151)*. Then,* (2.161) *becomes the famous Maxwell-Garnett (Clausius-Mossotti, Maxwell, Lorenz, Lorentz, Maxwell-Garnett) approximation [128]*

$$\varepsilon_e = \frac{1 + f\varrho}{1 - f\varrho} + O(f^3). \tag{2.165}$$

The precision is increased here according to [179] established for equal disks.

Some comments are needed to explain the formula (2.165). *For any positive f, it is easy to present such a location of perfectly conducting disks ($\varrho = 1$) that a collection C_M will contain two percolation chains along the axes. Hence, the averaged permittivity of C_M will be infinite. It cannot be reached by the approximation* (2.165). *Conversely, any macroscopically isotropic composite with circular inclusions satisfies* (2.165) *[179]. This assertion does not contradict the previous observation since the formula* (2.165) *is asymptotic with a positive infinitesimally small number f. In this sense, the asymptotic formula* (2.165) *is universal and holds for sufficiently small f.*

Example 2 *Let the ellipse L_1 be defined by equation $\frac{x_1^2}{(1+\alpha)^2}+\frac{x_2^2}{(1-\alpha)^2}=1$, where for definiteness $0<\alpha<1$. The semi-axes of the ellipse are equal to $(1+\alpha)$ and $(1-\alpha)$. The ellipse equation can be written in the complex form*

$$\overline{z} = z\frac{1+\alpha^2}{2\alpha} - \frac{1-\alpha^2}{2\alpha}\sqrt{z^2-4\alpha}, \quad z \in L_1, \tag{2.166}$$

where the square root is defined on the complex plane without the slit $(-2\sqrt{\alpha}, 2\sqrt{\alpha})$ on the x_1-axis. Moreover, the square root tends to $+\infty$ as $z = x_1 > 0$ tends to infinity along the real axis.

Consider the $\mathbb{R}$-linear problem (2.146)–(2.147). *One can check that the functions*

$$\varphi_1(z) = \frac{z}{1-\varrho\alpha}, \; \varphi(z) = \frac{1}{1-\varrho\alpha}\left[z\left(1-\varrho\frac{1+\alpha^2}{2\alpha}\right)+\varrho\frac{1-\alpha^2}{2\alpha}\sqrt{z^2-4\alpha}\right] \tag{2.167}$$

satisfy the problem (2.146)–(2.147). *The dipole moment of the ellipse has the form*

$$M_1 = \varrho\frac{1-\alpha^2}{1-\varrho\alpha}. \tag{2.168}$$

Consider equal non-overlapping elliptic inclusions G_k obtained from G_1 by translation and rotation by an angle θ_k. The complex potentials for the inclusion G_k can be obtained from (2.167) *by the transformation $z \mapsto z\exp(\mathrm{i}\theta_k)+a_k$. The dipole moment has the form*

$$M_k = \varrho\frac{1-\alpha^2}{1-\varrho^2\alpha^2}\left[1+\exp(\mathrm{i}\theta_k)\varrho\alpha\right]. \tag{2.169}$$

Let inclination angle θ be considered a random variable uniformly distributed on $(0,\pi)$. Then, the mathematical expectation $\mathbf{E}[\theta]=0$ and

$$\lim_{n\to\infty}\frac{1}{n}\sum_{k=1}^{n} M_k = \frac{\varrho|G_1|}{\pi}\frac{1}{1-\varrho^2\alpha^2}, \tag{2.170}$$

since the area of ellipse holds $|G_1| = \pi(1-\alpha^2)$. Using this formula, we obtain from (2.159) *and* (2.160)

$$\varepsilon_e = \frac{1+\frac{\varrho f}{1-\varrho^2\alpha^2}}{1-\frac{\varrho f}{1-\varrho^2\alpha^2}} + O(f^2) = 1+\frac{2\varrho f}{1-\varrho^2\alpha^2}+O(f^2). \tag{2.171}$$

It is worth comparing (2.171) *with the extended formula* (4.148).

We return to ellipses in Sections 4.2, 4.3 and extend the formulas to the complex permittivity of components.

2.7 EFFECTIVE PERMITTIVITY TENSOR FOR MULTI-PHASE COMPOSITES

Above in this chapter, we derived formulas for real-valued permittivity. We now begin to explore complex values of the normalized permittivity of components ε_k. Consequently, the contrast parameter ϱ_k also becomes complex. Building upon the foundations laid out, we extend these formulas to the complex domain through analytic continuation in terms of ϱ_k. To ensure the validity of this continuation, it is imperative that we establish a groundwork for understanding the interplay between complex permittivity and its matrix representation. This will be our main goal as we move forward.

The effective permittivity tensor with complex components will be denoted in the same way

$$\boldsymbol{\varepsilon}_\perp = \begin{pmatrix} \varepsilon_{11} & \varepsilon_{12} \\ \varepsilon_{21} & \varepsilon_{22} \end{pmatrix}, \tag{2.172}$$

where $\varepsilon_{12} = \varepsilon_{21}$. Besides it, the designations of Section 2.4 for the vector-complex potentials and other values will be used here.

Introduce the averaged value of a function $F(z)$ over the unit periodicity cell Q

$$\langle F(z)\rangle = \int_Q F(x_1 + ix_2)\, \mathrm{d}x_1 \mathrm{d}x_2. \tag{2.173}$$

Here, the real plane coordinates $\mathbf{x} = (x_1, x_2)$ are identified with the complex coordinate $z = x_1 + \mathrm{i}x_2$. The complex effective permittivity tensor can be defined through the following relation [113]

$$\langle \varepsilon(z)\nabla u(z)\rangle = \boldsymbol{\varepsilon}_\perp \langle \nabla u(z)\rangle, \tag{2.174}$$

where $\varepsilon(z) = 1$ in G and $\varepsilon(z) = \varepsilon_k$ in G_k. The value $\nabla u_{ext} := \langle \nabla u(z)\rangle$ determines the complex external flux applied to the unit periodicity cell Q. It follows from the theory of homogenization [22, 113] that in order to determine the tensor $\boldsymbol{\varepsilon}_\perp$, it is sufficient to find two local fields $\nabla u(z)$ for two external fluxes

$$\nabla u_{ext[1]} = \begin{pmatrix} 1 \\ 0 \end{pmatrix} \quad \text{and} \quad \nabla u_{ext[2]} = \begin{pmatrix} 0 \\ 1 \end{pmatrix}. \tag{2.175}$$

Let us fix the first external flux from (2.175). Consider the vector equality with complex components

$$\langle \nabla u\rangle = \int_G \nabla u\, \mathrm{d}x_1 \mathrm{d}x_2 + \sum_{k=1}^{N} \int_{G_k} \nabla u_k \mathrm{d}x_1 \mathrm{d}x_2. \tag{2.176}$$

Transform the double integrals by the divergence theorem associated to the Ostrogradsky-Gauss theorem of real vector analysis

$$\int_G \nabla u \, \mathrm{d}x_1 \mathrm{d}x_2 = \int_{\partial Q} u\mathbf{n} \, \mathrm{d}s + \sum_{k=1}^{N} \int_{\partial G_k} (u_k - u)\mathbf{n} \, \mathrm{d}s, \tag{2.177}$$

where $\mathbf{n}$ denotes the outward normal unit vector to the boundary $\partial G = \partial Q - \sum_{k=1}^{N} \partial G_k$. It follows from the first relation (2.20) that the integrals over ∂G_k vanish. The integral over ∂Q is calculated by the jumps of u per unit periodicity cell Q given by (2.25), see for details Chapter 3 and Remark 5 in Section 2.2.2 of [93]. Ultimately, we have

$$\langle \nabla u(z) \rangle = \begin{pmatrix} 1 \\ 0 \end{pmatrix} \text{ and } \boldsymbol{\varepsilon}_{\perp} \langle \nabla u(z) \rangle = \begin{pmatrix} \varepsilon_{11} \\ \varepsilon_{21} \end{pmatrix}. \tag{2.178}$$

We now proceed to calculate the left part of (2.174) which has to be equal to the second vector of (2.178)

$$\begin{pmatrix} \varepsilon_{11} \\ \varepsilon_{21} \end{pmatrix} = \int_G \nabla u \, \mathrm{d}x_1 \mathrm{d}x_2 + \sum_{k=1}^{N} \varepsilon_k \int_{G_k} \nabla u_k \mathrm{d}x_1 \mathrm{d}x_2. \tag{2.179}$$

Green's formula will be applied to the components of (2.179)

$$\int_S \left(\frac{\partial F}{\partial x_1} - \frac{\partial G}{\partial x_2} \right) \mathrm{d}x_1 \mathrm{d}x_2 = \int_{\partial S} G \, \mathrm{d}x_1 + F \, \mathrm{d}x_2. \tag{2.180}$$

The components of (2.179) can be written in the form

$$\varepsilon_{11} = \int_{\partial Q} u \, \mathrm{d}x_2 + \sum_{k=1}^{N} (\varepsilon_k - 1) \int_{\partial G_k} u_k \, \mathrm{d}x_2, \tag{2.181}$$

$$-\varepsilon_{21} = \int_{\partial Q} u \, \mathrm{d}x_1 + \sum_{k=1}^{N} (\varepsilon_k - 1) \int_{\partial G_k} u_k \, \mathrm{d}x_1. \tag{2.182}$$

Using the jump conditions per unit periodicity cell Q (2.25) we calculate the integrals over ∂Q

$$\int_{\partial Q} u \, \mathrm{d}x_1 = - \int_{-\frac{1}{2}}^{\frac{1}{2}} \left[u\left(x_1 + \frac{\mathrm{i}}{2}\right) - u\left(x_1 - \frac{\mathrm{i}}{2}\right) \right] \mathrm{d}x_1 = 0, \tag{2.183}$$

$$\int_{\partial Q} u \, \mathrm{d}x_2 = \int_{-\frac{1}{2}}^{\frac{1}{2}} \left[u\left(\frac{1}{2} + \mathrm{i}x_2\right) - u\left(-\frac{1}{2} + \mathrm{i}x_2\right) \right] \mathrm{d}x_2 = 1. \tag{2.184}$$

Again, applying Green's formula for G_k we obtain

$$\begin{pmatrix} \varepsilon_{11} \\ \varepsilon_{21} \end{pmatrix} = \begin{pmatrix} 1 \\ 0 \end{pmatrix} + \sum_{k=1}^{N} (\varepsilon_k - 1) \int_{G_k} \nabla u_k \, \mathrm{d}x_1 \mathrm{d}x_2. \tag{2.185}$$

Below, the formulas (2.27)–(2.36) will be used including the representation

$$\psi_k(z) = \frac{1}{2}(I+\boldsymbol{\alpha}_k)\begin{pmatrix} \dfrac{\partial u_k'}{\partial x_1} - \mathrm{i}\dfrac{\partial u_k'}{\partial x_2} \\ \\ \dfrac{\partial u_k''}{\partial x_1} - \mathrm{i}\dfrac{\partial u_k''}{\partial x_2} \end{pmatrix}, \quad z \in G_k, \tag{2.186}$$

which can be written in the equivalent form

$$\begin{pmatrix} \dfrac{\partial u_k'}{\partial x_1} - \mathrm{i}\dfrac{\partial u_k'}{\partial x_2} \\ \\ \dfrac{\partial u_k''}{\partial x_1} - \mathrm{i}\dfrac{\partial u_k''}{\partial x_2} \end{pmatrix} = \frac{2}{|1+\varepsilon_k|^2}\begin{pmatrix} 1+\varepsilon_k' & \varepsilon_k'' \\ -\varepsilon_k'' & 1+\varepsilon_k' \end{pmatrix}\psi_k(z), \quad z \in G_k. \tag{2.187}$$

Then, the gradient ∇u_k can be written in terms of the complex potentials

$$\nabla u_k \equiv \begin{pmatrix} \dfrac{\partial u_k'}{\partial x_1} + \mathrm{i}\dfrac{\partial u_k''}{\partial x_1} \\ \\ \dfrac{\partial u_k'}{\partial x_2} + \mathrm{i}\dfrac{\partial u_k''}{\partial x_2} \end{pmatrix} = \frac{2}{\varepsilon_k+1}\begin{pmatrix} \mathrm{Re}\,\psi_{1k} + \mathrm{iRe}\,\psi_{2k} \\ \\ -\mathrm{Im}\,\psi_{1k} - \mathrm{iIm}\,\psi_{2k} \end{pmatrix}. \tag{2.188}$$

where $\psi_{1k}(z)$ and $\psi_{2k}(z)$ are coordinate of the vector function $\psi_k(z)$.

Substitute (2.188) into (2.185) taking into account (2.32)

$$\begin{pmatrix} \varepsilon_{11} \\ \varepsilon_{21} \end{pmatrix} = \begin{pmatrix} 1 \\ 0 \end{pmatrix} + 2\sum_{k=1}^{N} \varrho_k \int_{G_k} \begin{pmatrix} \mathrm{Re}\,\psi_{1k} + \mathrm{iRe}\,\psi_{2k} \\ -\mathrm{Im}\,\psi_{1k} - \mathrm{iIm}\,\psi_{2k} \end{pmatrix} \mathrm{d}x_1 \mathrm{d}x_2, \tag{2.189}$$

where the scalar value ϱ_k has the form (2.36). Therefore, in order to calculate (2.189) we have to find the vector function $\frac{d}{dz}\varphi_k(z) = \psi_k(z) = (\psi_{1k}(z), \psi_{2k}(z))^\top$ satisfying the boundary value problem (2.34)–(2.35).

The formula (2.189) becomes (2.54) for the real-valued permittivity because $\psi_{1k} = \frac{d\varphi_k}{dz}$ and $\psi_{2k} = 0$. The eigenvalue ϱ_k of the contrast parameter matrix becomes the scalar real contrast parameter ϱ_k.

2.8 INTEGRAL EQUATIONS AND THEIR APPROXIMATE SOLUTION FOR MULTI-PHASE COMPOSITES

The scalar $\mathbb{R}$-linear problem was reduced to the system of scalar integral equations (2.51) in Section 2.5. Using the same arguments, we reduce the vector $\mathbb{R}$-linear problem (2.34)–(2.35) to the vector system of integral equations up to an arbitrary additive constant vector

$$\boldsymbol{\varphi}_k(z) = \sum_{m=1}^{N} \frac{1}{2\pi i}\int_{L_m} \boldsymbol{\beta}_m \overline{\boldsymbol{\varphi}_m(t)} E_1(t-z)\,\mathrm{d}t + \begin{pmatrix} z \\ 0 \end{pmatrix}, \; z \in G_k \; (k = 1,2,\ldots,N). \tag{2.190}$$

In the scalar case, the contrast approximations were defined by the constant $\varrho_0 = \max_{k=1,2,\ldots,N} |\varrho_k|$. In the vector-matrix case, introduce the spectral norm $\|\boldsymbol{\beta}_m\|$ of the matrix $\boldsymbol{\beta}_m$ as the square root of the maximum eigenvalue of $\boldsymbol{\beta}_m^\top \boldsymbol{\beta}_m$. We have $\boldsymbol{\beta}_m^\top \boldsymbol{\beta}_m = |\varrho_m|^2 I$, where ϱ_m is the eigenvalue of $\boldsymbol{\beta}_m$ given by (2.36). Hence, the contrast approximations can be defined by the parameter

$$\varrho_0 = \max_{m=1,2,\ldots,N} \left| \frac{\varepsilon_m - 1}{\varepsilon_m + 1} \right|. \tag{2.191}$$

Following Section 2.6.3, we find an approximate solution to equations (2.190) by two iterations. Similar to (2.83) the first-order term has the form

$$\boldsymbol{\varphi}_k^{(1)}(z) = \begin{pmatrix} z \\ 0 \end{pmatrix} + O(1), \quad z \in G_k. \tag{2.192}$$

The second-order approximation is found similar to (2.85)

$$\boldsymbol{\varphi}_k^{(2)}(z) = \sum_{m=1}^{N} \frac{\boldsymbol{\beta}_m}{2\pi i} \int_{L_m} \begin{pmatrix} \bar{t} \\ 0 \end{pmatrix} E_1(t-z)\, \mathrm{d}t + \begin{pmatrix} z \\ 0 \end{pmatrix} + O(\varrho_0^2), \quad z \in G_k. \tag{2.193}$$

Below, we omit the precision in ϱ_0 for shortness and keep track of the precision in f. Differentiate equation (2.193)

$$\psi_k^{(2)}(z) = \frac{d}{dz} \boldsymbol{\varphi}_k^{(2)}(z) = \begin{pmatrix} 1 \\ 0 \end{pmatrix} + \sum_{m=1}^{N} \frac{\boldsymbol{\beta}_m}{2\pi i} \int_{L_m} \begin{pmatrix} 1 \\ 0 \end{pmatrix} \bar{t}\, E_2(t-z)\, \mathrm{d}t, \tag{2.194}$$

Using the relation

$$\boldsymbol{\beta}_m \begin{pmatrix} 1 \\ 0 \end{pmatrix} = \frac{1}{|1+\varepsilon_m|^2} \begin{pmatrix} |\varepsilon_m|^2 - 1 \\ 2\varepsilon_m'' \end{pmatrix} \tag{2.195}$$

write (2.194) in the form

$$\psi_k^{(2)}(z) = \begin{pmatrix} 1 \\ 0 \end{pmatrix} + \frac{1}{\pi} \sum_{m=1}^{N} \frac{1}{|1+\varepsilon_m|^2} \begin{pmatrix} |\varepsilon_m|^2 - 1 \\ 2\varepsilon_m'' \end{pmatrix} \frac{dg_m}{dz}(z),\ z \in G_k. \tag{2.196}$$

where $g_m(z)$ has the form (2.87), hence,

$$\frac{dg_m}{dz}(z) = \frac{1}{2i} \int_{L_m} \bar{t}\, E_2(t-z)\, \mathrm{d}t. \tag{2.197}$$

Write the relation (2.196) by coordinates

$$\begin{aligned} \psi_{1k}^{(2)}(z) &= 1 + \tfrac{1}{\pi} \textstyle\sum_{m=1}^{N} \frac{|\varepsilon_m|^2 - 1}{|1+\varepsilon_m|^2} \frac{dg_m}{dz}(z) \\ \psi_{2k}^{(2)}(z) &= \tfrac{1}{\pi} \textstyle\sum_{m=1}^{N} \frac{2\varepsilon_m''}{|1+\varepsilon_m|^2} \frac{dg_m}{dz}(z), \quad z \in G_k. \end{aligned} \tag{2.198}$$

Taking into account the identity

$$\varrho_m \equiv \frac{\varepsilon_m - 1}{\varepsilon_m + 1} = \frac{|\varepsilon_m|^2 - 1 + 2\mathrm{i}\varepsilon_m''}{|1+\varepsilon_m|^2}, \tag{2.199}$$

find the vector

$$\begin{pmatrix} \mathrm{Re}\,\psi_{1k}^{(2)} + \mathrm{iRe}\,\psi_{2k}^{(2)} \\ -\mathrm{Im}\,\psi_{1k}^{(2)} - \mathrm{iIm}\,\psi_{2k}^{(2)} \end{pmatrix} = \begin{pmatrix} 1 \\ 0 \end{pmatrix} + \frac{1}{\pi}\sum_{m=1}^{N} \varrho_m \begin{pmatrix} \mathrm{Re}\,\frac{dg_m}{dz} \\ -\mathrm{Im}\,\frac{dg_m}{dz} \end{pmatrix}. \tag{2.200}$$

Introduce the averaged scalar contrast parameter over inclusions similar to the case of real-valued permittivity (2.89)

$$\langle \varrho \rangle := \sum_{k=1}^{N} \varrho_k |G_k|. \tag{2.201}$$

Substitute (2.200) into (2.189)

$$\begin{pmatrix} \varepsilon_{11} \\ \varepsilon_{21} \end{pmatrix} = (1 + 2\langle \varrho \rangle) \begin{pmatrix} 1 \\ 0 \end{pmatrix}$$
$$+ \tfrac{2}{\pi} \textstyle\sum_{k=1}^{N} \sum_{m=1}^{N} \varrho_k \varrho_m \int_{G_k} \begin{pmatrix} \mathrm{Re}\, h_{km} \\ -\mathrm{Im}\, h_{km} \end{pmatrix} \mathrm{d}x_1 \mathrm{d}x_2 + O(\varrho_0^3 f^{7/2}), \tag{2.202}$$

where h_{km} has the form (2.91) and ϱ_0 is given by (2.191). The precision of (2.202) is taken as the precision of (2.90) due to the same application of Schwarz's scheme. One can see that the same values h_{km} are needed for the calculation of the permittivity tensor in the case of real-valued permittivity by (2.90) and of complex permittivity by (2.202). The formulas (2.90) and (2.202) are similar, and (2.202) has to include (2.90) as a particular case. The difference between (2.90) and (2.202) consists of the assumption of the contrast parameters ϱ_k which have to be real in (2.90) and can be complex in (2.202).

This observation allows us to state the following assertion

Rule $\mathbb{R} \to \mathbb{C}$:
Any formula for the real-valued permittivity of components is transformed into a formula for the complex permittivity.

This Rule is loosely stated and needs explanations since a formula for the real-valued permittivity must first be written in the corresponding form to be extended to the complex permittivity. The formulas from [93, 180] and others were obtained using the classic scalar complex potentials. For instance, the formula (2.90) for the real-valued permittivity was written in the form

$$\varepsilon_{11} - \mathrm{i}\varepsilon_{21} = 1 + 2\langle \varrho \rangle + \tfrac{2}{\pi} \textstyle\sum_{k=1}^{N} \sum_{m=1}^{N} \varrho_k \varrho_m \int_{G_k} h_{km}\, \mathrm{d}x_1 \mathrm{d}x_2 + O(\varrho_0^3 f^{7/2}). \tag{2.203}$$

Hence, two real equations (2.90) are written in the form of one complex equation. In order to extend (2.203) to the complex permittivity, first, one has to write (2.203) in the vector form (2.202). After this, one can extend the real-valued vector to the complex-valued vector, assuming that the contrast parameters ϱ_k become complex.

Now, we take the second external flux (2.175). Analogously the previous transformations, we arrive at the formula

$$\begin{pmatrix} \varepsilon_{22} \\ -\varepsilon_{12} \end{pmatrix} = \begin{pmatrix} 1 \\ 0 \end{pmatrix} + 2\sum_{k=1}^{N} \varrho_k \int_{G_k} \begin{pmatrix} \mathrm{Re}\,\psi_{1k} + i\mathrm{Re}\,\psi_{2k} \\ -\mathrm{Im}\,\psi_{1k} - i\mathrm{Im}\,\psi_{2k} \end{pmatrix} \mathrm{d}x_1 \mathrm{d}x_2, \tag{2.204}$$

where ψ_{1k} and ψ_{2k} differ from the values of (2.189) calculated for the first flux. Two vector equations (2.189) and (2.202) can be written in the form

$$\begin{pmatrix} \varepsilon_{11} & \varepsilon_{12} \\ \varepsilon_{21} & \varepsilon_{22} \end{pmatrix} = \begin{pmatrix} 1 & 0 \\ 0 & 1 \end{pmatrix} + 2\sum_{k=1}^{N} \varrho_k \times$$
$$\int_{G_k} \begin{pmatrix} \mathrm{Re}\,\psi_{1k[1]} + \mathrm{iRe}\,\psi_{2k[1]} & \mathrm{Im}\,\psi_{1k[2]} + \mathrm{iIm}\,\psi_{2k[2]} \\ -\mathrm{Im}\,\psi_{1k[1]} - \mathrm{iIm}\,\psi_{2k[1]} & \mathrm{Re}\,\psi_{1k[2]} + \mathrm{iRe}\,\psi_{2k[2]} \end{pmatrix} \mathrm{d}x_1 \mathrm{d}x_2, \tag{2.205}$$

where the subscripts [1] and [2] indicate the solutions constructed by the corresponding external fields (2.175).

The formula (2.124) was derived for the real-valued permittivity. Using **Rule** $\mathbb{R} \to \mathbb{C}$ one can apply this formula to the complex permittivity

$$\boldsymbol{\varepsilon}_{\perp} = (1 + 2\langle\varrho\rangle)I + 2f \begin{pmatrix} \mathrm{Re}\,\mathcal{J}_0 & -\mathrm{Im}\,\mathcal{J}_0 \\ -\mathrm{Im}\,\mathcal{J}_0 & -\mathrm{Re}\,\mathcal{J}_0 \end{pmatrix}$$
$$+ 2f^2 \begin{pmatrix} \mathrm{Re}\,\mathcal{L}_0 & -\mathrm{Im}\,\mathcal{L}_0 \\ -\mathrm{Im}\,\mathcal{L}_0 & 2 - \mathrm{Re}\,\mathcal{L}_0 \end{pmatrix} + 2f^{5/2} \begin{pmatrix} \mathrm{Re}\,\mathcal{V}_{01} & -\mathrm{Im}\,\mathcal{V}_{01} \\ -\mathrm{Im}\,\mathcal{V}_{01} & -\mathrm{Re}\,\mathcal{V}_{01} \end{pmatrix} \tag{2.206}$$
$$+ 2f^3 \begin{pmatrix} \mathrm{Re}\,\mathcal{V}_{02} & -\mathrm{Im}\,\mathcal{V}_{02} \\ -\mathrm{Im}\,\mathcal{V}_{02} & -\mathrm{Re}\,\mathcal{V}_{02} \end{pmatrix} + 2f^{7/2} \begin{pmatrix} \mathrm{Re}\,\mathcal{V}_{03} & -\mathrm{Im}\,\mathcal{V}_{03} \\ -\mathrm{Im}\,\mathcal{V}_{03} & -\mathrm{Re}\,\mathcal{V}_{03} \end{pmatrix}$$
$$+ O(\varrho_0^3 f^4),$$

where the values from (2.206) are calculated by the same formulas (2.117) and (2.110)–(2.115).

The first invariant of the tensor $\boldsymbol{\varepsilon}_{\perp}$ takes the form

$$\frac{1}{2}(\varepsilon_{11} + \varepsilon_{22}) = 1 + 2\langle\varrho\rangle + 2\langle\varrho\rangle^2 + O(\varrho_0^3). \tag{2.207}$$

One can see that it does not depend on the location of inclusions in the considered approximation.

The normalized longitudinal permittivity of the considered fibrous composite is calculated by the mean value, see formula (3.142) from [235]

$$\varepsilon_{\parallel} = 1 - f + \sum_{k=1}^{N} \varepsilon_k f_k, \tag{2.208}$$

where f_k stands for the concentration of the phase of permittivity ε_k.

3 Effective Medium Approximation

Much Ado About Nothing

William Shakespeare

3.1 CLUSTER MODIFICATION

Various modifications of the first-order iteration of implicit scheme (2.57) were proposed and applied in many publications. The precision up to $O(f^2)$ of the first-order iteration (2.134) is established in the previous section. Theorem 4 implies that various modifications of implicit scheme (2.57) cannot improve the precision. In particular, it was established in Section 2.6.6 that the classical Maxwell self-consistent method (SCM) cannot be applied to higher-order approximations in f, i.e., it may be used only for composites with non-interacting inclusions. Dilute dispersed media and composites with neutral inclusions [51, 155] refer to such a class of composites.

The present section demonstrates the same fact for a finite collection of inclusions (cluster). Figure 3.1 illustrates an attempt to replace a composite with a finite cluster. The proper result of such a replacement is discussed in [168], where it is demonstrated that the applied SCM leads to the effective permittivity of finite clusters diluted in the host shown in Figure 3.1c. The same result holds for an elastic problem [173] and is expected for all types of homogenization problems for dispersed composites discussed in [115].

Return to the example discussed in Remark 2 on page 52. In the framework of SCM, consider the finite collection of disks displayed in Figure 3.2 representing two half-planes comprised of infinite square and hexagonal arrays of disks. The effective permittivity of this homogenized structure does not exist. However, a self-consistent method will give another result. The correspondence $(a) \Leftrightarrow (c)$ in Figure 3.1 yields the effective permittivity of dilute clusters bounded by the dashed line in Figure 3.2.

A rigorous mathematical study can properly explain this effect of dilute clusters. Consider a collection C_M of inclusions G_m $(m = 1, 2, \ldots, M)$ displayed in Figure 3.1a. Other inclusions G_m $(m = M+1, M+2, \ldots, N)$ are not taken into account. Instead of

DOI: 10.1201/9781032646633-3

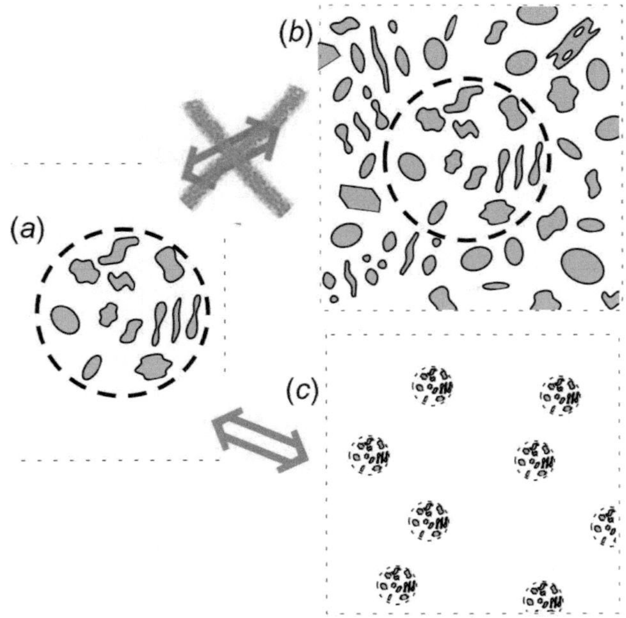

Figure 3.1 Illustration of the methodologically wrong and correct self-consistent concepts: a finite collection (a) embedded in the infinite medium and bounded by the dashed circle does not represent a composite (b) and does represent dilute clusters (c).

(2.134) consider the system of equations

$$\varphi_k^{(1)}(z) - \sum_{m=1}^{M} \frac{\varrho}{2\pi i} \int_{L_m} \frac{\overline{\varphi_k^{(1)}(t)}}{t-z} \mathrm{d}t = \sum_{m=1}^{M} \frac{\varrho}{2\pi i} \int_{L_m} \bar{t} E_0(t-z) \mathrm{d}t \tag{3.1}$$

$$+ \sum_{m=M+1}^{N} \frac{\varrho}{2\pi i} \int_{L_m} \bar{t} E_1(t-z) \mathrm{d}t + z, \quad z \in G_k \quad (k = 1, 2, \ldots, M),$$

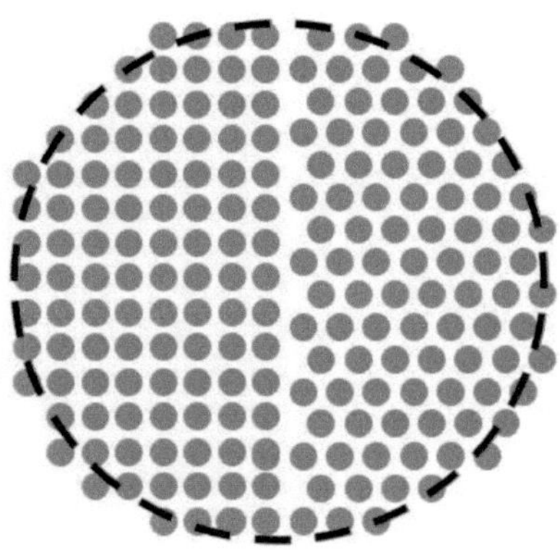

Figure 3.2 The finite collection of disks consisting of the halves of square and hexagonal arrays enclosed to a circle. The infinite arrays have the same concentration but different effective permittivity.

where the function $E_0(z) := E_1(z) - \frac{1}{z}$ is regular in Q. The concentration f now is the concentration of clusters C_M in the composite displayed in Figure 3.1c. The cluster C_M has its interior concentration defined as the ratio $f_C = |C_M|^{-1} \sum_{m=1}^{M} |G_m|$, where $|C_M|$ stands for the area of C_M, i.e., of the domain bounded by the dashed circle. The concentrations f and f_C are unrelated, and f is introduced as an independent parameter following Maxwell's self-consistent approach.

This modification of the first iteration considers interactions among the inclusions G_m $(m = 1, 2, \ldots, M)$ in the cluster C_M. The impact of other inclusions of composite onto the cluster C_M is in the right part of (3.1) found in the zero-order iteration, where the integral terms over L_m $(m \neq k)$ correspond to the periodic images of the cluster (with the kernel E_0) and to other doubly periodic inclusions (with the kernel E_1). The first iteration (3.1) yields the effective constant up to $O(f^2)$, where f is the cluster concentration.

The above mathematical consideration can be supported by the following physical arguments. The local field deep interior of the collection in Figure 3.1a can be accurately computed. However, the local field near the boundary of the cluster (dashed circle) is distorted by the induced charge P_{add} first explained in [146, Section 2.4]. Rayleigh [206] understood the effect of additional polarization charge arising on the boundary of the cluster since he used the proper Eisenstein summation for which P_{add} vanishes. Schwarz's method shows that the distortion P_{add} on the formal calculation of the effective constants has the second-order in f. Therefore, the geometrical transformation from (a) to (b) in Figure 3.1 is methodologically wrong and must be replaced by the transformation from (a) to the dilute structure (c).

Nevertheless, the scheme in Figure 3.1 can be corrected in some cases as shown in Figure 3.3. For definiteness, we consider a macroscopically isotropic composite. The double periodic problem is reduced to a boundary value problem for the quadruple cell $\mathcal{Q}$, which consists of the RVE Q and its three adjusted symmetrically with respect to the vertical and horizontal axes domains Q^*, Q_*, Q_*^*. The effective permittivity of $\mathcal{Q}$ coincides with the effective permittivity of the united quarters $Q \cup Q^* \cup Q_* \cup Q_*^*$.

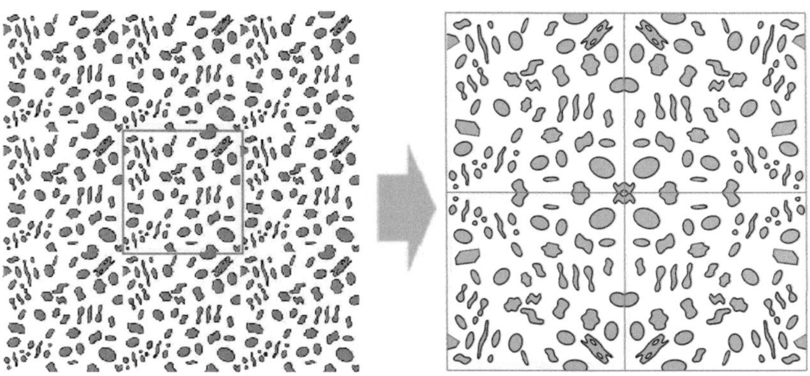

Figure 3.3 Illustration of the correct methodology contrary to the transition from (a) to (b) in Figure 3.1. A double periodic composite is replaced here by the quadruple cell $\mathcal{Q}$.

Due to symmetry, we have the mixed boundary value problem for the domain Q. The Dirichlet conditions $u(\pm 1, y) = \pm 1$ are given on the vertical lines of ∂Q and the Neumann conditions $\frac{\partial u}{\partial x_2}(x, \pm 1) = 0$ on the horizontal lines of ∂Q. The above construction of an auxiliary quadruple cell complements the study [22, p. 212-222] concerning a cuboid/rectangular cell with a symmetric shape inclusion. This scheme produces the accurate result because the induced polarization vanishes, $P_{add} = 0$, for the rectangular shape under consideration, taking into account the relevant boundary conditions.

3.2 EFFECTIVE MEDIUM APPROXIMATION AND OTHER PSEUDO-MODIFICATIONS

In the previous section, we demonstrate that SCM is not "very wrong" and gives the correct results for dilute clusters after the proper interpretation. This assertion coincides with Maxwell's SCM. Below, we consider some popular modifications of SCM. Their validity is based on intuition without any justification. In the present section, we discuss Effective Medium Approximation (EMA). It can be seen that the EMA worsens the result, and we cannot offer any reasonable interpretation of this method.

3.2.1 PRECISION OF EFFECTIVE MEDIUM APPROXIMATION

In the present section, we modify the scheme considered in the previous section following the widely used extensions of SCM, sometimes called Effective Medium Approximation. Consider a finite collection of inclusions C_M in a large disk $|z| < R$. As before, the inclusions of permittivity ε_1 are embedded in the medium of unit permittivity in $|z| < R$. Let this composite disk $|z| < R$ be embedded in the medium of the unknown effective permittivity ε_e. For simplicity of consideration, let the homogenized medium be macroscopically isotropic.

Introduce the "effective" contrast parameter

$$\varrho_e = \frac{\varepsilon - \varepsilon_e}{\varepsilon + \varepsilon_e}. \tag{3.2}$$

Instead of (3.1) we obtain the system of equations

$$\varphi_k^{(1)}(z) - \sum_{m=1}^{M} \frac{\varrho}{2\pi i} \int_{L_k} \overline{\varphi_k^{(1)}(t)} \frac{1}{t-z} \mathrm{d}t \tag{3.3}$$

$$= \frac{\varrho_e}{2\pi i} \int_{|t|=R} \overline{\varphi_e(t)} \frac{1}{t-z} \mathrm{d}t + z, \quad z \in G_k \quad (k = 1, 2, \ldots, M),$$

where $\varphi_e(z)$ denotes the complex potential in $|z| > R$. The system (3.3) has to be completed by the integral equation

$$\varphi_e(z) - \frac{\varrho_e}{2\pi i} \int_{|t|=R} \overline{\varphi_e(t)} \frac{1}{t-z} \mathrm{d}t = \sum_{m=1}^{M} \frac{\varrho}{2\pi i} \int_{L_k} \overline{\varphi_k^{(1)}(t)} \frac{1}{t-z} \mathrm{d}t + z, \; |z| > R. \tag{3.4}$$

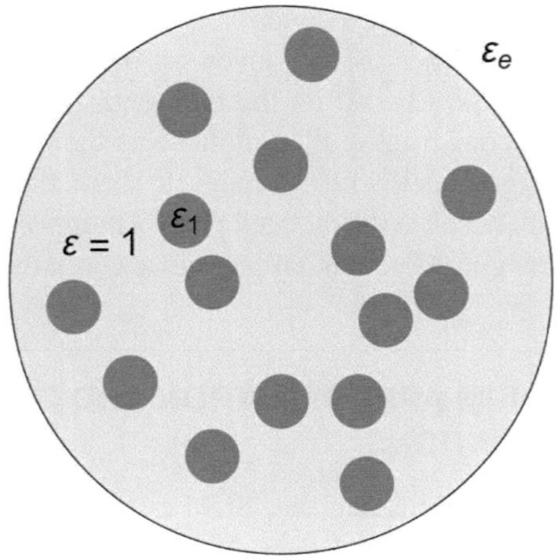

Figure 3.4 Modification of EMA by the introduction of the fictitious medium in $|z| > R$ of permittivity ε_e.

This equation (3.4) can be considered as an equation for the $(M+1)$th inclusion $|z| > R$ in Schwarz's method for $(M+1)$ inclusions.

Let the functions $\varphi_e(z)$ and $\varphi_k^{(1)}(z)$ be found. An additional relation on ϱ_e has to be considered to determine the effective permittivity ε_e. The result will be valid at most up to $O(f^2)$ since the first iteration of Schwarz's method cannot exceed this precision.

The discussed modification means that the representative cluster C_M interacts with the homogenized exterior medium with the unknown effective constant ε_e. Let the local field be found near C_M and the averaged effective permittivity ε_c over C_M. The value ε_C depends on ε_e. Ultimately, the following equation on ε_e has to be solved

$$\varepsilon_C(\varepsilon_e) = \varepsilon_e. \tag{3.5}$$

The precision of EMA was not analyzed in the previous publications. We now fill this gap, consider the local flux in circular inclusions up to $O(f^2)$ for simplicity, and analyze the result. In the framework of EMA, it is assumed that the cluster concentration f_C coincides with the global concentration f of inclusions.

The problem considered can be stated as follows: Let the large disk $|z| < R$ contains M mutually disjoint small circular inclusions $|z - a_k| < r$ $(k = 1, 2, \ldots, M)$ of permittivity ε_1 as displayed in Figure 3.4. Let the origin $z = 0$ lie out of inclusions. The permittivity of large disk without small disks is normalized to unity $\varepsilon = 1$. Let the domain $|z| > R$ be occupied by a material of permittivity ε_e. The large disk can be normalized so that its area holds unity, i.e., $R^2 = \frac{1}{\pi}$. The concentration of small disks is equal to $f = f_C = M\pi r^2$. The macroscopic isotropy of the considered cluster is suggested for simplicity. Moreover, it is supposed that inclusions in the disks are diluted, i.e., the averaged permittivity of the cluster C_M is calculated by (2.165).

In the framework of EMA, we have to find ε_C as a function of the exterior permittivity ε_e. Next, equation (3.5) has to be solved. In the considered case of dilute composites, this equation becomes $\varepsilon_e = \varepsilon_C$ with the given by (3.5) constant ε_C. Here, we do not follow the homogenization theory and make the assumptions that $f_C = f$ and $\varepsilon_e = \varepsilon_C$ following EMA. Below, we show that these assumptions are inconsistent. Using (3.2) we obtain from (3.5)

$$\varrho_e = \varrho f + O(f^3). \tag{3.6}$$

Introduce the complex flux analytic in the considered domains and continuous in their closures using equations (2.40)

$$\psi(z) = \varphi'(z) \equiv \frac{\partial u}{\partial \xi_1} - i\frac{\partial u}{\partial \xi_2}, \quad z = \xi_1 + i\xi_2 \in G, \tag{3.7}$$

$$\psi_k(z) = \varphi_k'(z) \equiv \frac{\varepsilon+1}{2}\left(\frac{\partial u}{\partial \xi_1} - i\frac{\partial u}{\partial \xi_2}\right), \ |z-a_k| \le r \ (k = 1,2,\ldots,M), \tag{3.8}$$

$$\psi_e(z) = \varphi_e'(z) \equiv \frac{\varepsilon_e+1}{2}\left(\frac{\partial u}{\partial \xi_1} - i\frac{\partial u}{\partial \xi_2}\right), \ |z| \ge R. \tag{3.9}$$

The external flux $(1,0)^\top$ is consistent with the condition $\psi_e(\infty) = 1$. The functions (3.7)–(3.9) satisfy the $\mathbb{R}$-linear problem [93] which express the perfect contact between the components

$$\psi(t) = \psi_k(t) + \varrho\left(\frac{r}{t-a_k}\right)^2 \overline{\psi_k(t)}, \quad |t-a_k| = r \quad (k = 1,2,\ldots,M), \tag{3.10}$$

$$\psi(t) = \psi_e(t) + \varrho_e\left(\frac{R}{t}\right)^2 \overline{\psi_e(t)}, \quad |t| = R. \tag{3.11}$$

It is shown in [93] that the $\mathbb{R}$-linear problem (3.10)–(3.11) is reduced to the system of functional equations

$$\psi_k(z) = \varrho r^2 \sum_{m\neq k}^{M} \frac{1}{(z-a_m)^2}\overline{\psi_m\left(\frac{r^2}{\overline{z-a_m}} + a_m\right)} + \varrho_e\left(\frac{R}{z}\right)^2 \overline{\psi_e\left(\frac{R^2}{\overline{z}}\right)} + 1, \tag{3.12}$$
$$|z-a_k| \le r \quad (k = 1,2,\ldots,M),$$

$$\psi_e(z) = \varrho r^2 \sum_{m=1}^{M} \frac{1}{(z-a_m)^2}\overline{\psi_m\left(\frac{r^2}{\overline{z-a_m}} + a_m\right)} + 1, \quad |z| \ge R. \tag{3.13}$$

The functional equations will be solved up to $O(f^2)$. It is assumed within the considered precision that $r^2 = O(f)$ when an arbitrarily number of inclusions M is fixed. The constant ϱ_e can be estimated by (3.6). In order to get the approximation of $\psi_k(z)$ in $|z-a_k| \le r$, it is sufficient to take $\psi_e(z) = 1 + O(f)$ and to use the zero approximation $\psi_k(z) = 1 + O(f)$

$$\psi_k(z) = 1 + \varrho r^2\left[\sum_{m\neq k}^{M} \frac{1}{(z-a_m)^2} + \frac{M}{z^2}\right] + O(f^2), \quad |z-a_k| \le r \tag{3.14}$$
$$(k = 1,2,\ldots,M).$$

Here, the relation $\varrho_e R^2 = \varrho r^2 M$ is used.

The formula (3.14) for the complex flux $\psi_k(z)$ in the inclusion $|z-a_k| \leq r$ is derived in the framework of EMA. Consider now the cluster modification discussed in the previous section for dilute disks. It differs from EMA by absence of the boundary $|z| = R$. EMA can be formally reduced to the cluster modification by the assumption $\varepsilon_e = 1$ in $|z| \geq R$. If we do it, instead of (3.14) we arrive at the formula

$$\psi_k(z) = 1 + \varrho r^2 \textstyle\sum_{m\neq k}^{M} \frac{1}{(z-a_m)^2} + O(f^2),\ |z-a_k| \leq r\ (k = 1,2,\ldots,M). \tag{3.15}$$

The formula (3.15) is correct. The formula (3.14) deduced by the EMA arguments is wrong.

The approximation (3.15) can be used for the calculation of the effective permittivity of the collection of disks

$$\varepsilon_e^{(M)} = 1 + 2\varrho \sum_{k=1}^{M} \int_{G_k} \psi_k(\xi_1 + i\xi_2)\, \mathrm{d}\xi_1 \mathrm{d}\xi_2. \tag{3.16}$$

Substitution of (3.15) into (3.16) yields

$$\varepsilon_e^{(M)} = 1 + 2\varrho f + 2\varrho^2 f^2 \frac{1}{\pi M} \sum_{k=1}^{M} \sum_{m\neq k}^{M} \frac{1}{(a_k - a_m)^2} + O(f^3). \tag{3.17}$$

It corresponds to the general formula [93, (4.2.26)–(4.2.27)] derived for a doubly periodic structure. The conditionally convergent limit $\varepsilon_e = \lim_{M\to\infty} \varepsilon_e^{(M)}$ has to be calculated by the Eisenstein summation [231] which yields for macroscopically isotropic composites [93] and

$$\lim_{M\to\infty} \frac{1}{M} \sum_{k=1}^{M} \sum_{m\neq k}^{M} \frac{1}{(a_k - a_m)^2} = \pi. \tag{3.18}$$

Therefore, the introduction of the fictitious effective medium to $|z| > R$ (see Figure 3.4) yields the additional term Mz^{-2} in the complex flux $\psi_k(z)$ in G_k calculated by (3.14). The scheme without usage of such a fictitious effective medium yields the correct expression (3.15) for $\psi_k(z)$.

3.2.2 CLUSTER EXTENSION TO INFINITY

In the present section, we consider a sequence of extending clusters C_M $(M = 1,2,\ldots)$ displayed in Figure 3.5. Here, every cluster C_M is bounded by a closed curve and contains M inclusions. The same investigation holds for the modified EMA illustrated in Figure 3.4 when the radius of large disk R tends to infinity. Some authors think that "one can systematically correct such dilute-limit formulas by taking into account interactions between pairs of spheres, triplets of spheres, and so on." Usually, one performs computations for a finite cluster, calls it by the way RVE, and declares that it approximates the whole composite. We consider the limit $M \to \infty$ in order to understand what such a scheme approximates.

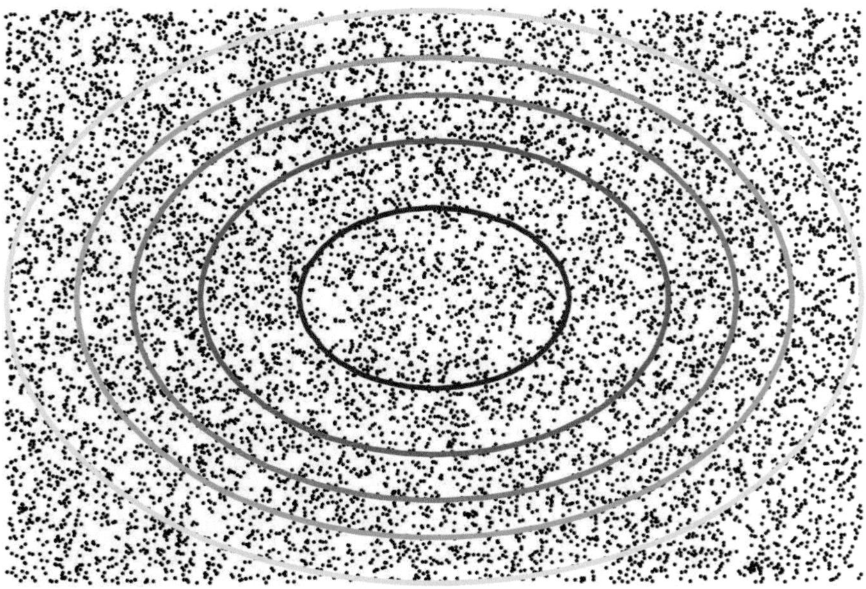

Figure 3.5 Extending cluster to the whole plane.

It is convenient to study the limit by Schwarz's method. In the considered case, the operator equation (2.52) takes the form

$$u_k^{(M)} = \varrho \sum_{m=1}^{M} F_m u_m^{(M)} + u_0, \quad z \in G_k \ (k = 1, 2, \ldots, M). \tag{3.19}$$

where M inclusions are considered on the plane. Equations (3.19) in the limit case become

$$u_k^{(\infty)} = \varrho \sum_{m=1}^{\infty} F_m u_m^{(\infty)} + u_0, \quad z \in G_k \ (k = 1, 2, \ldots). \tag{3.20}$$

One can formally obtain $u_k^{(\infty)}$ due to all the mutual interactions among inclusions. Let the solution $u_k^{(M)}$ of equations (3.19) be obtained by a numerical method. Therefore, all the interactions between the inclusions G_k $(k = 1, 2, \ldots, M)$ are included in $u_k^{(M)}$ considered as a series in the framework of Schwarz's method. It doesn't matter that we don't distinguish these interactions in the numerical result for $u_k^{(M)}$, but we know that all the interactions among M inclusions are considered. If one extends the cluster C_M to C_{M_1} $(M_1 > M)$ as shown in Figure 3.5, the solution $u_k^{(M_1)}$ will contain the additional interactions between the first M inclusions and the additional inclusions G_k $(k = M+1, M+2, \ldots, M_1)$, and between these additional inclusions. Let this process be continued numerically. As a result one obtains a finite set of solutions $u_k^{(M_j)}$ $(j = 1, 2, \ldots, L)$ and the corresponding effective conductivities $\varepsilon_e^{(M_j)}$ of the clusters. Ultimately, one has a finite set of numbers $\varepsilon_e^{(M)}$ obtained by a numerical procedure. At the same time, these numbers can be considered as the partial sums of the series arising by applying Schwarz's method. It is suggested that

$$\varepsilon_e^{(\infty)} = \lim_{M\to\infty} \varepsilon_e^{(M)}. \tag{3.21}$$

However, the limit (3.21) is conditionally convergent [161, 168, 169, 173, 180] and can lead to nothing, more precisely, to everything, since the corresponding series can be arranged in a permutation so that the resulting series will converge to any number including infinity. This false trick used in SCM manipulations leads to illusory different formulas called models[1].

The special order of summation must be used to get the correct limit (3.21). It is related to the direction of the external field. The external field is parallel to the x_1-axis in the considered case. Hence, the Eisenstein summation has to be used. The Eisenstein summation is reduced to the iterated limit when, first, the limit along the x_1-axis has to be found, i.e., for an infinite strip $-\ell < x_2 < \ell$, next, for $\ell \to \infty$.

Rayleigh [206] in 1892 first applied the proper Eisenstein summation. McPhedran and McKenzie [146] in 1978 gave the physical explanation of the Rayleigh method by the induced charge effect; see [169] for mathematical and historical details. These two fundamental works went unnoticed by specialists in material sciences, possibly due to the prevailing focus on conductivity-related terminology like transport, permeability, and permittivity rather than on the phenomenon of elasticity. However, Laplace's equation governs the antiplane elasticity, serving as a specific instance within the broader spectrum of elastic stress and strain fields elucidated by biharmonic functions. Consequently, the externally induced field that emerges in conductivity [146] logically manifests in elasticity [173] and other domains like viscous flow [67].

In order to explicitly demonstrate that (3.21) can fail in the framework of SCM, we consider the regular square array of equal disks. Consider ℓ^2 ($\ell = 2, 4, \ldots$) disks of radius $r < \frac{1}{2}$ centered at $a_{m_1,m_2} = m_1 + \mathrm{i}m_2 - \frac{(\ell+1)(1+\mathrm{i})}{2}$ ($m_1, m_2 = 1, 2, \ldots, \ell$). These disks belong to the finite square cell $Q^{(\ell)} = \{z = x + iy \in \mathbb{C} : -\frac{\ell}{2} < x, y < \frac{\ell}{2}\}$ of size $\ell \times \ell$, see Figure 3.6. The introduced double numeration of disks can be transformed into the linear numeration $k = 1, 2, \ldots, \ell^2$ used before. Let $G^{(\ell)}$ denote the square $Q^{(\ell)}$ without disks. We are interested in the effective permittivity $\varepsilon_e^{(\ell)}$ of the cluster of disks $Q^{(\ell)}$, as $\ell \to \infty$. Following the SCM approach, we define the effective permittivity by the relation

$$\varepsilon_e^{(\ell)} e^{(\ell)} = j^{(\ell)}, \tag{3.22}$$

where

$$e^{(\ell)} = \int_{G^{(\ell)}} \frac{\partial u}{\partial x}\, \mathrm{d}x\mathrm{d}y + \sum_{k=1}^{\ell^2} \int_{G_k} \frac{\partial u_k}{\partial x}\, \mathrm{d}x\mathrm{d}y \tag{3.23}$$

[1]A discussion of the notion *model* can be found in [176, Chapter 9].

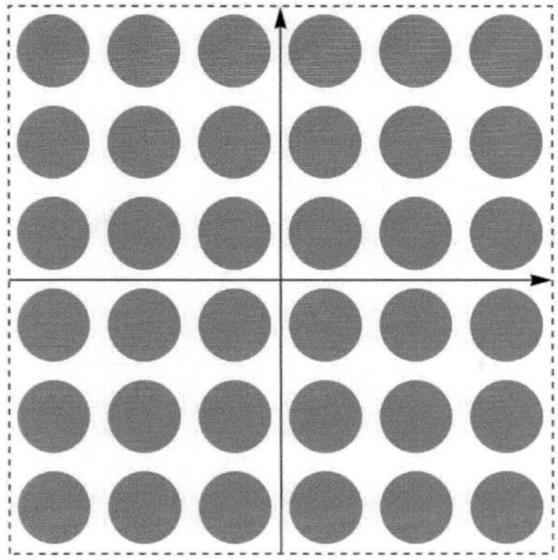

Figure 3.6 Square cell $Q^{(\ell)}$ with ℓ^2 disks. The cell is displayed for $\ell = 6$.

and

$$j^{(\ell)} = \int_{G^{(\ell)}} \frac{\partial u}{\partial x}\, \mathrm{d}x\mathrm{d}y + \varepsilon_1 \sum_{k=1}^{\ell^2} \int_{G_k} \frac{\partial u_k}{\partial x}\, \mathrm{d}x\mathrm{d}y. \tag{3.24}$$

The integrals (3.22)–(3.24) can be transformed by Green's formulas following [93, Chapter 3, Sec. 2.2.2]. Symmetry of the considered problem and the condition $u = u_k$ on ∂G_k yield

$$e^{(\ell)} = \int_{\partial Q^{(\ell)}} u\, \mathrm{d}y = 4 \int_0^{\frac{\ell}{2}} u\left(\frac{\ell}{2}, y\right) \mathrm{d}y \tag{3.25}$$

and

$$j^{(\ell)} = \int_{\partial Q^{(\ell)}} u\, \mathrm{d}y + (\varepsilon_1 - 1) \sum_{k=1}^{\ell^2} \int_{G_k} \frac{\partial u_k}{\partial x}\, \mathrm{d}x\mathrm{d}y, \tag{3.26}$$

where $\partial G^{(\ell)} = \partial Q^{(\ell)} - \cup_{k=1}^{\ell^2} \partial G_k$. Using (2.40) and the mean value theorem for a harmonic function, we arrive at the formula

$$\varepsilon_e^{(\ell)} = 1 + \frac{2\varrho\pi r^2}{e^{(\ell)}} \sum_{k=1}^{\ell^2} \mathrm{Re}\, \psi_k^{(\ell)}(a_k). \tag{3.27}$$

Here, $\psi_k^{(\ell)}(z) = \varphi_k'(z)$ denotes the complex flux in G_k of the considered finite cluster cell $Q^{(\ell)}$ caused by the external gradient $\nabla u = (1,0)^\top$ at infinity. It is worth noting that the value $\varepsilon_e^{(\ell)}$ can be calculated by a numeric method but the result must coincide with the exact formula (3.27).

The local field, in particular $\psi_k^{(\ell)}(a_k)$, can be calculated by exact formulas written in [93]. In order to establish the discrepancy in the limit (3.21) when $M = \ell^2$, it is sufficient to consider the expressions up to $O(r^4)$. Using the asymptotic formula from [93] similar to (3.13) we obtain the estimation $e^{(\ell)} = \pi\ell^2 + O(r^4)$. The latter formula is suggested after numerical computations for $\ell \leq 1000$. Let $f_C = \pi r^2$ stand for the concentration of disks in the cluster. Substitute (3.15) into (3.27)

$$\varepsilon_e^{(\ell)} = 1 + 2\varrho f_C + \frac{2\varrho^2 f_C^2}{\pi} \mathrm{Re}\, S_2^{(\ell)} + O(f_C^3) = 1 + 2\varrho f_C + O(f_C^3), \tag{3.28}$$

where the double sum

$$S_2^{(\ell)} = \sum_{k_1,k_2=1}^{\ell} \sum_{\substack{m_1,m_2=1 \\ (m_1,m_2)\neq(k_1,k_2)}}^{\ell} \frac{1}{\left(a_{k_1,k_2} - a_{m_1,m_2}\right)^2}. \tag{3.29}$$

vanishes due to the symmetry of the square cluster $Q^{(\ell)}$, when a_{km} can be replaced by $\mathrm{i}a_{km}$ without change of $S_2^{(\ell)}$.

Consider now the infinite square array of disks. Its effective permittivity is given by the exact formula (4.2.28) from [93]. Its approximation has the form

$$\varepsilon_e^{(\infty)} = 1 + 2\varrho f + 2\varrho^2 f^2 + O(f^3), \tag{3.30}$$

where $f = \pi r^2$. One can see a discrepancy in the limit (3.21) where $\varepsilon_e^{(\ell)}$ is calculated by (3.28) and $\varepsilon_e^{(\infty)}$ by (3.30).

The cluster $Q^{(\ell)}$ has a square shape. A change of shape yields a change of result. The shape dependence and the corrector, induced charge on the boundary, were investigated in [146, 157].

A proper numerical scheme based on the homogenization theory has to be applied in the periodic statement. Its consistency with the Eisenstein summation and the theory of random composites is demonstrated in [67, 93] and papers cited therein.

3.2.3 EMPIRICAL FORMULAS

Engineers often use empirical and fitted formulas for a set of experiments. However, a variation of parameters can significantly change the results. Therefore, it is essential to know the validity domain of suggested formulas. In the present section, we analyze the precision of some known approximations of the effective permittivity based on empirical arguments.

Bruggeman's equation for disks of permittivity ε_1 embedded in the host of unit permittivity has the form [45]

$$f\frac{\varepsilon_1 - \varepsilon_e}{\varepsilon_1 + \varepsilon_e} + (1-f)\frac{1-\varepsilon_e}{1+\varepsilon_e} = 0. \tag{3.31}$$

Here, a macroscopically isotropic composite of the scalar permittivity ε_e is considered for shortness. The Bruggeman formula can be regarded as a two-point Taylor expansion of the effective permittivity around two reference points $f = 0$ and $f = 1$.

The particular property characterizes this Bruggeman formula for which the inclusions and the environment are treated symmetrically. The physical meaning of (3.31)

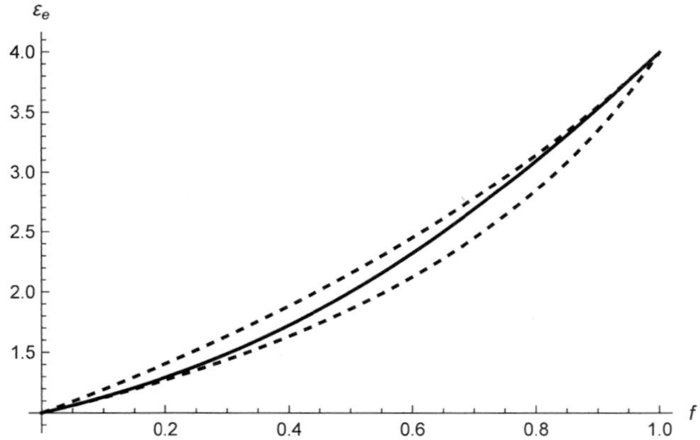

Figure 3.7 Hashin-Shtrikman bounds (dashed) and Bruggeman's formula (solid) for the effective permittivity when $\varepsilon_1 = 4$.

is that the formula balances both mixing components with respect to the unknown effective medium, using the volume fraction of each component as weight (f for the inclusions and $(1-f)$ for the environment).

Equation (3.31) was analyzed in [179, 180] for real permittivity. First, it was noted that the formula (3.31) was mistakenly treated in [55] as a theoretical improvement of the Maxwell-Garnett formula (2.13). The formula (3.31) is based on the empirical hypothesis of the zero flux deviation [55, eq. (1.31)] which is valid only up to $O(f)$. Equation (3.31) yields the asymptotic formula

$$\varepsilon_e = 1 + 2\varrho f + 2\varrho^2 f^2 (1+\varrho) + O(f^3 \varrho^4). \tag{3.32}$$

The proper formula (2.13) within the considered precision was proved in [179]. Hence, the coefficient in f^2 is wrong in (3.32). There are other cases when (3.31) does not hold, and one needs to be cautious in its application [143].

Empirical formulas are based on the correct conviction that everybody may sketch any curve between the Hashin-Shtrikman bounds (2.14)–(2.16) displayed in Figures 3.7 and 3.8. It doesn't matter which curve is taken in the case $\varepsilon_1 = 4$; see Figure 3.7. But it does in the case $\varepsilon_1 = 10000$, see Figure 3.8.

Figure 3.7 displays the Hashin-Shtrikman bounds for not high $|\varrho| = 0.6$. In the considered case of tight bounds, one may take any curve between the bounds with almost the same numerical values. Perhaps equation (3.31) yields the simplest curve, which lies between the Hashin-Shtrikman bounds. This is the main and unique reason why it is frequently used in engineering estimations. Figure 5.14 from Chapter 5 shows that (3.31) does not satisfy the higher-order bounds.

The differential scheme was introduced by Bruggeman [45] and considered as a refinement of EMA [198, 199] [155, Sections 10.7-10.8]. It leads to the relation

$$\frac{\varepsilon - \varepsilon_e}{\varepsilon - 1} \frac{1}{\sqrt{\varepsilon_e}} \approx 1 - f. \tag{3.33}$$

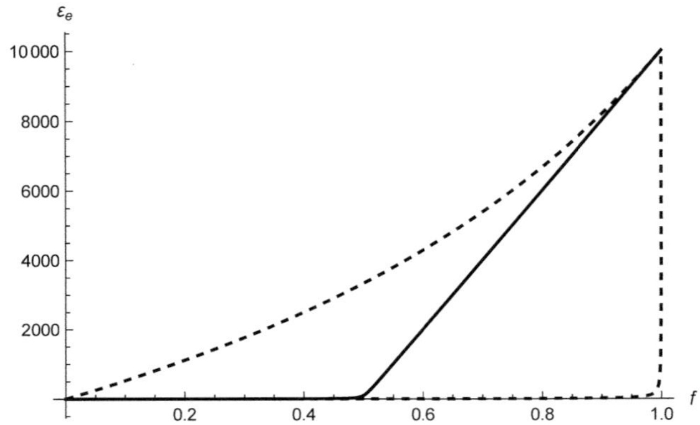

Figure 3.8 Hashin-Shtrikman bounds (dashed) and Bruggeman's formula (solid) for the effective permittivity when $\varepsilon = 10000$.

Equation (3.33) also yields the wrong second-order approximation in f

$$\varepsilon_e = 1 + 2\varrho f + 2\varrho^2 f^2 \left(1 + \frac{1}{2}\varrho\right) + O(f^3 \varrho^4). \tag{3.34}$$

The same applies to the Mori-Tanaka method commonly used in elasticity, where calculations are accurate up to $O(f^2)$. Although the second-order term is not evident in pure numerical computations, if it is relatively small, the resulting numerical outcomes may misleadingly suggest an improvement in the effective constants.

We end this section with some general arguments. One might argue that the effective permittivity is represented by a function $\varepsilon_e = \varepsilon_e(f, \varrho)$, which has different forms in different subdomains of the rectangle $[0,1] \times [-1,1]$, where $f \in [0,1]$ and $\varrho \in [-1,1]$. However, the function $\varepsilon_e = \varepsilon_e(f, \varrho)$ can be considered a unified analytic function. This follows from the expansion justified in [40], see also Theorem 3 from Section 2.6.1,

$$\varepsilon_e(f, \varrho) = \sum_{\ell=0}^{\infty} c_\ell(f^{1/2}) \varrho^\ell. \tag{3.35}$$

The series (3.35) converges absolutely and uniformly in the complex unit disk $|\varrho| < 1$. Therefore, $\varepsilon_e(f, \varrho)$ is given by the analytic function (3.35) whose analyticity may be disturbed in the complex domain $|\varrho| \geq 1$.

Further, we follow the theory [41,64,65] and the computational method [62,67,93] in the special aspect of expansion in $f^{1/2}$. In the case of deterministic composites, we consider the set of non-overlapping inclusions δG_k. Here, δ is the coefficient shrinking the domain G_k $(0 < \delta < \delta_c)$. The inclusions become touching for $\delta = \delta_c$. This parameter determines the critical concentration f_c and the interval $|f| < f_c$. A random composite is considered as a random set of non-overlapping domains $\{G_1, G_2, \ldots, G_N\}$ satisfying a probabilistic distribution and investigated using computer simulations [62,67,93]. The critical concentration f_c is determined as the mathematical expectation of the considered class of composites. It is worth noting that a constructive description of a class of dispersed composites, i.e., its RVE, can be performed through the structural sums [62,67,93]. The behavior of $\varepsilon_e(f,\varrho)$, as $f \to f_c$ and $|\varrho| \to 1$ requires a separate discussion due to percolation effects [36,67,93,119].

We end this section with a remark concerning the Hashin-Shtrikman bounds and Bruggeman's heuristic equation (3.31) for complex permittivity of components. The Hashin-Shtrikman bounds for 2D macroscopically isotropic composites for complex permittivity were derived in [153]. They can be represented on the complex plane ε_e for fixed f and permittivity of phases as the domain Ω bounded by two circular arcs. The complex parametric equations of arcs have the form

$$Z_1(x) = f\varepsilon_1 + (1-f)\varepsilon_2 - \frac{f(1-f)(\varepsilon_2-\varepsilon_1)^2}{2(x\varepsilon_1+(1-x)\varepsilon_2)},$$

$$Z_2(x) = \left[\frac{f}{\varepsilon_1} + \frac{1-f}{\varepsilon_2} - \frac{f(1-f)\left(\frac{1}{\varepsilon_2}-\frac{1}{\varepsilon_1}\right)^2}{2\left(\frac{x}{\varepsilon_1}+\frac{1-x}{\varepsilon_2}\right)}\right]^{-1}, \quad x \geq \frac{1-f}{2}. \tag{3.36}$$

Here, it is convenient to consider the complex permittivity ε_1 and ε_2 of inclusions and matrix, respectively, without normalization.

Figure 3.9 displays the complex Hashin-Shtrikman domain Ω for the fixed $f = 0.25$ and $\varepsilon_1 = -58-36.5\mathrm{i}$ and $\varepsilon_2 = 60$. The permittivity is taken from [96] for the frequency 10^{11}. The calculated by (3.31) Bruggeman's point $\varepsilon_B = 39.8-49.6\mathrm{i}$ is just one of the infinitely many points which one may take in Ω if the location of inclusions is not determined.

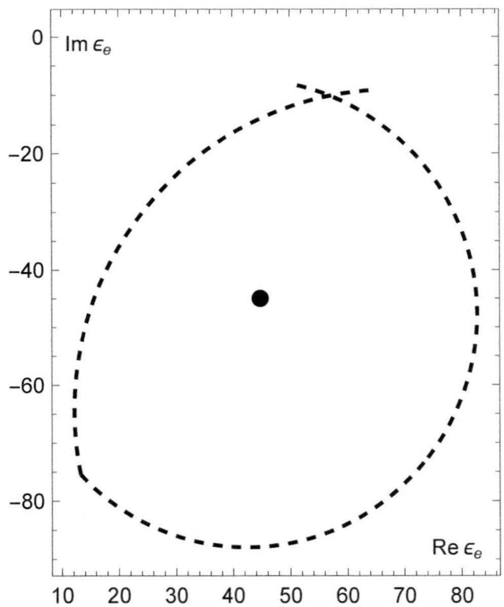

Figure 3.9 Bruggeman's approximation (3.31) on the complex plane ε_e for $f = 0.2$, $\varepsilon_1 = -58-36.5\mathrm{i}$ and $\varepsilon_2 = 60$ (point $\varepsilon_B = 49.4-39.5\mathrm{i}$) and the Hashin-Shtrikman domain bounded by the circle arcs (dashed lines).

3.2.4 CONCLUSION TO CHAPTER

In the present chapter, the generalized alternating method of Schwarz is outlined in general form and realized in the case of real-valued permittivity of components. Approximate analytical formulas for the effective properties of dispersed composites with the exactly derived precision of their validity are deduced. It is demonstrated that the Maxwell-Garnett (Clausius-Mossotti) formula plays a central role in approximations used in applications. In particular, the domain of its validity is established; it is higher than other popular approximations. The Hashin-Shtrikman bounds (2.14)–(2.16) and (3.36) are helpful in the estimation of the effective permittivity ε_e. If the bounds are tight, the Maxwell-Garnett and Brugemann approximations yield barely distinguishable numerical values.

In the previous publications, many other formulas were given without analysis of their precision. This led, at most, to formally different but asymptotically equivalent formulas for the same classes of composites. We outlined above typical methods and tricks used in SCM and their modifications. Analyzing manipulations applied in thousands of papers concerning such an approximation is difficult, but our study suggests revising some methods that lead to doubtful results. The precision analysis of SCM must be performed before a declaration of its advantages is made.

A shortcoming of SCM, EMA, is related to neglecting the external charge induced by a finite cluster of inclusions. Regardless of the solution method applied to a sequence of clusters, the conditionally convergent series (3.18) arises when clusters extend to infinity. The convergence problem was resolved by the Eisenstein summation method applied to composites, first, by Rayleigh [206]. The exhaustive physical explanation of the induced charge effect is given in [146] and in the mathematical papers [157, 161, 168, 169, 173, 180]. This induced charge is lost in the SCM methodology and can lead to incorrect formulas.

Instead of over-optimistic SCM treated as "alive and well Maxwell's methodology," the following Maxwell remark is pertinent here. "This result may be obtained in other ways, but that here given involves only the repetition of the result already obtained for a single sphere" [139, p. 365]. The principal shortage of SCM is ignoring precision on the parameters ϱ and f. "Although the comparison with experiment was very favorable in many cases, it is clear from theoretical grounds that also the self-consistent effective moduli are not rigorous quantities" (Kröner [123, p. 72]). "A study of these examples makes it plain that the theory is unreliable under extreme conditions, *except when the dispersed phase is sufficiently dilute*" (R. Hill [106]).

Despite these cautions, various SCMs and applications arose without analysis of their validity. Many of them contain empirical arguments intertwined with mathematical manipulations. Nevertheless, the first iteration step of Schwarz's method is visible in SCM as the leading approximation. Sometimes, it is hidden in the statement of boundary value problem for RVE or in an integral estimation up to $O(f^2)$. Bergman's note [35] serves as an excellent treatment of SCM based on physical intuition and asymptotically reasonable manipulations.

Numerical comparisons of various SCM and other formulas can demonstrate their similarities and significant differences. It is unsurprising since different dispersed random composites can have different effective properties. Figure 3.7 show that one may use an approximate universal formula for small $|\varrho|$. Hashin-Shtrikman bounds in Figure 3.8 show the opposite fact for higher $|\varrho|$. The dependence of effective permittivity on the concentration for $|\varrho| = 1$ was discussed in [93] for disks where the square array (Chapter 6), the hexagonal array (Chapter 7) and a class of random composites (Chapter 9) were studied using renormalization approach. It was shown that the critical exponent s near the percolation threshold $\varepsilon_e \sim (f_c - f)^{-s}$ takes the values $0.5 \leq s \leq 1.3$. This result is not attainable by SCM.

The present study shows how to overcome the shortcomings of SCM through Schwarz's method. It concerns an analytical approach that consists of the symbolically and numerically implemented blocks related by iterations in ϱ and f [67,93]. In the chapter, we present such a block as integral equations developed in [98], [105]. One can apply a FEM or another method to the block problem. The corresponding computations must be based on homogenization principles and the proper definition of RVE.

4 Circular and Elliptic Inclusions

4.1 FUNCTIONAL EQUATIONS

In Section 2.4, the local fields and effective permittivity were determined by solution to the $\mathbb{R}$-linear problem (2.34)–(2.35) for the periodic unit square Q. In the present section, we consider this problem when the inclusions G_k are non-overlapping disks $|z-a_k| = r_k$ $(k = 1,2,\ldots,N)$ occupied by a material with the normalized complex permittivity $\varepsilon_k = \varepsilon_k' + \mathrm{i}\varepsilon_k''$. The general unit periodicity cell Q shown in Figure 4.1 will be taken instead of the unit square cell. The unit cell Q is formed by the fundamental translation vectors expressed by complex numbers $\boldsymbol{\omega}_1$ and $\boldsymbol{\omega}_2$, where $\boldsymbol{\omega}_1 > 0$ and $|Q| = \boldsymbol{\omega}_1 \mathrm{Im}\, \boldsymbol{\omega}_2 = 1$.

The permittivity of the host is normalized to unity. The contrast parameter matrix has the form (2.33). The concentration of each phase of inclusions is equal to $f_k = \pi r_k^2$. The total concentration $f = \sum_{k=1}^{N} \pi r_k^2$ will be also used.

We develop the general case of complex permittivity ε_k following the real case discussed in [93, 156] by functional equations. The $\mathbb{R}$-linear problem (2.34)–(2.35) can be written in terms of the complex vector fluxes $\boldsymbol{\psi}(z) = \frac{d}{dz}\boldsymbol{\varphi}(z)$ in $G \cup L$ and $\boldsymbol{\psi}_{\boldsymbol{k}}(z) = \frac{d}{dz}\boldsymbol{\varphi}_{\boldsymbol{k}}(z)$ in $|z-a_k| \le r_k$ [93, 156, 185]

$$\boldsymbol{\psi}(t) = \psi_k(t) + \boldsymbol{\beta}_k \left(\frac{r_k}{t-a_k}\right)^2 \overline{\boldsymbol{\psi}_k(t)}, \quad t \in L_k \quad (k = 1,2,\ldots,N). \tag{4.1}$$

The vector-function $\boldsymbol{\psi}(z)$ is double periodic

$$\boldsymbol{\psi}(z+\boldsymbol{\omega}_j) = \boldsymbol{\psi}(z), \quad j = 1,2. \tag{4.2}$$

The given external field $\nabla u_{ext} = (\xi_1, \xi_2)^\top$ is determined by the conditions (2.22) [163]

$$\int_z^{z+\omega_j} \boldsymbol{\psi}(t)\, \mathrm{d}t = \boldsymbol{\varphi}(z+\omega_j) - \boldsymbol{\varphi}(z), \quad j = 1,2. \tag{4.3}$$

For definiteness, let the external field be applied along the x_1-axis

$$\boldsymbol{\psi}_{ext}(z) = \begin{pmatrix} 1 \\ 0 \end{pmatrix}. \tag{4.4}$$

Introduce the inversion (reflection) in the circle $|z-a_k| = r_k$

$$z_{(k)}^* = a_k + \frac{r_k^2}{\overline{z-a_k}}. \tag{4.5}$$

DOI: 10.1201/9781032646633-4

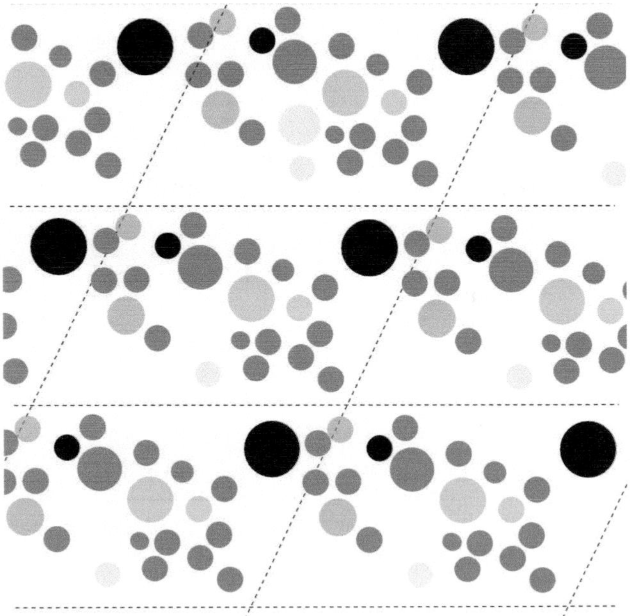

Figure 4.1 Unit periodicity cell Q with different circular inclusions.

By Schwarz' symmetry principle, if a function $F(z)$ is analytic in $|z-a_k| < r_k$, the function $\overline{F\left(z^*_{(k)}\right)}$ is analytic in $|z-a_k| > r_k$.

The components of the vector-function $\boldsymbol{\psi}_k(z)$ belong to the space $\mathcal{H}(G^+)$ introduced on page 33. Consider in $\mathcal{H}(G^+)$ the linear operator which transforms the vector-function $\boldsymbol{\psi}_k(z)$

$$W_{k\mathbf{m}}\boldsymbol{\psi}_k(z) = \left(\frac{r_k^2}{z-a_k-\mathbf{m}}\right)^2 \overline{\boldsymbol{\psi}_k\left(a_k + \frac{r_k^2}{\overline{z-a_k-\mathbf{m}}}\right)}, \tag{4.6}$$

where $|z-a_k| \leq r_k$ $(k = 1,2,\ldots,N)$ and

$$\mathbf{m} = \boldsymbol{\omega}_1 m_1 + \boldsymbol{\omega}_2 m_2 \quad (m_1, m_2 \in \mathbb{Z}). \tag{4.7}$$

The notation $\mathbb{Z}^2_{\boldsymbol{\omega}}$ will be used for the lattice (4.7) . The set $\mathbb{Z}^2_{\boldsymbol{\omega}}$ with the excluded one term $m_1 = m_2 = 0$ will be also considered and denoted by $\mathbb{Z}^2_{\boldsymbol{\omega}} \backslash (0,0)$. It is convenient to consider the double periodic vector function $\boldsymbol{\psi}_k(z)$ in the array of separate disks $\cup_{m_1,m_2 \in \mathbb{Z}}(G_k + \mathbf{m})$. The shift $a_k + \frac{r_k^2}{\overline{z-a_k-\mathbf{m}}}$ of the operator $W_{k\mathbf{m}}$ for $\mathbf{m} \neq (0,0)$ maps the closed disk $|z-a_k| \leq r_k$ onto the interior of the open disk $|z-a_k| < r_k$, since the shift is composed of the inversion of the disk $|z-a_k| \leq r_k$ in the circle $|z-a_k-\mathbf{m}| = r_k$ and of the return translation along the vector $(-\mathbf{m})$ into $|z-a_k| < r_k$ as displayed in Figure 4.2.

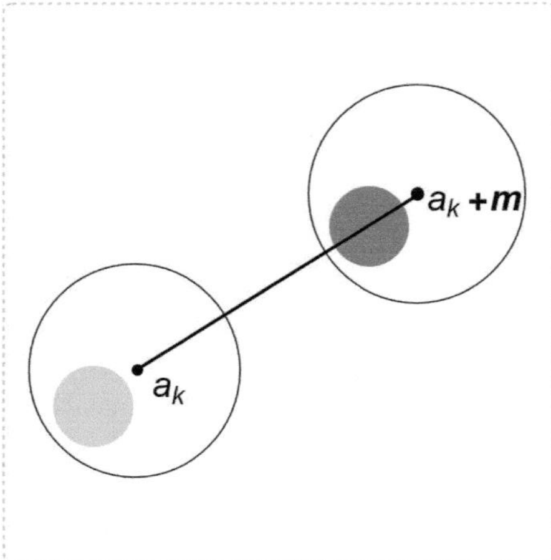

Figure 4.2 The inversion of the disk $|z-a_k| \le r_k$ in the circle $|z-a_k-\mathbf{m}| = r_k$ is shown by gray. The translation of the gray disk is shown by light gray. Therefore, the shift maps $|z-a_k| \le r_k$ onto $|z-a_k| < r_k$, the light gray disk.

Represent the vector function $\boldsymbol{\psi}_k(z)$ in the form of the Taylor series

$$\boldsymbol{\psi}_k(z) = \sum_{p=0}^{\infty} \boldsymbol{\psi}_{kp}(z-a_k)^p, \quad |z-a_k| < r, \tag{4.8}$$

with the vector coefficients $\boldsymbol{\psi}_{kp}$. Using the Eisenstein summation and the series (A.14) defined in Appendix A, calculate

$$\begin{aligned}
&\sum_{\mathbf{m}\in\mathbb{Z}^2_\omega} W_{k\mathbf{m}}\boldsymbol{\psi}_k(z) = \sum_{\mathbf{m}\in\mathbb{Z}^2_\omega} \sum_{p=0}^{\infty} \frac{r_k^{2(p+1)}\overline{\boldsymbol{\psi}_{kp}}}{(z-a_k-\mathbf{m})^{p+2}} = \\
&r_k^2\overline{\boldsymbol{\psi}_{k2}} \sum^{(e)}_{\mathbf{m}\in\mathbb{Z}^2_\omega} \frac{1}{(z-a_k-\mathbf{m})^2} + \sum_{\mathbf{m}\in\mathbb{Z}^2_\omega} \sum_{p=1}^{\infty} \frac{r_k^{2(p+1)}\overline{\boldsymbol{\psi}_{kp}}}{(z-a_k-\mathbf{m})^{p+2}} = \\
&r_k^2\overline{\boldsymbol{\psi}_{k2}}E_2(z-a_k) + \sum_{p=1}^{\infty} r_k^{2(p+1)}\overline{\boldsymbol{\psi}_{kp}}E_{p+2}(z-a_k), \quad |z-a_k| > r_k.
\end{aligned} \tag{4.9}$$

The function $E_2(z)$ is constructed by means of the conditionally convergent series $\sum^{(e)}$, the functions $E_{p+2}(z)$ $(p = 1, 2, \ldots)$ by absolutely convergent series. The Eisenstein functions are constructed for a fixed lattice $\mathbb{Z}^2_\omega$.

Theorem 5 ([157, 162]) •

- *The series* $\sum_{\mathbf{m}\in\mathbb{Z}^2_\omega\setminus(0,0)} W_{k\mathbf{m}}\boldsymbol{\psi}_k(z)$ *represented by the last line in* (4.9) *converges absolutely and almost uniformly in* $Q \cup \partial Q \setminus \{a_k\}$.
- *The function* $\sum_{\mathbf{m}\in\mathbb{Z}^2_\omega} W_{k\mathbf{m}}\boldsymbol{\psi}_k(z)$ *is analytic in the domain* $Q\setminus(G_k \cup L_k)$, *Hölder continuous in its closure and doubly periodic.*
- *The linear operator* $\sum_{\mathbf{m}\in\mathbb{Z}^2_\omega\setminus(0,0)} W_{k\mathbf{m}}\boldsymbol{\psi}_k(z)$ *is compact in* $\mathcal{H}(G^+)$.

Using this Theorem we introduce the function

$$\Phi(z) = \begin{cases} \boldsymbol{\psi}_l(z) - \sum_{k=1}^{\prime N} \boldsymbol{\beta}_k \sum_{\mathbf{m}\in\mathbb{Z}_\omega^2} W_{k\mathbf{m}} \boldsymbol{\psi}_k(z), \ |z-a_l| \le r_l \ (l = 1,2,\ldots,N), \\ \boldsymbol{\psi}(z) - \sum_{k=1}^{N} \boldsymbol{\beta}_k \sum_{\mathbf{m}\in\mathbb{Z}_\omega^2} W_{km} \boldsymbol{\psi}_k(z), \quad z \in G, \end{cases} \tag{4.10}$$

where the term with $l = k$ and $\mathbf{m} = (0,0)$ is skipped in the sum $\sum'$. It follows from Theorem 5 that the vector function $\Phi(z)$ is analytic in all the domains G, G_l $(l = 1,2,\ldots,N)$. One can check that the increment of $\Phi(z)$ across every circle $|t-a_l| = r_l$ vanishes due to (4.1)

$$\lim_{\substack{z\to t \\ |z-a_k|<r_k}} \Phi(z) - \lim_{\substack{z\to t \\ z\in G}} \Phi(z) = \boldsymbol{\psi}_l(t) - \boldsymbol{\psi}(t) + \boldsymbol{\beta}_l \left(\frac{r_l}{z-a_l}\right)^2 \overline{\psi_l(t)} = \begin{pmatrix} 0 \\ 0 \end{pmatrix}. \tag{4.11}$$

By the Principle of Analytic Continuation and the generalized Liouville's theorem on torus [10] $\Phi(z)$ is constant. It will be proved later that this constant vector is determined by (4.3) and

$$\Phi(z) = \begin{pmatrix} 1 \\ 0 \end{pmatrix}. \tag{4.12}$$

Writing the latter equation by using the definition of $\Phi(z)$ in G_k and (4.6) we obtain the system of functional equations

$$\boldsymbol{\psi}_l(z) = \sum_{k=1}^{\prime N} \boldsymbol{\beta}_k \sum_{\mathbf{m}\in\mathbb{Z}_\omega^2} \left(\frac{r_k}{z-a_k-\mathbf{m}}\right)^2 \overline{\boldsymbol{\psi}_k\left(a_k + \frac{r_k^2}{\overline{z-a_k-\mathbf{m}}}\right)} + \begin{pmatrix} 1 \\ 0 \end{pmatrix}, \tag{4.13}$$
$$|z-a_l| \le r_l \quad (l = 1,2,\ldots,N).$$

A discrete form of the functional equations (4.13) is obtained by the substitution of (4.9) into (4.13)

$$\boldsymbol{\psi}_l(z) = \sum_{k=1}^{N} \boldsymbol{\beta}_k \sum_{p=0}^{\infty} r_l^{2p+2} \hat{E}_{p+2}(z-a_k;l) \overline{\boldsymbol{\psi}_{kp}} + \begin{pmatrix} 1 \\ 0 \end{pmatrix}, \tag{4.14}$$
$$|z-a_l| \le r_l \quad (l = 1,2,\ldots,N).$$

Here, the function $\hat{E}_{p+2}(z-a_k;l)$ is defined in the fixed disk $|z-a_l| \le r_l$ as follows

$$\hat{E}_{p+2}(z-a_k,l) = \begin{cases} E_{p+2}(z-a_k), & k \ne l, \\ E_{p+2}(z-a_l) - (z-a_l)^{-p-2}, & k = l. \end{cases} \tag{4.15}$$

In particular, if $a_k = a_l$, then $\hat{E}_{p+2}(a_l - a_k, l) \equiv \hat{E}_{p+2}(0,l) := S_{p+2}$.

When the vector functions $\boldsymbol{\psi}_l(z)$ are determined, the vector function can be found from (4.10) in G

$$\boldsymbol{\psi}(z) = \sum_{k=1}^{N} \boldsymbol{\beta}_k \sum_{\mathbf{m}\in\mathbb{Z}_\omega^2} W_{km} \boldsymbol{\psi}_k(z) + \begin{pmatrix} 1 \\ 0 \end{pmatrix}, \quad z \in G. \tag{4.16}$$

The latter equation justifies the choice of the constant vector (4.12) because the external flux in the not-perturbed case $\boldsymbol{\beta}_k = \mathbf{0}$ becomes the required (4.4).

Remark 3 *An asymptotic analysis was developed in [142] for similar boundary value problems for a multiply connected domain, where* $\boldsymbol{\psi}_k(z)$ *are approximated by constants* C_k *and the system of functional equations* (4.13) *becomes the system of linear algebraic equations (9.22) from [142].*

The method of successive approximations can be applied to (4.13) or equivalently to (4.14). Consider the vector function $\Psi(z) = \psi_k(z)$ defined in $G^+ = \cup_{k=1}^N G_k$ as an element of the space $\mathcal{H}^+(G^+)$ introduced on page 33. Equations (4.13) or (4.14) can be shortly presented in the operator form

$$\Psi(z) = \begin{pmatrix} 1 \\ 0 \end{pmatrix} + \sum_{k=1}^{N} (\mathcal{A}_k \boldsymbol{\beta}_k \overline{\Psi})(z), \quad z \in G^+ \cup \partial G^+, \tag{4.17}$$

where the operator $\mathcal{A}_k$ is written in the discrete form

$$\mathcal{A}_k F = \sum_{p=0}^{\infty} r^{2p+2} \hat{E}_{p+2}(z - a_k; l) F_{kp}, \quad F \in \mathcal{H}^+(G^+). \tag{4.18}$$

The vector function $F(z)$ is analytic in $|z - a_k| < r_k$, its Taylor coefficient in $(z - a_k)^p$ is denoted by F_{kp}.

Remark 4 *The operator* (4.18) *has such a structure that it can be defined on the scalar functions* $F \in \mathcal{H}^+(G^+)$.

The vector equation (4.17) can be written in the shortened form

$$\Psi(z) = \begin{pmatrix} 1 \\ 0 \end{pmatrix} + \sum_{k=1}^{N} (\mathcal{B}_k \Psi)(z), \quad z \in G^+ \cup \partial G^+, \tag{4.19}$$

where the operator $\mathcal{B}_k \Psi := \mathcal{A}_k \boldsymbol{\beta}_k \overline{\Psi}$ is compact in the space $\mathcal{H}^+(G^+)$ [156, 162, 185]. The action of the operator $\mathcal{B}_k$ can be described more precisely as follows. First, the Taylor coefficients $\boldsymbol{\psi}_{kp}$ of $\Psi(z)$ in $|z - a_k| < r_k$ are found and the complex conjugation is taken. Next, the matrix $\boldsymbol{\beta}_k$ is multiplied by the vectors $\overline{\boldsymbol{\psi}_{kp}}$ for all $p = 0, 1, \ldots$, the result is denoted by F_{kp}. The final step consists of the linear combination (4.18).

The method of successive approximations applied to (4.19) yields the series

$$\Psi(z) = \begin{pmatrix} 1 \\ 0 \end{pmatrix} + \sum_{s=1}^{\infty} \sum_{k_1, \ldots, k_s} \mathcal{B}_{k_1} \mathcal{B}_{k_2} \ldots \mathcal{B}_{k_s} \begin{pmatrix} 1 \\ 0 \end{pmatrix}, \tag{4.20}$$

where $k_1, k_2, \ldots, k_s$ run over $1, 2, \ldots, N$. The series (4.20) can be written in the form

$$\Psi(z) = \begin{pmatrix} 1 \\ 0 \end{pmatrix} + \sum_{s=1}^{\infty} \sum_{k_1, \ldots, k_s} \mathcal{A}_{k_1} \boldsymbol{\beta}_{k_1} \mathbf{C} \mathcal{A}_{k_2} \boldsymbol{\beta}_{k_2} \mathbf{C} \ldots \mathcal{A}_{k_s} \boldsymbol{\beta}_{k_s} \mathbf{C} \begin{pmatrix} 1 \\ 0 \end{pmatrix}, \tag{4.21}$$

where $\mathbf{C}$ denotes the operator of complex conjugation.

It follows from (4.18) that the operator $\mathcal{A}_k$ consists of the multiplications of the vectors F_{kp} by scalar complex functions. Moreover, the matrices $\boldsymbol{\beta}_k$ are real, and their multiplication commutes with the scalar multiplier. Therefore, (4.21) can be written in the form

$$\Psi(z) = \begin{pmatrix} 1 \\ 0 \end{pmatrix} + \sum_{s=1}^{\infty} \sum_{k_1,\ldots,k_s} \mathcal{A}_{k_1} \mathbf{C} \mathcal{A}_{k_2} \ldots \mathbf{C} \mathcal{A}_{k_s} \upsilon_s, \tag{4.22}$$

where

$$\upsilon_s \equiv \begin{pmatrix} \upsilon_{1s} \\ \upsilon_{2s} \end{pmatrix} = \boldsymbol{\beta}_{k_1} \boldsymbol{\beta}_{k_2} \ldots \boldsymbol{\beta}_{k_s} \begin{pmatrix} 1 \\ 0 \end{pmatrix}. \tag{4.23}$$

It follows from (2.37) to (2.39) that the coordinates of the real vector (4.23) satisfy the equality

$$\begin{pmatrix} \upsilon_{1s} \\ \upsilon_{2s} \end{pmatrix} = \begin{pmatrix} \mathrm{Re}\, [\varrho_{k_1} \varrho_{k_2} \cdots \varrho_{k_s}] \\ \mathrm{Im}\, [\varrho_{k_1} \varrho_{k_2} \cdots \varrho_{k_s}] \end{pmatrix}. \tag{4.24}$$

This two-dimensional vector equality yields the scalar complex relation

$$\upsilon_{1s} + i\upsilon_{2s} = \varrho_{k_1} \varrho_{k_2} \cdots \varrho_{k_s}. \tag{4.25}$$

4.1.1 GENERAL FORMULAS FOR THE EFFECTIVE PERMITTIVITY TENSOR AND LOCAL FIELD

In the case of circular inclusions $G_k = \{z \in \mathbb{C} : |z - a_k| < r_k\}$, equation (2.189) takes the simple form

$$\begin{pmatrix} \varepsilon_{11} \\ \varepsilon_{21} \end{pmatrix} = \begin{pmatrix} 1 \\ 0 \end{pmatrix} + 2\sum_{k=1}^{N} \varrho_k \pi r_k^2 \begin{pmatrix} \mathrm{Re}\, \psi_{1k}(a_k) + i\mathrm{Re}\, \psi_{2k}(a_k) \\ -\mathrm{Im}\, \psi_{1k}(a_k) - i\mathrm{Im}\, \psi_{2k}(a_k) \end{pmatrix}. \tag{4.26}$$

due to the mean value theorem for a function harmonic in a disk.

Consider the case of real coefficients ε_k. The matrix (2.33) takes the form

$$\boldsymbol{\beta}_k = \varrho_k I, \tag{4.27}$$

where the scalar contrast parameter ϱ_k given by (2.36) is real and $|\varrho_k| \leq 1$. In this case, the system (4.13) breaks up into two scalar systems, which can be solved by the method of successive approximations convergent for any $|\varrho_k| \leq 1$ and for any non-overlapping circular inclusions $|z - a_k| \leq r_k$ [93]. In the case of touching disks, the convergence holds at least for $|\varrho_k| < 1$.

Due to (4.27) the vector-matrix formula (4.22) is reduced to the scalar one

$$\Psi_1(z) = 1 + \sum_{s=1}^{\infty} \sum_{k_1,\ldots,k_s} \mathcal{A}_{k_1} \mathbf{C} \mathcal{A}_{k_2} \ldots \mathbf{C} \mathcal{A}_{k_s} \upsilon_{1s}, \tag{4.28}$$

where $\Psi_1(z) = \psi_k(z)$ defined in $|z - a_k| \leq r$ $(k = 1, 2, \ldots, N)$ is a scalar analytic function now. It follows from (4.25) that $\upsilon_{1s} = \varrho_{k_1} \varrho_{k_2} \cdots \varrho_{k_s}$ with the vanishing imaginary part $\upsilon_{2s} = 0$. The formula (4.26) becomes

$$\begin{pmatrix} \varepsilon_{11} \\ \varepsilon_{21} \end{pmatrix} = \begin{pmatrix} 1 \\ 0 \end{pmatrix} + 2\sum_{k=1}^{N} \varrho_k \pi r_k^2 \begin{pmatrix} \mathrm{Re}\, \psi_{1k}(a_k) \\ -\mathrm{Im}\, \psi_{1k}(a_k) \end{pmatrix}. \tag{4.29}$$

The vector relation (4.29) was written in [93] as a complex relation on $\varepsilon_{11} - \mathrm{i}\varepsilon_{12}$, where $\varepsilon_{12} = \varepsilon_{21}$. The series (4.28) was used to compute $\psi_k(a_k) = \Psi_1(a_k)$. The analogous formula was derived for the component $\varepsilon_{22} + \mathrm{i}\varepsilon_{21}$.

We proceed to demonstrate that the same scheme can be repeated for the complex eigenvalues ϱ_k. Using (4.22) and (4.24) find the vector-function

$$\Psi(z) = \begin{pmatrix} 1 \\ 0 \end{pmatrix} + \sum_{s=1}^{\infty} \sum_{k_1,\ldots,k_s} \mathcal{A}_{k_1} \mathbf{C} \mathcal{A}_{k_2} \ldots \mathbf{C} \mathcal{A}_{k_s} \begin{pmatrix} \mathrm{Re}\,[\varrho_{k_1} \varrho_{k_2} \cdots \varrho_{k_s}] \\ \mathrm{Im}\,[\varrho_{k_1} \varrho_{k_2} \cdots \varrho_{k_s}] \end{pmatrix}, \qquad (4.30)$$

where in general $v_{2s} \neq 0$. It is worth noting that the relation (4.26) can be used now only in the vector form since this vector contains complex components contrary to the real vector (4.29). Therefore, the vector $\begin{pmatrix} \varepsilon_{11} \\ \varepsilon_{21} \end{pmatrix}$ for real and complex ε_k is calculated by the same formula (4.26) where $\Psi(z)$ is found by (4.30).

Using the scalar form of the operator $\mathcal{A}_k$ described in Remark 4 on page 80 find $\psi_{1k}(a_k)$ and $\psi_{2k}(a_k)$ from (4.30)

$$\psi_{1k}(a_k) = 1 + \textstyle\sum_{s=1}^{\infty} \sum_{k_1,\ldots,k_s} \mathcal{A}_{k_1} \mathbf{C} \mathcal{A}_{k_2} \ldots \mathbf{C} \mathcal{A}_{k_s}(a_k)\, \mathrm{Re}\,[\varrho_{k_1} \varrho_{k_2} \cdots \varrho_{k_s}],$$

$$\psi_{2k}(a_k) = \textstyle\sum_{s=1}^{\infty} \sum_{k_1,\ldots,k_s} \mathcal{A}_{k_1} \mathbf{C} \mathcal{A}_{k_2} \ldots \mathbf{C} \mathcal{A}_{k_s}(a_k)\, \mathrm{Im}\,[\varrho_{k_1} \varrho_{k_2} \cdots \varrho_{k_s}]. \qquad (4.31)$$

Introduce the contrast average $\langle \varrho \rangle := \sum_{k=1}^{N} \varrho_k \pi r_k^2$. Substitute (4.31) into (4.26)

$$\begin{pmatrix} \varepsilon_{11} \\ \varepsilon_{21} \end{pmatrix} = (1 + 2\langle \varrho \rangle) \begin{pmatrix} 1 \\ 0 \end{pmatrix} + \\ +2 \textstyle\sum_{s=1}^{\infty} \sum_{k_1,\ldots,k_s} \varrho_{k_1} \varrho_{k_2} \cdots \varrho_{k_s} \begin{pmatrix} \mathrm{Re}\,[\mathcal{A}_{k_1} \mathbf{C} \mathcal{A}_{k_2} \ldots \mathbf{C} \mathcal{A}_{k_s}(a_k)] \\ -\mathrm{Im}\,[\mathcal{A}_{k_1} \mathbf{C} \mathcal{A}_{k_2} \ldots \mathbf{C} \mathcal{A}_{k_s}(a_k)] \end{pmatrix}. \qquad (4.32)$$

The analogous formula (2.202) holds for the vector $\begin{pmatrix} \varepsilon_{22} \\ \varepsilon_{12} \end{pmatrix}$. Then, the whole tensor can be calculated by (2.205).

In particular, the formula (4.32) holds for the real-valued permittivity when all ϱ_k are real. In this case, equations (4.31) become

$$\psi_{1k}(a_k) = 1 + \textstyle\sum_{s=1}^{\infty} \sum_{k_1,\ldots,k_s} \varrho_{k_1} \varrho_{k_2} \cdots \varrho_{k_s} \mathcal{A}_{k_1} \mathbf{C} \mathcal{A}_{k_2} \ldots \mathbf{C} \mathcal{A}_{k_s}(a_k), \\ \psi_{2k}(a_k) = 0. \qquad (4.33)$$

Substitution of (4.33) into (4.29) yields the same formula (4.32). This observation confirms **Rule** $\mathbb{R} \to \mathbb{C}$ stated on 58.

Remark 5 *According to* **Rule** $\mathbb{R} \to \mathbb{C}$, *one can apply the formulas derived for real-valued permittivity to the complex permittivity for composites. The formulas summarized in [93] were derived for deterministic and random composites. These formulas* *are the known most general formulas established for the effective properties of composites with circular, non-overlapping inclusions. The formulas from [93] need a slight modification to be formally applied for the complex permittivity since they were obtained using the classic scalar complex potentials. For instance, for real* ϱ_k *two real coordinate equations* (4.32) *were written in the form of one complex relation*

$$\varepsilon_{11} - \mathrm{i}\varepsilon_{21} = 1 + 2\langle\varrho\rangle + 2\textstyle\sum_{s=1}^{\infty}\sum_{k_1,\dots,k_s} \varrho_{k_1}\cdots\varrho_{k_s}\mathcal{A}_{k_1}\mathbf{C}\mathcal{A}_{k_2}\cdots\mathbf{C}\mathcal{A}_{k_s}(a_k). \tag{4.34}$$

4.1.2 APPROXIMATE ANALYTICAL FORMULAS FOR THE LOCAL FIELD

The expressions for the local fields and tensors are presented in the previous Section in the form of an operator infinite series. In the present Section, we describe a constructive symbolic computational scheme for the local field. It extends an algorithm developed in [38,93] for real-valued permittivity.

For simplicity, consider the system of functional equations (4.14) for equal disks when $\boldsymbol{\beta}_k = \boldsymbol{\beta}$ and $r_k = r$ $(k = 1,2,\dots,N)$

$$\begin{gathered}\boldsymbol{\psi}_l(z) = \boldsymbol{\beta}\textstyle\sum_{k=1}^{N}\sum_{p=0}^{\infty} r^{2p+2}\hat{E}_{p+2}(z-a_k;l)\overline{\boldsymbol{\psi}_{kp}} + \begin{pmatrix}1\\0\end{pmatrix},\\ |z-a_l| \le r_l \quad (l = 1,2,\dots,N),\end{gathered} \tag{4.35}$$

where $\boldsymbol{\psi}_{kp}$ are the coefficient in the Taylor expansions (4.8). We are looking for the vector-functions $\boldsymbol{\psi}_l(z)$ in the form of the approximation up to $O(f^{n+1})$

$$\boldsymbol{\psi}_l(z) = \sum_{s=0}^{n}\boldsymbol{\psi}_l^{(s)}(z)r^{2s} + O(r^{2(n+1)}), \quad |z-a_l| \le r \quad (l = 1,2,\dots,N). \tag{4.36}$$

Let the vector-function $\boldsymbol{\psi}_l^{(s)}(z)$ be expanded into its Taylor series

$$\boldsymbol{\psi}_l^{(s)}(z) = \sum_{p=0}^{\infty}\boldsymbol{\psi}_{lp}^{(s)}(z-a_l)^p, \quad |z-a_l| \le r \quad (l = 1,2,\dots,N). \tag{4.37}$$

Therefore,

$$\boldsymbol{\psi}_l(z) = \sum_{s=0}^{n}\sum_{p=0}^{\infty}\boldsymbol{\psi}_{lp}^{(s)}r^{2s}(z-a_l)^p, \quad |z-a_l| \le r \quad (l = 1,2,\dots,N) \tag{4.38}$$

and

$$\boldsymbol{\psi}_{kp} = \sum_{q=0}^{n-1}\boldsymbol{\psi}_{kp}^{(q)}r^{2q} + O(r^{2n}) \quad (k = 1,2,\dots,N;\ n = 1,2,\dots). \tag{4.39}$$

Substitute (4.36) and (4.39) into the left and right parts of (4.35), respectively, keeping the terms at most $O(r^{2n})$

$$\sum_{s=0}^{n}\boldsymbol{\psi}_l^{(s)}(z)r^{2s} = \boldsymbol{\beta}\sum_{k=1}^{N}\sum_{0\le p+q+1\le n} r^{2(p+q+1)}\hat{E}_{p+2}(z-a_k;l)\overline{\boldsymbol{\psi}_{kp}^{(q)}} + \begin{pmatrix}1\\0\end{pmatrix} + O(r^{2(n+1)}), \tag{4.40}$$

where $p, q = 0, 1, \ldots$. Selection of the coefficients in the same powers of r^2 yields the following recurrence scheme to write explicitly the terms of (4.36)

$$\begin{aligned}
&\boldsymbol{\psi}_l^{(0)}(z) = \begin{pmatrix} 1 \\ 0 \end{pmatrix}, \\
&\boldsymbol{\psi}_l^{(s)}(z) = \boldsymbol{\beta} \sum_{k=1}^N \sum_{p+q+1=s} \hat{E}_{p+2}(z - a_k; l) \overline{\boldsymbol{\psi}_{kp}^{(q)}} = \\
&\boldsymbol{\beta} \sum_{k=1}^N \sum_{p=0}^{s-1} \left[\hat{E}_2(z - a_k; l) \overline{\boldsymbol{\psi}_{k0}^{(s-1)}} + \hat{E}_3(z - a_k; l) \overline{\boldsymbol{\psi}_{k1}^{(s-2)}} + \ldots \right. \\
&\left. + \hat{E}_{s+1}(z - a_k; l) \overline{\boldsymbol{\psi}_{k,s-1}^{(0)}} \right], \quad s = 1, 2, \ldots, n.
\end{aligned} \tag{4.41}$$

The first equation in (4.41) gives the zeroth term. The next term $\boldsymbol{\psi}_l^{(s)}(z)$ is calculated explicitly by (4.41) through the Taylor coefficients $\boldsymbol{\psi}_{kp}^{(q)}$ of the vector-functions $\boldsymbol{\psi}_k^{(q)}(z)$ $(q = 0, 1, \ldots, s-1)$ found in the previous steps of algorithm. Therefore, the algorithm uses the Taylor coefficients of vector functions found in the previous steps and computes their linear combinations with the Eisenstein functions. This algorithm can be easily implemented by a symbolic computation package.

Below, we write a few terms

$$\begin{aligned}
&\boldsymbol{\psi}_l^{(0)}(z) = \begin{pmatrix} 1 \\ 0 \end{pmatrix}, \quad \boldsymbol{\psi}_l^{(1)}(z) = \boldsymbol{\beta} \sum_{k=1}^N \hat{E}_2(z - a_k; l) \begin{pmatrix} 1 \\ 0 \end{pmatrix}, \\
&\boldsymbol{\psi}_l^{(2)}(z) = \boldsymbol{\beta}^2 \sum_{k,k_1=1}^N \overline{E_2(a_k - a_{k_1})} \hat{E}_2(z - a_k; l) \begin{pmatrix} 1 \\ 0 \end{pmatrix}, \\
&\boldsymbol{\psi}_l^{(3)}(z) = \boldsymbol{\beta}^3 \sum_{k,k_1,k_2=1}^N \overline{E_2(a_k - a_{k_1})} E_2(a_{k_1} - a_{k_2}) \hat{E}_2(z - a_k; l) \begin{pmatrix} 1 \\ 0 \end{pmatrix} \\
&-2\boldsymbol{\beta}^2 \sum_{k,k_1=1}^N \overline{E_3(a_k - a_{k_1})} \hat{E}_3(z - a_k; l) \begin{pmatrix} 1 \\ 0 \end{pmatrix}.
\end{aligned} \tag{4.42}$$

Here, it is assumed that the coefficient $E_p(a_k - a_\ell)$ is equal to the lattice sum S_p if $a_k = a_\ell$. Otherwise, it is calculated by the standard Eisenstein function.

The expressions (4.42) can be simplified by using (4.23) and (4.24) for equal $\varrho_k = \varrho$ $(k = 1, 2, \ldots, N)$

$$\boldsymbol{\beta}^s \begin{pmatrix} 1 \\ 0 \end{pmatrix} = \begin{pmatrix} \operatorname{Re} \varrho^s \\ \operatorname{Im} \varrho^s \end{pmatrix}. \tag{4.43}$$

Therefore, (4.42) becomes

$$\boldsymbol{\psi}_l^{(1)}(z) = \sum_{k=1}^N \hat{E}_2(z-a_k;l)\begin{pmatrix} \operatorname{Re}\varrho \\ \operatorname{Im}\varrho \end{pmatrix},$$

$$\boldsymbol{\psi}_l^{(2)}(z) = \sum_{k,k_1=1}^N \overline{E_2(a_k-a_{k_1})}\hat{E}_2(z-a_k;l)\begin{pmatrix} \operatorname{Re}\varrho^2 \\ \operatorname{Im}\varrho^2 \end{pmatrix},$$

$$\boldsymbol{\psi}_l^{(3)}(z) = \sum_{k,k_1,k_2=1}^N \overline{E_2(a_k-a_{k_1})}E_2(a_{k_1}-a_{k_2})\hat{E}_2(z-a_k;l)\begin{pmatrix} \operatorname{Re}\varrho^3 \\ \operatorname{Im}\varrho^3 \end{pmatrix}$$

$$-2\sum_{k,k_1=1}^N \overline{E_3(a_k-a_{k_1})}\hat{E}_3(z-a_k;l)\begin{pmatrix} \operatorname{Re}\varrho^2 \\ \operatorname{Im}\varrho^2 \end{pmatrix}. \tag{4.44}$$

The approximations (4.41) are substituted into (4.36). This yields the approximation of the complex potential $\boldsymbol{\psi}_l(z)$ in the disk $|z-a_l| \le r$ up to $O(r^{2n})$ with $n=4$

$$\boldsymbol{\psi}_l(z) = \begin{pmatrix} 1 \\ 0 \end{pmatrix} + r^2\sum_{k=1}^N \hat{E}_2(z-a_k;l)\begin{pmatrix} \operatorname{Re}\varrho \\ \operatorname{Im}\varrho \end{pmatrix}$$

$$+r^4\sum_{k,k_1=1}^N \overline{E_2(a_k-a_{k_1})}\hat{E}_2(z-a_k;l)\begin{pmatrix} \operatorname{Re}\varrho^2 \\ \operatorname{Im}\varrho^2 \end{pmatrix}$$

$$+r^6\left[\sum_{k,k_1,k_2=1}^N \overline{E_2(a_k-a_{k_1})}E_2(a_{k_1}-a_{k_2})\hat{E}_2(z-a_k;l)\begin{pmatrix} \operatorname{Re}\varrho^3 \\ \operatorname{Im}\varrho^3 \end{pmatrix}\right. \tag{4.45}$$

$$\left.-2\sum_{k,k_1=1}^N \overline{E_3(a_k-a_{k_1})}\hat{E}_3(z-a_k;l)\begin{pmatrix} \operatorname{Re}\varrho^2 \\ \operatorname{Im}\varrho^2 \end{pmatrix}\right] + O(r^{2n}),$$

$$|z-a_l| \le r \quad (l=1,2,\ldots,N).$$

Using the relation (2.31) we determine the flux in the inclusion

$$\begin{pmatrix} \dfrac{\partial u_l'}{\partial x_1} + \mathrm{i}\dfrac{\partial u_l'}{\partial x_2} \\ \\ \dfrac{\partial u_l''}{\partial x_1} + \mathrm{i}\dfrac{\partial u_l''}{\partial x_2} \end{pmatrix} = 2(I+\boldsymbol{\alpha}_l)^{-1}\overline{\boldsymbol{\psi}_l(z)}, \quad |z-a_l| \le r. \tag{4.46}$$

Applying (2.32) to (4.46) we get

$$\begin{pmatrix} \dfrac{\partial u_l'}{\partial x_1} + \mathrm{i}\dfrac{\partial u_l'}{\partial x_2} \\ \\ \dfrac{\partial u_l''}{\partial x_1} + \mathrm{i}\dfrac{\partial u_l''}{\partial x_2} \end{pmatrix} = \frac{2}{|1+\varepsilon_k|^2}\begin{pmatrix} 1+\varepsilon_k' & \varepsilon_k'' \\ -\varepsilon_k'' & 1+\varepsilon_k' \end{pmatrix}\overline{\boldsymbol{\psi}_l(z)}, \quad |z-a_l| \le r. \tag{4.47}$$

The complex potential $\boldsymbol{\psi}(z)$ in the domain G can be found by the formula

$$\boldsymbol{\psi}(z) = \boldsymbol{\beta}\sum_{k=1}^N \sum_{0 \le p+q+1 \le n} r^{2(p+q+1)}E_{p+2}(z-a_k;l)\overline{\boldsymbol{\psi}_{kp}^{(q)}} + \begin{pmatrix} 1 \\ 0 \end{pmatrix} + O(r^{2(n+1)}), \tag{4.48}$$

where $n = 4$ and $\boldsymbol{\psi}_{kp}^{(q)}$ is the pth coefficient in the Taylor expansion of $\boldsymbol{\psi}_k^{(q)}(z)$ given by (4.36) in the disk $|z-a_k| < r$. Due to (2.30), we obtain the flux in the domain G

$$\begin{pmatrix} \frac{\partial u'}{\partial x_1} + \mathrm{i}\frac{\partial u'}{\partial x_2} \\ \\ \frac{\partial u''}{\partial x_1} + \mathrm{i}\frac{\partial u''}{\partial x_2} \end{pmatrix} = \overline{\boldsymbol{\psi}(z)}, \quad z \in G. \tag{4.49}$$

Example 3 *Consider a regular square array when $S_2 = \pi$ and $S_3 = 0$ [93]. Hence,* (4.42) *becomes*

$$\begin{aligned} &\boldsymbol{\psi}_1^{(1)}(z) = \hat{E}_2(z)\begin{pmatrix} \mathrm{Re}\,\varrho \\ \mathrm{Im}\,\varrho \end{pmatrix}, \quad \boldsymbol{\psi}_1^{(2)}(z) = \pi\hat{E}_2(z)\begin{pmatrix} \mathrm{Re}\,\varrho^2 \\ \mathrm{Im}\,\varrho^2 \end{pmatrix}, \\ &\boldsymbol{\psi}_1^{(3)}(z) = \pi^2\hat{E}_2(z)\begin{pmatrix} \mathrm{Re}\,\varrho^3 \\ \mathrm{Im}\,\varrho^3 \end{pmatrix}, \quad |z| < r, \end{aligned} \tag{4.50}$$

where $\hat{E}_2(z) = E_2(z) - z^{-2}$. Substitution of (4.50) *into* (4.36) *yields the asymptotic formula for the local field*

$$\boldsymbol{\psi}_1(z) = \begin{pmatrix} 1 \\ 0 \end{pmatrix} + \frac{f}{\pi}\hat{E}_2(z)\begin{pmatrix} \mathrm{Re}\,\frac{\varrho}{1-f\varrho} \\ \mathrm{Im}\,\frac{\varrho}{1-f\varrho} \end{pmatrix} + O(f^4), \quad |z| \le r, \tag{4.51}$$

where the formula $f = \pi r^2$ is used. Here, the following asymptotic formula is applied

$$f\varrho + f^2\varrho^2 + f^3\varrho^3 = \frac{f\varrho}{1-f\varrho} + O(f^4). \tag{4.52}$$

The local flux in the inclusion can be found by (4.47)

$$\begin{pmatrix} \frac{\partial u_1'}{\partial x_1} + \mathrm{i}\frac{\partial u_1'}{\partial x_2} \\ \\ \frac{\partial u_1''}{\partial x_1} + \mathrm{i}\frac{\partial u_1''}{\partial x_2} \end{pmatrix} = \frac{2}{|1+\varepsilon_1|^2}\begin{pmatrix} 1+\varepsilon_1' & \varepsilon_1'' \\ -\varepsilon_1'' & 1+\varepsilon_1' \end{pmatrix}\overline{\boldsymbol{\psi}_1(z)}, \quad |z| \le r, \tag{4.53}$$

where $\boldsymbol{\psi}_1(z)$ is given by (4.51).

Using (4.16) *and* (4.49), *we obtain the asymptotic formula for the local flux in the matrix*

$$\begin{pmatrix} \frac{\partial u'}{\partial x_1} + \mathrm{i}\frac{\partial u'}{\partial x_2} \\ \\ \frac{\partial u''}{\partial x_1} + \mathrm{i}\frac{\partial u''}{\partial x_2} \end{pmatrix} = \begin{pmatrix} 1 \\ 0 \end{pmatrix} + \frac{f}{\pi}\overline{E_2(z)}\begin{pmatrix} \mathrm{Re}\,\frac{\varrho}{1-f\varrho} \\ \mathrm{Im}\,\frac{\varrho}{1-f\varrho} \end{pmatrix} + O(f^4), \quad z \in G. \tag{4.54}$$

The ratio of the permittivity of disks occupied by glioma to the host medium is given by the equation [96]

$$\varepsilon_1 = -\frac{3.5\,(2.45678\cdot 10^{10}\mathrm{i}\omega + \omega^2 + 2.52773\cdot 10^{19})}{\omega(3\omega - 10^{11}\mathrm{i})}, \tag{4.55}$$

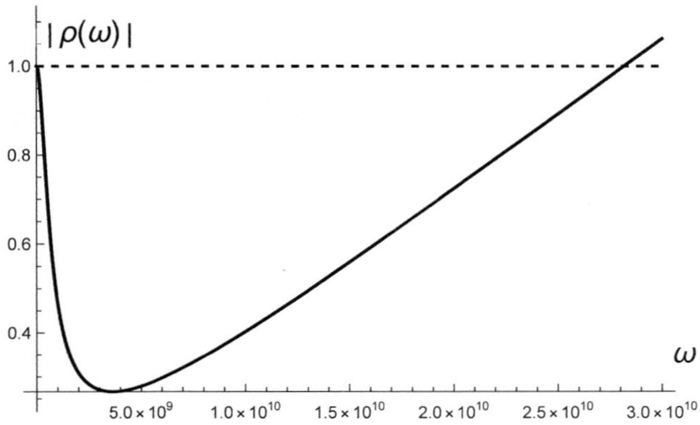

Figure 4.3 Dependence of $|\varrho|$ on ω for $0 < \omega < 3 \cdot 10^{10}$.

where ω denotes the frequency. The contrast parameter can be written as the following function on ω

$$\varrho = \frac{-2.80251 \cdot 10^{10} \mathrm{i}\omega + 13\omega^2 + 1.76941 \cdot 10^{20}}{3.71975 \cdot 10^{11} \mathrm{i}\omega + \omega^2 + 1.76941 \cdot 10^{20}}. \tag{4.56}$$

The dependence of absolute value $|\varrho(\omega)|$ on the frequency ω is displayed in Figures 4.3 and 4.4, where $|\varrho(\omega)|$ tends to 13 as ω tends to infinity. The convergence of Schwarz's method was proved for $|\varrho(\omega)| < 1$, i.e., for $\omega < 2.81718 \cdot 10^{10}$.

It is suggested that the region of convergence increases at least up to $\frac{f}{f_c}|\varrho(\omega)| < 1$, where f_c stands for the critical concentration when the disks become touching. In the first considered numerical example, $r = 0.35$, $\omega = 10^{10}$, $\frac{f}{f_c} = 0.49$ and $|\varrho(10^{10})| = 0.403$, where $\varepsilon_1 = 0.668 - 0.639\mathrm{i}$ and $\varrho = -0.045 - 0.4\mathrm{i}$ are calculated by (4.55) and (4.56), respectively. In the second example, $r = 0.25$, $\omega = 10^{11}$, $\varepsilon_1 = -0.967 - 0.609\mathrm{i}$, $\varrho = 0.821 - 3.275\mathrm{i}$, $\frac{f}{f_c} = 0.25$ and $|\varrho(10^{11})| = 3.376$, hence $\frac{f}{f_c}|\varrho(\omega)| = 0.844$.

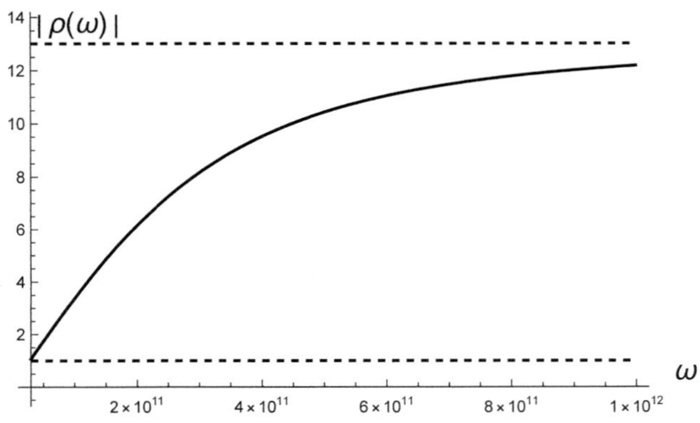

Figure 4.4 Dependence of $|\varrho|$ on ω for $\omega > 3 \cdot 10^{10}$.

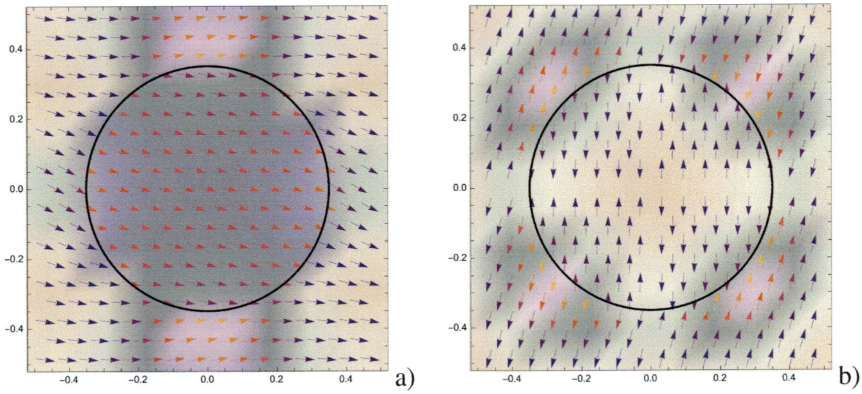

Figure 4.5 The local flux calculated with (4.51) and (4.53) in the unit periodicity cell for $\omega = 10^{10}$, $\varepsilon_1 = 0.668 - 0.639\mathrm{i}$, $\varrho = -0.045 - 0.4\mathrm{i}$ and $r = 0.35$ ($f = 0.38$). Data are for (a) the real part of the flux and (b) the imaginary part of the flux. The colors correspond to the intensity of the flux shown in Figure 4.7.

Numerical examples of the local flux for the regular unit square array of disks of radius $r = 0.35$ are presented in Figures 4.5, 4.6, 4.7 and 4.8.

4.1.3 REGULAR ARRAY OF EQUAL DISKS

In the present subsection, we consider a regular array of equal disks of radius r in the unit periodicity cell Q generated by two fundamental translation vectors $\boldsymbol{\omega}_1$ and $\boldsymbol{\omega}_2$. The periodicity cell is displayed in Figure 4.9. The complex contrast parameter ϱ for a two-phase medium and the concentration $f = \pi r^2$ will be used in the present section.

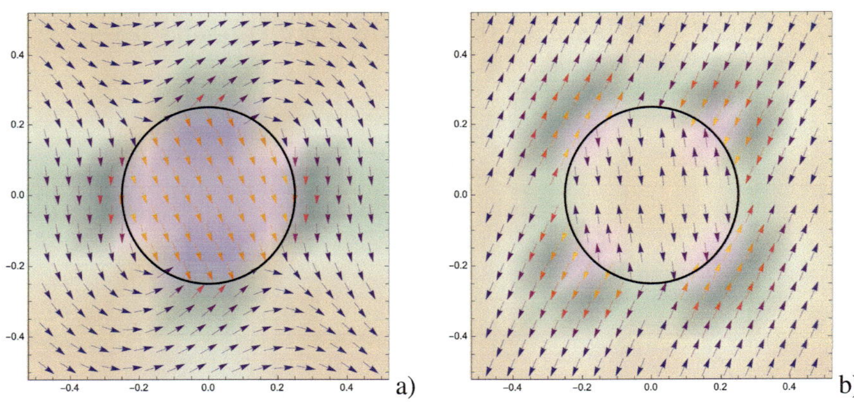

Figure 4.6 The local flux calculated for $\omega = 10^{11}$, $\varepsilon_1 = -0.967 - 0.609\mathrm{i}$, $\varrho = 0.821 - 3.275\mathrm{i}$ and $r = 0.25$ ($f = 0.2$). Data are for (a) the real part of the flux and (b) the imaginary part of the flux. The colors correspond to the intensity of the flux shown in Figure 4.8.

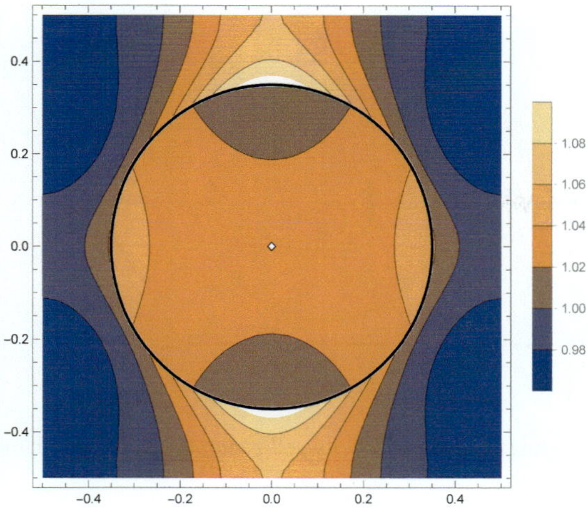

Figure 4.7 The local intensity of flux for $\omega = 10^{10}$, expressed by the absolute value of the flux represented by the real and imaginary parts in Figure 4.5.

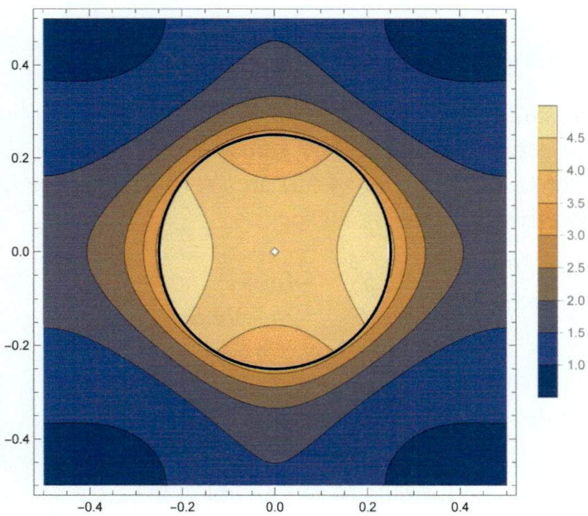

Figure 4.8 The local intensity of flux for $\omega = 10^{11}$ corresponding Figure 4.6.

Introduce the complex constants

$$s_k^{(m)} = \frac{(2m+2k-3)!}{(2m-1)!(2k-2)!} S_{2(m+k-1)}, \quad k, m = 1, 2, \ldots, \tag{4.57}$$

$$s = s_{l_1}^{(1)} s_{l_2}^{(l_1)} \ldots s_{l_k}^{(l_{k-1})} s_1^{l_k}, \tag{4.58}$$

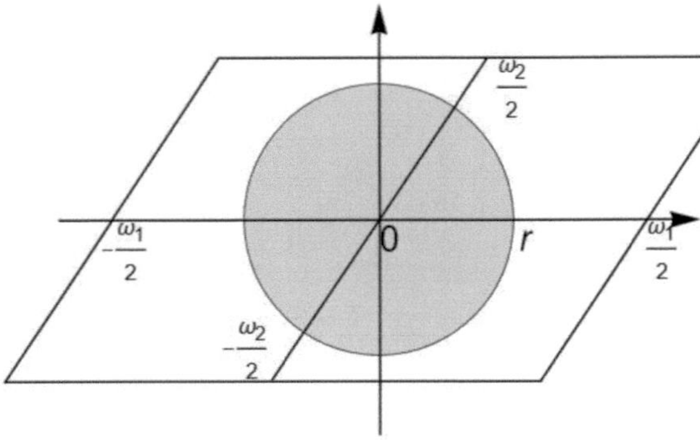

Figure 4.9 A disk of radius r in the periodicity cell generated by two fundamental translation vectors $\boldsymbol{\omega}_1$ and $\boldsymbol{\omega}_2$.

where the Eisenstein-Rayleigh lattice sums are used

$$S_{2p} = \sum_{m_1,m_2 \in \mathbb{Z}^2 \setminus (0,0)} \frac{1}{(m_1\boldsymbol{\omega}_1 + m_2\boldsymbol{\omega}_2)^{2p}}, \quad p = 1,2,\ldots. \tag{4.59}$$

Here, the Eisenstein summation described in Appendix A is applied to the conditionally convergent series S_2. The lattice sums of odd order S_{2p-1} vanish. The computationally effective formulas for (4.59) are (A.6)–(A.8). The following exact formula gives the effective permittivity tensor

$$\boldsymbol{\varepsilon}_\perp = (1+2\varrho f)I + \frac{2\varrho^2 f^2}{\pi} \begin{pmatrix} \operatorname{Re} S_2 & -\operatorname{Im} S_2 \\ -\operatorname{Im} S_2 & 2\pi - \operatorname{Re} S_2 \end{pmatrix}$$
$$+\frac{2\varrho^2 f^2}{\pi} \sum_{k=1}^{\infty} \varrho^k \sum_{l_1,l_2,\ldots,l_k=1}^{\infty} \left(\frac{f}{\pi}\right)^{2(l_1+\ldots+l_k)-k} \begin{pmatrix} \operatorname{Re} s & -\operatorname{Im} s \\ -\operatorname{Im} s & \operatorname{Re} s^* \end{pmatrix}, \tag{4.60}$$

where s^* is given by the same formula (4.58) except at the term $s_1^{(1)} = S_2$ which has to be replaced by $2\pi - S_2$. Though the formula (4.60) expresses $\boldsymbol{\varepsilon}_\perp$ in terms of the power series in ϱ and f, all its coefficients are exactly written.

Remark 6 *The formula* (4.60) *for complex* ϱ *can be derived from the corresponding exact formula for real* ϱ *established in [165], [93, Chapter 4] by analytic continuation of the absolutely convergent power series in* $-1 < \varrho < 1$ *onto the disk* $|\varrho| < 1$ *on the complex plane. This corresponds to* **Rule** $\mathbb{R} \to \mathbb{C}$ *on page 58. It is worth noting that the exact formula was first derived in [160] for a regular array of circular holes.*

Remark 7 *The formula* (4.60) *is a particular case of the Decomposition theorem 1 on page 25.*

The local real flux in the domain G for the real contrast parameter ϱ has the form [93, the exact formula (4.2.25)]

$$\begin{pmatrix} \frac{\partial u'}{\partial x_1}(z) \\ \\ \frac{\partial u'}{\partial x_2}(z) \end{pmatrix} = \begin{pmatrix} 1 \\ 0 \end{pmatrix} + \varrho r^2 \begin{pmatrix} \mathrm{Re}\, E_2(z) \\ -\mathrm{Im}\, E_2(z) \end{pmatrix}$$
$$+ \sum_{k=2}^{\infty} \varrho^k \sum_{l_1,l_2,\ldots,l_k=1}^{\infty} r^{4(l_1+\ldots+l_k)-2k} \begin{pmatrix} \mathrm{Re}\, sE_{l_k}(z) \\ -\mathrm{Im}\, sE_{l_k}(z) \end{pmatrix}, \quad z \in G, \tag{4.61}$$

where s is given by (4.58) and (4.57).

Example 4 *Consider a rectangular array of disks when the fundamental cell is defined by $\boldsymbol{\omega}_1 > 0$ and $\boldsymbol{\omega}_2 = \mathrm{i}\boldsymbol{\omega}_1^{-1}$. Then, the lattice sums S_{2p} are real and* (4.60) *up to $O(f^{11}|\varrho|^{11})$ gives*

$$\varepsilon_{11} = 1 + 2f\varrho + \frac{2f^2\varrho^2 S_2}{\pi} + \frac{2f^3\varrho^3 S_2^2}{\pi^2} + \frac{2f^4\varrho^4 S_2^3}{\pi^3} + \frac{2f^5\varrho^3}{\pi^4}\left(\varrho^2 S_2^4 + 3S_4^2\right)$$
$$+\frac{2f^6\varrho^4}{\pi^5}\left(\varrho^2 S_2^5 + 6S_2S_4^2\right) + \frac{2f^7\varrho^3}{\pi^6}\left(\varrho^4 S_2^6 + 9\varrho^2 S_2^2 S_4^2 + 5S_6^2\right)$$
$$+\frac{2f^8\varrho^4}{\pi^7}\left(\varrho^4 S_2^7 + 12\varrho^2 S_2^3 S_4^2 + 10S_2S_6^2 + 30S_4^2S_6\right)$$
$$+\frac{2f^9\varrho^3}{\pi^8}\left(\varrho^6 S_2^8 + 15\varrho^4 S_4^2 S_2^4 + 15\varrho^2 S_6^2 S_2^2 + 60\varrho^2 S_4^2 S_6 S_2 + 9\varrho^2 S_4^4 + 7S_8^2\right)$$
$$+\frac{2f^{10}\varrho^4}{\pi^9}(\varrho^6 S_2^9 + 18\varrho^4 S_4^2 S_2^5 + 20\varrho^2 S_6^2 S_2^3 + 90\varrho^2 S_4^2 S_6 S_2^2 + 27\varrho^2 S_2 S_4^4$$
$$+14S_2S_8^2 + 210S_4S_6S_8) + O(f^{11}|\varrho|^{11}). \tag{4.62}$$

The components $\varepsilon_{12} = \varepsilon_{21} = 0$ and ε_{22} coincides with (4.62) *where S_2 is replaced by $2\pi - S_2$.*

Example 5 *Consider a square array of disks when $\boldsymbol{\omega}_1 = 1$ and $\boldsymbol{\omega}_2 = \mathrm{i}$. The exact and numerical values for the lattice sums are given in Appendix A.*

First, we analyze the asymptotic formula (4.62) *for the square array when $S_2 = \pi$, $S_4 \approx 3.151211$, $S_6 = 0$, $S_8 = \frac{5}{7}S_4^2 \approx 4.2557732$*

$$\varepsilon_e = 1 + 2f\varrho + 2f^2\varrho^2 + 2f^3\varrho^3 + 2f^4\varrho^4 + 2f^5\varrho^5 + 2f^6\varrho^6 + 2f^7\varrho^7$$
$$+2f^8\varrho^8 + 2f^9\varrho^9 + 2f^{10}\varrho^{10} + \frac{6S_4^2 f^5\varrho^3}{\pi^4} + \frac{12S_4^2 f^6\varrho^4}{\pi^4} + \frac{18S_4^2 f^7\varrho^5}{\pi^4} + \frac{24S_4^2 f^8\varrho^6}{\pi^4}$$
$$+\frac{30S_4^2 f^9\varrho^7}{\pi^4} + \frac{36S_4^2 f^{10}\varrho^8}{\pi^4} + \frac{18S_4^4 f^9\varrho^5}{\pi^8} + \frac{54S_4^4 f^{10}\varrho^8}{\pi^8} + \frac{14S_8^2 f^9\varrho^3}{\pi^8}$$
$$+\frac{28S_8^2 f^{10}\varrho^4}{\pi^8} + O(f^{11}|\varrho|^{11}) = \frac{1+f\varrho}{1-f\varrho} + \frac{6S_4^2 f^5\varrho^3}{\pi^4(1-f\varrho)^2} + \frac{18S_4^4 f^9\varrho^5}{\pi^8}(1+3f\varrho)$$
$$+\frac{14S_8^2 f^9\varrho^3}{\pi^8}(1+2f\varrho) + O(f^{11}|\varrho|^{11}). \tag{4.63}$$

This formula explicitly demonstrates the precision of the Maxwell-Garnett formula up to $O(f^5|\varrho|^3)$ for the square array. Moreover, one can see that the Maxwell-Garnett rational approximation $\varepsilon_e \approx \frac{1+f\varrho}{1-f\varrho}$ is better than the polynomial approximation $\varepsilon_e \approx 1+2f\varrho+2f^2\varrho^2+2f^3\varrho^3+2f^4\varrho^4$ within the same precision since the rational approximation contains all the term $2f^p\varrho^p$ $(p = 1,2,\ldots)$ which are in the exact expansion (4.60).

The series (4.60) *converges in the disks $|\varrho| < 1$. The normalization method was developed in [93] for $\varrho = \pm 1$ to determine the critical behavior of $\varepsilon_e(f)$, as $f \to f_c = \frac{\pi}{4}$, when the disk become touching. The case $|\varrho| > 1$ has not yet been studied.*

The formula (4.63) *can be transformed to another asymptotically equivalent approximation for which the phase-interchange Keller relation [116] $\varepsilon_e(\varrho, f)\varepsilon_e(-\varrho, f) = 1$ is fulfilled. It can be done by the transformation of approximation $\varepsilon'_e(\varrho, f)$ to the expression [93]*

$$\varepsilon_e(\varrho, f) \approx \sqrt{\frac{\varepsilon'_e(\varrho, f)}{\varepsilon'_e(-\varrho, f)}}. \tag{4.64}$$

Note on a regular array of circular inclusions

The study of a regular array of circular inclusions displayed in Figure 4.9 can be summarized as follows. The method of functional equations applied in Section 4.1.3 yields the exact formula (4.60) for the effective permittivity tensor $\boldsymbol{\varepsilon}_\perp$. The term *exact formula* means that $\boldsymbol{\varepsilon}_\perp$ is given by the expression in the right part of (4.60) explicitly including the lattice sums S_{2p}, the eigenvalue ϱ of the contrast matrix and the concentration f. The lattice sums are explicitly given using the periods $\boldsymbol{\omega}_1$ and $\boldsymbol{\omega}_2$ in Appendix A. The formula (4.60) does not conceal any hidden parameter calculated by a numerical method like an infinite system of equations or an integral equation.

Equation (4.60) can be viewed as a double power series in both ϱ and f convergent within the Cartesian product of the unit disk $|\varrho| < 1$, and the interval $0 \le f \le f_c$, where f_c is the critical concentration of touching disks. For a fixed $f < f_c$, it is possible to extend the series (4.60) into the disk to $|\varrho| < r_0$, where $r_0 > 1$. When considering any desired precision $O(|\varrho|^{\ell_1} f^{\ell_2})$ in (4.60), the result becomes a polynomial $P_{\ell_1,\ell_2}(\varrho, f)$, as exemplified in (4.62).

We may stop here and say that the problem for a regular periodic array of disks has been solved. However, comparing (4.60) with the previous approximate formulas and numerical results deserves attention. We come back to this in Sections 4.1.5 and 4.1.6 devoted to the method of Rayleigh (1892) and Filshtinsky (1964). Further insights into their relevance to Natanzon's method (1935) for addressing plane elastic problems and historical context can be found in [172].

4.1.4 ARBITRARY LOCATION OF INCLUSIONS IN THE PERIODIC SQUARE UNIT CELL

Consider arbitrarily located equal non-overlapping disks of the same permittivity ε_1 in the periodic square unit cell as displayed in Figure 4.10. Let the radius of disks be r, and the centers of disks be $a_1, a_2, \ldots, a_N$. The considered composite is often called

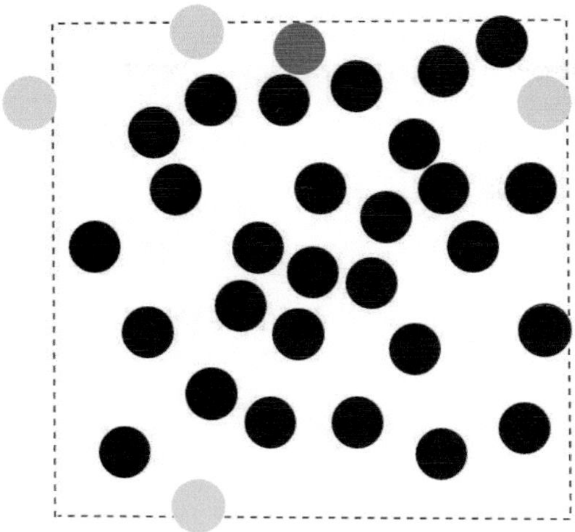

Figure 4.10 The periodic square unit cell with equal circular inclusions. The gray disks are identified by pairs in the double periodic structure.

a random composite in engineering. It is correct to say that this is a deterministic composite because a probabilistic distribution of points $a_1, a_2, \ldots, a_N$ has not been mentioned yet. The proper random approach will be discussed in the next section.

One can use the formula (4.32) to calculate the complex effective permittivity tensor. In order to derive higher-order constructive formulas convenient in computation, we have to transform the composition of operators $\mathcal{A}_k$ to a constructive form. This work has been done in [93] and publications cited therein. Therefore, we can rewrite approximate analytical formulas for the effective permittivity using the established correspondence between the formulas for the real-valued and complex permittivity. The high-order terms can be written by the algorithm described in [67,93]. The main object of the theory of such a composite is the set of structural sums, sometimes called basic sums and e-sums in previous publications.

Following [164], [93, Chapter 4] introduce the *structural sums*

$$e_{m_1,\ldots,m_q} = \frac{1}{N^{1+\frac{1}{2}(m_1+\cdots+m_q)}} \times$$
$$\sum_{k_0,k_1,\ldots,k_n} E_{m_1}(a_{k_0} - a_{k_1}) \mathbf{C} E_{m_2}(a_{k_1} - a_{k_2}) \ldots \mathbf{C}^{q+1} E_{m_q}(a_{k_{q-1}} - a_{k_q}), \tag{4.65}$$

where $E_p(z)$ are the Eisenstein functions. It is assumed for shortness that $E_p(a_k - a_m) =: S_p$, if $a_k = a_m$. For instance, e_2 and e_{22} take the following form

$$\begin{aligned} e_2 &= \frac{1}{N^2} \sum_{k_0=1}^{N} \sum_{k_1=1}^{N} E_2(a_{k_0} - a_{k_1}), \\ e_{22} &= \frac{1}{N^3} \sum_{k_0=1}^{N} \sum_{k_1=1}^{N} \sum_{k_2=1}^{N} E_2(a_{k_0} - a_{k_1}) \overline{E_2(a_{k_1} - a_{k_2})}. \end{aligned} \tag{4.66}$$

The structural sum (4.65) is a discrete multiple convolution of the Eisenstein functions. A fast, almost linear algorithm in N for its computation was developed in [67, Chapter 2]. This algorithm also includes the extended structural sum (4.65) to polydispersed 2D, 3D and elastic 2D composites.

Using the complex structural sums, we form the matrix structural sums

$$\mathbf{e}_{m_1,\ldots,m_q} = \begin{pmatrix} \mathrm{Re}\, e_{m_1,\ldots,m_q} & -\mathrm{Im}\, e_{m_1,\ldots,m_q} \\ -\mathrm{Im}\, e_{m_1,\ldots,m_q} & \mathrm{Re}\, e^*_{m_1,\ldots,m_q} \end{pmatrix}, \tag{4.67}$$

where $e^*_{m_1,\ldots,m_q}$ has the same form (4.65) except at the terms $E_2(a_k - a_m)$ which have to be replaced by the expressions $2\pi - E_2(a_k - a_m)$.

The effective permittivity tensor has the form

$$\boldsymbol{\varepsilon}_\perp = I + 2\varrho f(I + A_1 f + A_2 f^2 + \cdots), \tag{4.68}$$

where a few first matrices are written below

$$\begin{aligned}
A_1 &= \frac{\varrho}{\pi}\mathbf{e}_2, \\
A_2 &= \frac{\varrho^2}{\pi^2}\mathbf{e}_{22}, \\
A_3 &= \frac{\varrho^2}{\pi^3}\left[-2\mathbf{e}_{33} + \varrho\mathbf{e}_{222}\right], \\
A_4 &= \frac{\varrho^2}{\pi^4}\left[3\mathbf{e}_{44} - 4\varrho\mathbf{e}_{332} + \varrho^2\mathbf{e}_{2222}\right], \\
A_5 &= \frac{\varrho^2}{\pi^5}\left[-4\mathbf{e}_{55} + 6\varrho(\mathbf{e}_{442} + 2\mathbf{e}_{343}) - 2\varrho^2(\mathbf{e}_{2332} + 2\mathrm{Re}\,\mathbf{e}_{2233})\right. \\
&\quad \left. + \varrho^3\mathbf{e}_{22222}\right], \\
A_6 &= \frac{\varrho^2}{\pi^6}\left[5\mathbf{e}_{66} - 8\varrho(\mathbf{e}_{255} + 3\mathbf{e}_{354}) + \right. \\
&\quad + \varrho^2(6\mathrm{Re}\,\mathbf{e}_{2244} + 12\mathrm{Re}\,\mathbf{e}_{2343} + 4\mathbf{e}_{3333} + 3\mathbf{e}_{2442}) - \\
&\quad \left. -4\varrho^3(\mathbf{e}_{22233} + \mathbf{e}_{22332}) + \varrho^4\mathbf{e}_{222222}\right],
\end{aligned} \tag{4.69}$$

$$\begin{aligned}
A_7 &= \frac{\varrho^2}{\pi^7}\left[-6\mathbf{e}_{77} + 10\varrho\,(\mathbf{e}_{266} + 4\mathbf{e}_{365} + 3\mathbf{e}_{464})\right. \\
&\quad -2\varrho^2\,(2\mathbf{e}_{2255} + 6\mathbf{e}_{2354} + 6\mathbf{e}_{2453} + 2\mathbf{e}_{2552} + 3\mathbf{e}_{3344} + 9\mathbf{e}_{3443} \\
&\quad +6\mathbf{e}_{3542} + 3\mathbf{e}_{4433} + 6\mathbf{e}_{4532} + 2\mathbf{e}_{5522}) \\
&\quad +2\varrho^3\,(3\mathrm{Re}\,\mathbf{e}_{22244} + 6\mathrm{Re}\,\mathbf{e}_{22343} + 3\mathrm{Re}\,\mathbf{e}_{22442} + 4\mathrm{Re}\,\mathbf{e}_{23333} \\
&\quad +3\mathbf{e}_{23432} + 2\mathbf{e}_{33233}) - 2\varrho^4\,(2\mathbf{e}_{222233} + 2\mathbf{e}_{222332} + \mathbf{e}_{223322}) \\
&\quad \left. +\varrho^5\mathbf{e}_{2222222}\right].
\end{aligned}$$

$$
\begin{aligned}
A_8 \;=\; & \frac{\varrho^2}{\pi^8}[7\mathbf{e}_{88}+\varrho^4\mathrm{Re}\,(6\mathbf{e}_{244222}+8\mathbf{e}_{333322}+6\mathbf{e}_{442222})+16\varrho^2\mathrm{Re}\,\mathbf{e}_{5533}\\
& -2\varrho^5(\mathbf{e}_{2223322}+2\mathbf{e}_{2233222}+4\mathbf{e}_{2332222}+4\mathbf{e}_{3322222})+\varrho^4(6\mathbf{e}_{222343}\\
& +6\mathbf{e}_{223432}+3\mathbf{e}_{224422}+4\mathbf{e}_{233233}+4\mathbf{e}_{233332}+6\mathbf{e}_{234322}+4\mathbf{e}_{332233}\\
& +4\mathbf{e}_{332332}+6\mathbf{e}_{343222})-\varrho^3(12\mathbf{e}_{22354}+12\mathbf{e}_{22453}+6\varrho^3\mathbf{e}_{23344}+18\mathbf{e}_{23443}\\
& +12\mathbf{e}_{23542}+6\mathbf{e}_{24433}+12\mathbf{e}_{24532}+8\mathbf{e}_{25522}+12\mathbf{e}_{33343}+6\mathbf{e}_{33442}+12\mathbf{e}_{34333}\\
& +18\mathbf{e}_{34432}+12\mathbf{e}_{35422}+12\mathbf{e}_{44233}+6\mathbf{e}_{44332}+12\mathbf{e}_{45322}+8\mathbf{e}_{55222})\\
& +\varrho^2(20\mathbf{e}_{2365}+30\mathbf{e}_{2464}+20\mathbf{e}_{2563}+5\mathbf{e}_{2662}+36\mathbf{e}_{3454}+48\mathbf{e}_{3553}+20\mathbf{e}_{3652}\\
& +9\mathbf{e}_{4444}+36\mathbf{e}_{4543}+30\mathbf{e}_{4642}+20\mathbf{e}_{5632}+10\mathbf{e}_{6622})\\
& -\varrho(30\mathbf{e}_{376}+60\varrho\mathbf{e}_{475}+60\varrho\mathbf{e}_{574}+30\varrho\mathbf{e}_{673}+12\varrho\mathbf{e}_{772})+\varrho^6\mathbf{e}_{22222222}].
\end{aligned}
$$

The coefficients (4.69) are obtained from the corresponding coefficients written in the previous works for real-valued permittivity by application of **Rule** $\mathbb{R} \to \mathbb{C}$ on page 58 in the following way. Consider the matrix coefficient A_1 having the complex component for the complex contrast parameter ϱ

$$A_1 = \frac{\varrho}{\pi}\begin{pmatrix} \mathrm{Re}\, e_2 & -\mathrm{Im}\, e_2 \\ -\mathrm{Im}\, e_2 & 2\pi - \mathrm{Re}\, e_2 \end{pmatrix}. \tag{4.70}$$

In the case of real ϱ, the matrix (4.70) was written in the previous works through two complex numbers

$$\hat{A}_1 = \frac{\varrho}{\pi}e_2, \quad \hat{A}_1^* = \frac{\varrho}{\pi}(2\pi - e_2), \quad A_1 = \begin{pmatrix} \mathrm{Re}\, \hat{A}_1 & -\mathrm{Im}\, \hat{A}_1 \\ -\mathrm{Im}\, \hat{A}_1 & \mathrm{Re}\, \hat{A}_1^* \end{pmatrix}. \tag{4.71}$$

The matrix form A_1 of (4.71) was not used. Its components were written in terms of the real and imaginary parts of the complex numbers $\hat{A}_1$ and $\hat{A}_1^*$. It is clear that the matrix A_1 from (4.70) and (4.71) coincide for real ϱ. The formula (4.71) for complex ϱ is no longer true and must be replaced by (4.70). The same rule concerns other coefficients A_n.

The next coefficients A_n can be exactly computed by the symbolic-numerical algorithm [93, Chapter 2]. The formulas (4.68)–(4.69) hold for a regular array considered in Section 4.1.3, if we replace $\mathbf{e}_{m_1,m_2,\ldots,m_s}$ by $S_{m_1}S_{m_2}\ldots S_{m_s}$.

It is worth noting that a numerical result has to be reasonable for $|\varrho| < 1$ when the convergence is justified[1]. The latter inequality may be fulfilled not for all frequency oscillations ω. More precisely, the inequality $|\varrho(\omega)| \geq 1$ holds for glioma cells at high frequencies $\omega > 3 \cdot 10^{10}$ as shown in Figure 4.4. Such a dependence requires separate numerical and theoretical investigations of (4.68)–(4.69). The case of touching disks for $\varrho = \pm 1$ was comprehensively studied by an asymptotic renormalization method [91, 92] and by structural approximation [182].

As noted above, the case of real-valued permittivity was systematically discussed and summarized in [93]. The obtained results are extended by **Rule** $\mathbb{R} \to \mathbb{C}$ on page

[1]It is suggested on page 87 that the convergence holds for $\frac{f}{f_c}|\varrho| < 1$.

58 when the complex contrast parameter lies in the unit disk $|\varrho| \leq 1$ on the complex plane. The effective permittivity tensor (4.68) is presented as an analytic function in the concentration f. It can also be considered as an analytic function in $\boldsymbol{\varepsilon}_{\perp}(\varrho)$ for a fixed $0 \leq f \leq f_c$ in a disk represented by its Taylor series. Here, f_c denotes the critical concentration when the disks with the fixed center extend up to touching. The convergence in ϱ takes place at least in the unit disk $|\varrho| < 1$. The radius of convergence R of the disk $|\varrho| < R$ can exceed 1 for a fixed $f < f_c$ and even for $f = f_c$ if the touching disks do not form a percolation chain.

The main question can be stated as follows. Whether the function $\boldsymbol{\varepsilon}_{\perp}(\varrho)$ is analytically continued out of the disk $|\varrho| < R$ and how to construct its continuation. This is a classical problem of the theory of analytic functions. Padé approximants can be applied to constructively solve this problem [220].

Analytical properties of the complex effective dielectric constant were systematically studied by David Bergman [31, 34]. Besides contribution to the theory of bounds, we want to pay attention to the rational approximations discussed therein.

Keller's relation $\varepsilon_e(\varrho, f)\varepsilon_e(-\varrho, f) = 1$ [116] can be applied following Example 5 to transform an approximation $\varepsilon_e(\varrho, f)$ to the expression (4.64).

4.1.5 INFINITE SYSTEM OF LINEAR ALGEBRAIC EQUATIONS

The compactness of the operator in the right part of equation (4.13) in the space $\mathcal{H}(G^+)$ can be useful in the considered question. Introduce the matrices $\boldsymbol{\beta}'_k = \varrho_0^{-1}\boldsymbol{\beta}_k$ where ϱ_0 is given by (2.191) and $\varrho_0 > 1$. One can see that the spectral norm of the matrix $\boldsymbol{\beta}'_k$ does not exceed 1.

Write equation (4.13) in the form

$$\boldsymbol{\psi}_l(z) = \varrho_0 {\sum}'^{N}_{k=1}\boldsymbol{\beta}'_k \sum_{\mathbf{m}\in\mathbb{Z}^2_\omega} \left(\frac{r_k}{z-a_k-\mathbf{m}}\right)^2 \overline{\boldsymbol{\psi}_k\left(a_k + \frac{r_k^2}{\overline{z-a_k-\mathbf{m}}}\right)} + \begin{pmatrix} 1 \\ 0 \end{pmatrix}, \quad (4.72)$$
$$|z-a_l| \leq r_l \quad (l = 1,2,\ldots,N).$$

Let the vector-function $\boldsymbol{\psi}_l(z)$ be expanded into its Taylor series

$$\boldsymbol{\psi}_l(z) = \sum_{p=0}^{\infty} \boldsymbol{\psi}_{lp}(z-a_l)^p, \quad |z-a_l| < r_l \quad (l = 1,2,\ldots,N). \quad (4.73)$$

Let the disk $|z-a_l| \leq r_l$ be fixed and $k \neq l$. Transform the expression

$$\frac{1}{z-a_k-\mathbf{m}} = \sum_{s=0}^{\infty}(-1)^s \frac{(z-a_l)^s}{(a_l-a_k-\mathbf{m})^{s+1}}, \quad (4.74)$$

where $\frac{|z-a_l|}{|a_l-a_k-\mathbf{m}|} \leq \frac{r_l}{r_l+r_k} < 1$. We have

$$\overline{\boldsymbol{\psi}_k\left(a_k + \frac{r_k^2}{\overline{z-a_k-\mathbf{m}}}\right)} = \sum_{p=0}^{\infty} \frac{r_k^{2p}\overline{\boldsymbol{\psi}_{kp}}}{(z-a_k-\mathbf{m})^p}. \quad (4.75)$$

Substitute (4.74) into (4.75) and find

$$
\begin{aligned}
y_{kl\mathbf{m}} &= \left(\frac{r_k}{z-a_k-\mathbf{m}}\right)^2 \overline{\boldsymbol{\psi}_k\left(a_k+\frac{r_k^2}{\overline{z-a_k-\mathbf{m}}}\right)} = \sum_{p=0}^{\infty} \frac{r_k^{2(p+1)}\overline{\boldsymbol{\psi}_{kp}}}{(z-a_k-\mathbf{m})^{p+2}} = \\
&\sum_{p=0}^{\infty} r_k^{2(p+1)}\overline{\boldsymbol{\psi}_{kp}} \left[\sum_{s=0}^{\infty} \frac{(-1)^s (z-a_l)^s}{(a_l-a_k-\mathbf{m})^{s+1}}\right]^{p+2} = \\
&\left\{ r_k^2\overline{\boldsymbol{\psi}_{k0}} \left[\sum_{s=0}^{\infty} \frac{(-1)^s (z-a_l)^s}{(a_l-a_k-\mathbf{m})^{s+1}}\right]^2 + r_k^4\overline{\boldsymbol{\psi}_{k1}} \left[\sum_{s=0}^{\infty} \frac{(-1)^s (z-a_l)^s}{(a_l-a_k-\mathbf{m})^{s+1}}\right]^3 \right. \\
&\left. + r_k^6\overline{\boldsymbol{\psi}_{k2}} \left[\sum_{s=0}^{\infty} \frac{(-1)^s (z-a_l)^s}{(a_l-a_k-\mathbf{m})^{s+1}}\right]^4 + \ldots \right\}.
\end{aligned} \tag{4.76}
$$

Expand the square brackets and find the sum over **m** from (4.73)

$$
\begin{aligned}
Y_{kl} &= \sum_{\mathbf{m}\in\mathbb{Z}_\omega^2} y_{kl\mathbf{m}} = \sum_{\mathbf{m}\in\mathbb{Z}_\omega^2} \left\{ r_k^2\overline{\boldsymbol{\psi}_{k0}} \sum_{s_1,s_2=0}^{\infty} \frac{(-1)^{s_1+s_2}(z-a_l)^{s_1+s_2}}{(a_l-a_k-\mathbf{m})^{s_1+s_2+2}} \right. \\
&+ r_k^4\overline{\boldsymbol{\psi}_{k1}} \sum_{s_1,s_2,s_3=0}^{\infty} \frac{(-1)^{s_1+s_2+s_3}(z-a_l)^{s_1+s_2+s_3}}{(a_l-a_k-\mathbf{m})^{s_1+s_2+s_3+3}} \\
&\left. + r_k^6\overline{\boldsymbol{\psi}_{k2}} \sum_{s_1,s_2,s_3,s_4=0}^{\infty} \frac{(-1)^{s_1+s_2+s_3+s_4}(z-a_l)^{s_1+s_2+s_3+s_4}}{(a_l-a_k-\mathbf{m})^{s_1+s_2+s_3+s_4+4}} + \ldots \right\}
\end{aligned} \tag{4.77}
$$

Using the Eisenstein functions, transform the above equations

$$
\begin{aligned}
Y_{kl} &\equiv Y_{kl}(z) = r_k^2\overline{\boldsymbol{\psi}_{k0}} \sum_{s_1,s_2=0}^{\infty} (-1)^{s_1+s_2}(z-a_l)^{s_1+s_2} E_{s_1+s_2+2}(a_l-a_k) \\
&+ r_k^4\overline{\boldsymbol{\psi}_{k1}} \sum_{s_1,s_2,s_3=0}^{\infty} (-1)^{s_1+s_2+s_3}(z-a_l)^{s_1+s_2+s_3} E_{s_1+s_2+s_3+3}(a_l-a_k) \\
&+ r_k^6\overline{\boldsymbol{\psi}_{k2}} \sum_{s_1,\ldots,s_4=0}^{\infty} (-1)^{s_1+\ldots+s_4}(z-a_l)^{s_1+\ldots+s_4} E_{s_1+\ldots+s_4+4}(a_l-a_k) + \ldots
\end{aligned} \tag{4.78}
$$

The expressions $Y_{kl}(z)$ are represented as power series in $(z-a_l)$. Analogous formulas can be obtained for $k=l$. Let Y_{klp} denote the Taylor pth coefficient of $Y_{kl}(z)$ in the disk $|z-a_l|<r_l$. Substitute (4.73) and (4.78) into (4.72) and select the coefficients on $(z-a_l)^p$

$$
\boldsymbol{\psi}_{lp} = \varrho_0 \sum_{k=1}^{N} \boldsymbol{\beta}'_k Y_{klp} + \delta_{k0}\begin{pmatrix} 1 \\ 0 \end{pmatrix}, \quad l=1,2,\ldots,N; \quad p=0,1,\ldots, \tag{4.79}
$$

where δ_{k0} denotes the Kronecker symbol. One can see that Y_{klp} is an infinite linear combination of $\overline{\boldsymbol{\psi}_{kj}}$. Therefore, equations (4.79) can be considered as an infinite $\mathbb{R}$-linear algebraic system on the coefficients $\boldsymbol{\psi}_{lp}$.

Equation (4.79) is considered in a discrete space $h^+(G^+)$ isomorphic to the continuous space $\mathcal{H}^+(G^+)$. The compactness of the discrete operator defined by the right part of (4.79) was considered in the space of sequences $h^2(G^+)$ (Taylor coefficients of the vector-functions $\boldsymbol{\psi}_l(z)$) isomorphic to the classic Hardy type space $H^2(G^+)$).

It is worth noting that $H^2(G^+)$ is a Hilbert space. Similar to Theorem 5, the compactness of the continuous operator in $H^2(G^+)$ was proved in [185]. This yields the compactness of the discrete operator in $h^2(G^+)$. The Pumping Principle [185] implies that every solution from $H_2(G^+)$ belongs to $\mathcal{H}^+(G^+)$. The same holds for the space $h^2(G^+)$ and $h^+(G^+)$.

It is noted above that the formulas for $k = l$ can be obtained by analogous calculations. We now proceed to study a regular double periodic array of disks, i.e., the case $N = 1$, which contains the necessary formulas for $k = l$.

Consider the unit periodicity cell displayed in Figure 4.9. Equations (4.72) for $N = 1$ become

$$\boldsymbol{\psi}_1(z) = \boldsymbol{\beta}_1 \sum_{\mathbf{m}\in\mathbb{Z}^2_{\omega}\setminus(0,0)} \overline{\left(\frac{r}{z-\mathbf{m}}\right)^2 \boldsymbol{\psi}_1\left(\frac{r^2}{z-\mathbf{m}}\right)} + \begin{pmatrix} 1 \\ 0 \end{pmatrix}, \quad |z| \le r, \tag{4.80}$$

where the set $\mathbb{Z}^2_{\omega}\setminus(0,0)$ consists of the points $\boldsymbol{\omega}_1 m_1 + \boldsymbol{\omega}_2 m_2$ for $m_1, m_2 \in \mathbb{Z}$ except $m_1 = m_2 = 0$. The vector-function $\boldsymbol{\psi}_1(z)$ can be expanded into its Taylor series

$$\boldsymbol{\psi}_1(z) = \sum_{p=0}^{\infty} \boldsymbol{\alpha}_p z^p, \quad |z| < r, \tag{4.81}$$

where $\boldsymbol{\alpha}_p = \boldsymbol{\psi}_{1p}$ is introduced for shortness. Then,

$$Y(z) := \sum_{\mathbf{m}\in\mathbb{Z}^2_{\omega}\setminus(0,0)} \overline{\left(\frac{r}{z-\mathbf{m}}\right)^2 \boldsymbol{\psi}\left(\frac{r^2}{z-\mathbf{m}}\right)} = \sum_{\mathbf{m}\in\mathbb{Z}^2_{\omega}\setminus(0,0)} \sum_{p=0}^{\infty} \frac{r^{2(p+1)}\overline{\boldsymbol{\alpha}_p}}{(z-\mathbf{m})^{p+2}}. \tag{4.82}$$

Using the Eisenstein summation, we obtain

$$Y(z) = \sum_{n=0}^{\infty} r^{2(p+1)} \overline{\boldsymbol{\alpha}_p} \hat{E}_{p+2}(z). \tag{4.83}$$

It should be noted that in the considered case, $k = l = 1$, hence, the function $\hat{E}_{p+2}(z)$ defined by (4.15) is used.

Substitute (4.83) into (4.80)

$$\boldsymbol{\psi}_1(z) = \boldsymbol{\beta}_1 \sum_{p=0}^{\infty} r^{2(p+1)} \overline{\boldsymbol{\alpha}_p} \hat{E}_{p+2}(z) + \begin{pmatrix} 1 \\ 0 \end{pmatrix}, \quad |z| \le r. \tag{4.84}$$

Application of the method of successive approximations yields the exact and approximate formula from Section 4.1.3. The method absolutely converges if the eigenvalue of the matrix $\boldsymbol{\beta}_1$ satisfies the inequality $|\varrho_1| < 1$ for any r including the touching disks when $r = r_c = \frac{\pi}{4}$. In the case $|\varrho_1| = 1$, the method of successive approximations converges uniformly, at least for all $r < r_0$ where r_0 is a fixed radius less than r_c.

In order to compare the results with other methods, consider a discrete analog of the continuous equation (4.84). Select the coefficients on the powers of z in (4.84) using equations (A.29) and (A.24) we arrive at the infinite system of $\mathbb{R}$-linear equations on the Taylor vector coefficients $\boldsymbol{\alpha}_l$

$$\boldsymbol{\alpha}_l = \boldsymbol{\beta}_1 \sum_{n=0}^{\infty} (-1)^n r^{2(n+1)} C_{l+n+1}^{n+1} S_{l+n+2} \overline{\boldsymbol{\alpha}_n} + \delta_{l0} \begin{pmatrix} 1 \\ 0 \end{pmatrix}, \; l = 0, 1, \dots \tag{4.85}$$

Due to the compactness of the operator from the right part of (4.80) in the space $\mathcal{H}^+(|z| \le r)$ the operator from the right part of (4.85) is compact in the discrete space $h^+(|z| \le r)$. Therefore, the truncation method can be applied to the equation (4.85).

4.1.6 METHODS OF RAYLEIGH AND OF NATANZON-FILSHTINSKY

In 1892, Rayleigh [206] proposed a method based on the representations of the unknown scalar real potentials $u_1(z)$ in the disk $|z| < r$ and $u(z)$ in G by the Laurent and Taylor series

$$u(z) = A_0 + \mathrm{Re} \sum_{l=1}^{\infty} \left(A_l z^l + B_l z^{-l} \right), \; z \in G, \quad u_1(z) = \mathrm{Re} \sum_{l=0}^{\infty} C_l z^l, \; |z| < r. \tag{4.86}$$

Rayleigh used the trigonometric form of the series (4.86) and the symmetry of rectangular lattice when the unknown coefficients are real. Using the scalar boundary conditions (2.20) for $N = 1$, Rayleigh established linear relations between A_l, B_l, C_l and reduced the problem to an infinite system of linear algebraic equations.

Rayleigh's system is usually written for the coefficients $B_l = \frac{1-\sigma_1}{2l} r^{2l} \overline{\alpha_l}$, sometimes separately for the real and imaginary parts of B_l. Rayleigh's system is presented here in the form (4.85). McPhedran, with coworkers, had developed the method of Rayleigh and extended it to some other fundamental physical problems [144–146, 148].

In 1935, Natanzon [194] proposed a method for the plain-strain problem with a doubly periodic array of circular holes. Filshtinsky [73] adapted Natanzon's method in 1964 to the conductivity problems. The method of Natanzon-Filshtinsky is based on the representation of complex potentials by series

$$\varphi(z) = Az - \sum_{k=0}^{\infty} A_k \frac{\zeta^{(k)}(z)}{k!}, \; z \in G, \quad \varphi_1(z) = \sum_{k=0}^{\infty} C_k z^k, \; |z| < r. \tag{4.87}$$

where A, A_k and C_k are undetermined constants, $\zeta(z)$ the Weierstrass elliptic function. Substitution of the series (4.87) to the scalar boundary condition (2.34) leads to the same infinite linear algebraic system (4.85). Analytical approximate lower-order formulas were derived up to $O(r^4)$. It corresponds to the terms $k = 0$ and neglects the terms $k > 0$ in (4.87). Rayleigh (1892) and Filshtinsky (1964) approaches lead to the same infinite system of linear algebraic equations (4.85).

A finite system can be obtained by truncating the infinite Rayleigh/Filshtinsky system, producing the polynomial $P_{\ell_1,\ell_2}(\varrho, f)$ discussed on page 92. Consequently, both Rayleigh and Filshtinsky's methods are reduced to numerical procedures for calculating the coefficients of the exact series (4.60), represented symbolically for any lattice configuration. This clarification is essential, as Rayleigh, Filshtinsky, McPhedran, and other researchers acknowledged that their solutions were numeric despite presenting lower-order approximate formulas. It is worth mentioning that numerous publications continue to refer to analogous numerical polynomial approximations obtained through infinite systems as "exact," even after the exact formulas [17, 160, 165] derived in 1998 and 2007.

The functional equation (4.80) explicitly written for the vector field $\boldsymbol{\psi}_1(z)$ in the disk can be solved by the method of successive approximations. A few first approximations are given by (4.51) and (4.54). The exact formula (4.61) for the complex potentials $\psi_1(z)$ and $\psi(z)$ in the case of real permittivity was derived in [160, 165]. **Rule** $\mathbb{R} \to \mathbb{C}$ concerns the effective tensor, not the local fields. Hence, a separate investigation should be done for an extension of the local field (4.61) to the complex permittivity. The discrete form of functional equation (4.80) shows its equivalence to the infinite systems that Rayleigh and Filshtinsky derive. One can see that the analytical results presented in Section 4.1.4 in symbolic form are preferable to the numerical results that might be obtained from the infinite system (4.85).

4.1.7 RANDOM COMPOSITES AND RVE

As it was noted in Section 2.1, randomness in composites is strictly related to the measure theory [94, 113, 223]. In the considered case of non-overlapping disks, the measure theory is applied to the characteristic set equal to ε_1 in inclusions and to 1 in the matrix. Theoretically, any such a measurable set is determined by the infinite set of n-point correlation functions ($n = 1, 2, \ldots$) [30, 123, 226].

We highlight the constructive part of the general homogenization theory of random composites in order to discuss a classification of various dispersed composites. Modern engineering implementations are often reduced to a picture taken from somewhere like the illustrating Figure 4.10, next, computation by a numerical method (FEM) of the effective properties, and the declaration that the deal is done. In the best case, it can be explained that the considered dispersed random composite is a composite with uniform independent identically distributed (i.i.d.) inclusions. Occasionally, a few numerical examples are presented with a small number of inclusions per RVE. However, this somewhat simplistic approach overlooks the rich diversity of composites and their intricate classification [122, 195], including clustering and other geometrical factors.

Implementing multiple correlation functions for random composites is usually reduced to the well-studied spatial two-point correlation functions. Three- and four-point correlation functions were described through the Torquato–Milton parameters [155, 226]. Two-point correlation functions yield a rough differentiation of random composites since it is restricted to second-order approximations of the effective properties; see [67, 93] and discussion below.

Theoretically, the infinite set of multiple correlation functions completely describes a random composite. However, a virtual impossibility of computing the correlation functions of higher-orders restricts their applications. A popular opinion among authors can be represented by the following phrase [71]: "Except for some special cases, the infinite set of correlation functions is not known, and hence an exact solution is not possible." We agree with this opinion up to the second comma and disagree with the subsequent statement since the method of multiple correlation functions is not a unique analytical method that can be applied to dispersed composites. A much more computationally effective method of structural sums was proposed in [38, 162] and developed in [62, 67, 93, 164, 174, 180]. This method is based on the generalized Schwarz alternating method discussed in Sections 2.1- 2.2 and the solution to the Riemann-Hilbert and $\mathbb{R}$-linear problems [185] for an arbitrary multiply connected domain.

We now outline the theoretical results concerning RVE, which can be extended to the complex case. Equations (4.68)–(4.69) can be considered as a decomposition of $\boldsymbol{\varepsilon}_\perp$ on physical and geometrical terms for circular inclusions. One can see that $\boldsymbol{\varepsilon}_\perp$ is a power series in f with the coefficients A_k written as linear combinations of the structural sums. The coefficients in structural sums contain ϱ^l $(l \leq k)$.

We now proceed to explain how to apply the deterministic equations (4.68)–(4.69) to random structures in the framework of stochastic homogenization. Until the probabilistic distributions of centers and radii are introduced, one cannot say about a random composite. If one shows a special picture that looks random, it is just a deterministic picture since a picture must be described in terms of probabilistic distribution. It can be a set of pictures representing a simple statistical sample, i.e., a subset of the general population of composites in which each composite has an equal probability of being chosen.

Let a doubly periodic cell Q represent a random statistically homogeneous 2D composite [93, 113, 223]. We restrict ourselves now by equal disks. The centers $\mathbf{a} = \{a_1, a_2, \ldots, a_N\}$ form a random pattern of points satisfying a joint probabilistic distribution $\mathcal{P}$. It follows from the decomposition of $\boldsymbol{\varepsilon}_\perp$ that all the information that we need about $\mathcal{P}$ in order to determine $\boldsymbol{\varepsilon}_\perp$ is contained in structural sums. The structural sums can be considered as an alternative computationally effective method to use spatial multi-point correlation functions [30, 226]. The formulas (4.68)–(4.69) are obtained as a result of 8 iterations applied to functional equations explicitly calculated up to $O(f^{10})$. The same result could be obtained after the application of l-order correlation functions $(l = 1, 2, \ldots, 9)$ up to $O(f^{10})$. One can find A_k for $k \leq 14$ in [170]. A general fast constructive algorithm to write any A_k is described in [67, Chapter 2].

Equations (4.68)–(4.69) implies that $\boldsymbol{\varepsilon}_\perp = \boldsymbol{\varepsilon}_\perp(\mathbf{a})$ can be considered as a random function. The mathematical expectation $\mathbb{E}[\boldsymbol{\varepsilon}_\perp]$ over the fixed distribution $\mathcal{P}$ yields the effective permittivity tensor of the considered class of random composites. Different results for $\mathbb{E}[\boldsymbol{\varepsilon}_\perp]$ naturally correspond to different spatial distributions $\mathcal{P}$ of inclusions. Of course, the standard statistical analysis includes the determination of variance $Var[\boldsymbol{\varepsilon}_\perp]$, investigations of estimators, statistical hypothesis, etc. It is worth noting that it is difficult to analytically describe a distribution $\mathcal{P}$, but it is unnecessary. Usually, information is given by a set of pictures or by one large picture. Such a picture can be used as a statistical sample.

The described above approach leads to the theory of *analytical* Representative Volume Element (*a*RVE) proposed in [164]. It can be considered as a constructive application of the Decomposition Theorem 1 on page 25. One of *a*RVE applications is outlined below. Let us have at our disposal two plane pictures of dispersed composites and their digital treatment (see Chapter 6 and, for instance, [214]) in the form of two sets of the centers of inclusions $\mathbf{a} = \{a_1, a_2, \ldots, a_N\}$ and $\mathbf{a}' = \{a'_1, a'_2, \ldots, a'_{N'}\}$ scaled to the periodic unit square cell. We want to know whether these two composites belong to the same class of random composites.

First, we must calculate all the structural sums $e_{m_1,\ldots,m_s}$ for two sets of points $\mathbf{a}$ and $\mathbf{a}'$. It was established in [93, Chapter 2] that the coefficient A_k on f^{k+1} from (4.68) is a linear combination of the structural sums $e_{m_1,\ldots,m_s}$ with $m_1 + \ldots + m_s = 2k$. Therefore, we have to compare the set of complex values $\mathcal{E} = \{\mathcal{E}_1, \mathcal{E}_2, \ldots\}$ where

$$\mathcal{E}_1 = \{e_2\},\ \mathcal{E}_2 = \{e_{22}\},\ \mathcal{E}_3 = \{e_{33}, e_{222}\},\ \mathcal{E}_4 = \{e_{44}, e_{332}, e_{233}, e_{2222}\}, \ldots. \tag{4.88}$$

constructed for $\mathbf{a}$ and the set $\mathcal{E}'$ for $\mathbf{a}'$. Such a comparison of $\mathcal{E}$ and $\mathcal{E}'$ may be performed by elements or by elements of $\mathcal{E}_k$ multiplied by the weight f^{k+1}.

The set $\mathcal{E}$ is infinite and may be truncated by the concentration tolerance. Let it be equal to $O(f^8)$ for definiteness. Then, we have to calculate the finite set of structural sums $e_{m_1,\ldots,m_s}$ from (4.69)

$$\mathcal{E}_1 = \{e_2\},\ \mathcal{E}_2 = \{e_{22}\}, \ldots, \mathcal{E}_7 = \{e_{77}, e_{266}, \ldots, e_{2222222}\}. \tag{4.89}$$

It is worth noting that the 2-point correlation functions yield the same result as only one constant e_2; the 3-point correlation functions yield the same result as the infinitely many elements e_{pp} $(p = 2, 3, \ldots)$; the 4-point correlation functions yield the same result as the infinitely many triple elements e_{m_1,m_2,m_3} from $\mathcal{E}$ and so forth.

Using *a*RVE is equivalent to the usage of the multi-point correlation functions. We leave aside other characteristics of dispersed random composites based on the heuristic observations since they are based on one or a few numbers. It is impossible to completely describe a dispersed random composite by a heuristic parameter.

The main advantage of *a*RVE compared to other methods can be summarized in the following points.

- A class of random composites can be directly determined by a set of structural sums (4.65) without computation of its effective properties.
- In the framework of *a*RVE, random clustering composites can be theoretically simulated [195] as well as an observed composite can be investigated [214].
- The macroscopic isotropy can be quickly verified by means of structural sums [62,174], more accurately than by two-point correlation functions. For details see [174, equations (3.2)] including the isotropy condition $e_2 = \pi$.
- The number of inclusions per periodicity cell is practically not restricted due to the algorithm [93, Chapter 2] of complexity not exceeding $O(sN^2)$ in the computation of $e_{m_1,\ldots,m_s}$.

- aRVE is extended to 2D elastic and viscous problems and to 3D conductivity problems [93, 212] by application of the Eisenstein–Natanzon-Filshtinsky functions.
- aRVE can be applied to any distribution of disks on the plane, not only to the uniformly distributed inclusions tacitly considered in the majority of engineering papers.
- The method doesn't use virtually impossible computation of higher-order correlation functions.

4.2 ELLIPTIC INCLUSIONS. APPLICATION OF GENERAL FORMULAS

Consider non-overlapping equal ellipses with the semi-axes $r(1+\alpha)$ and $r(1-\alpha)$ $(0<\alpha_m<1)$. The components of the effective permittivity tensor $\boldsymbol{\varepsilon}_{\perp}$ aligned with the coordinate axes ellipses were estimated by Galeener [80]

$$\varepsilon_{11} \approx 1+\frac{2\varrho f}{1-\varrho(f+\alpha)}, \quad \varepsilon_{22} \approx 1+\frac{2\varrho f}{1-\varrho(f-\alpha)} \tag{4.90}$$

and by Cohen et al. [58]

$$\varepsilon_{11} \approx 1+\frac{2\varrho f}{1-\varrho(\alpha+f(1-\alpha))}, \quad \varepsilon_{22} \approx 1+\frac{2\varrho f}{1+\varrho(\alpha-f(1+\alpha))}. \tag{4.91}$$

Zimmerman [238] derived the following formula for the effective scalar permittivity of randomly oriented equal ellipses

$$\varepsilon_e \approx \frac{1+f\varrho-\alpha^2\varrho^2}{1-f\varrho-\alpha^2\varrho^2}. \tag{4.92}$$

Approximate equality is used in the above relations since the authors did not indicate regions of their validity. The transformation of (4.90) and (4.91) up to the third orders of f, $|\varrho|$ and up to $O(f^2\alpha)$ yields the same result

$$\varepsilon_{11} \approx 1+2f\varrho+2f(f+\alpha)\varrho^2, \quad \varepsilon_{22} \approx 1+2f\varrho+2f(f-\alpha)\varrho^2. \tag{4.93}$$

The invariant $\frac{1}{2}(\varepsilon_{11}+\varepsilon_{22}) \approx 1+2f\varrho+2f^2\varrho^2$ coincides with the transformed formula (4.92). The expression (4.92) coincides with the approximation (2.171) derived by Maxwell's approach is valid up to $O(f^2)$.

At the end of the present section, we apply the general formula (2.206) to equal elliptic inclusions uniformly distributed in the unit periodicity cell Q with the uniformly distributed inclinations to the x_1-axis. The concentrations of ellipses equals $f=N\pi r^2(1-\alpha^2)$.

One can check that the static moments s_{1k} of ellipses vanish. The value $\mathcal{J}_k$ for disks is equal to zero. The values $\mathcal{J}_k$ for ellipses are calculated by the formula

$$\mathcal{J}_k=\frac{1}{4\pi}\int_{L_k} d\bar{t} \int_{L_k} \frac{\overline{\tau}}{\tau-t} d\tau=\frac{1}{4\pi}\int_{L_k} d\bar{t} \int_{L_k} \frac{\overline{\tau-t}}{\tau-t} d\tau, \tag{4.94}$$

since [79]

$$\int_{L_k} \frac{d\tau}{\tau - t} = \pi i, \text{ hence } \frac{1}{4\pi}\int_{L_k} d\bar{t}\int_{L_k} \frac{\bar{t}}{\tau - t} d\tau = 0. \tag{4.95}$$

Write the complex equation of ellipse L_k in the local polar coordinates (r, θ) centered at a_k

$$t(r,\theta) = r\exp(i\varphi_k)[\exp(i\theta) + \alpha\exp(-i\theta)]. \tag{4.96}$$

It follows from (4.94) and (4.96) that it is sufficient to calculate $\mathcal{J}_k$ for $r = 1$, $\varphi_k = 0$ and the result multiply on $r^2\exp(-2i\varphi_k)$ to obtain

$$\mathcal{J}_k = \exp(-2i\varphi_k)\pi r^2\alpha(1-\alpha^2) = \exp(-2i\varphi_k)\frac{f\alpha}{N}. \tag{4.97}$$

Therefore, the shape form term becomes

$$\begin{aligned} \mathcal{J} = 2\varrho^2 \textstyle\sum_{k=1}^N \begin{pmatrix} \mathrm{Re}\,\mathcal{J}_k & -\mathrm{Im}\,\mathcal{J}_k \\ -\mathrm{Im}\,\mathcal{J}_k & -\mathrm{Re}\,\mathcal{J}_k \end{pmatrix} = \\ 2\varrho^2 f\alpha \tfrac{1}{N}\textstyle\sum_{k=1}^N \begin{pmatrix} \cos 2\varphi_k & \sin 2\varphi_k \\ \sin 2\varphi_k & -\cos 2\varphi_k \end{pmatrix}. \end{aligned} \tag{4.98}$$

Ultimately,

$$\begin{aligned} \boldsymbol{\varepsilon}_\perp = (1 + 2\varrho f + 2\varrho^2 f^2)\, I + 2\varrho^2 f\alpha \tfrac{1}{N}\textstyle\sum_{k=1}^N \begin{pmatrix} \cos 2\varphi_k & \sin 2\varphi_k \\ \sin 2\varphi_k & -\cos 2\varphi_k \end{pmatrix} \\ + \tfrac{2f^2\varrho^2}{\pi N^2}\textstyle\sum_{k,m=1}^N \begin{pmatrix} \mathrm{Re}\,\wp(a_m - a_k) & -\mathrm{Im}\,\wp(a_m - a_k) \\ -\mathrm{Im}\,\wp(a_m - a_k) & -\mathrm{Re}\,\wp(a_m - a_k) \end{pmatrix} + O(f^3\varrho^3). \end{aligned} \tag{4.99}$$

Equations (4.98)–(4.99) can be applied to deterministic and random inclinations of ellipses. Let the inclination of ellipses φ_k be random and uniformly distributed in $0 < \varphi_k < \pi$. Then, the mathematical expectation of $\cos 2\varphi$ and $\sin 2\varphi$ vanish; hence, the expectation $\mathbb{E}[\mathcal{J}]$ has zero components in the considered case.

Let the inclinations of ellipses φ_k be uniformly distributed in the sector $0 < \varphi < \varphi_0$. Then, using the integral $\frac{1}{\varphi_0}\int_0^{\varphi_0} \cdots d\varphi$ instead of the sum (4.98) we obtain

$$\mathbb{E}[\mathcal{J}] = 2\varrho^2 f\alpha \frac{\sin\varphi_0}{\varphi_0}\begin{pmatrix} \cos\varphi_0 & \sin\varphi_0 \\ \sin\varphi_0 & -\cos\varphi_0 \end{pmatrix}. \tag{4.100}$$

In the case of unidirectional ellipses, $\varphi_0 = 0$ and (4.100) becomes

$$\mathbb{E}[\mathcal{J}] = 2\varrho^2 f\alpha \begin{pmatrix} 1 & 0 \\ 0 & -1 \end{pmatrix}. \tag{4.101}$$

Figure 4.11 Elliptic inclusions on the plane

4.3 ELLIPTIC INCLUSIONS. HIGHER-ORDER FORMULAS

This section follows [166, 181] and is devoted to higher-order analytical approximate formulas for the effective permittivity of randomly distributed elliptic inclusions.

Let non-overlapping elliptical inclusions $G_m (m = 1, 2, \ldots, n)$ be located in the plane (x_1, x_2), see Figure 4.11. Let $r_m(1 \pm \alpha)$ denote the semi-axes of ellipses ($0 \leq \alpha < 1$ and $r_m > 0$) of permittivity ε_1 embedded in the host of the normalized unit permittivity. It follows from **Rule** $\mathbb{R} \to \mathbb{C}$ on page 58 that it is sufficient to consider the real permittivity ε_1.

Initially, let the number of inclusions n be finite on the plane. The limit $n \to \infty$ will be considered at the end of our constructions according to the Eisenstein summation. For convenience put the semi-axes of the mth ellipse equal to $r_m(1+\alpha_m)$ and $r_m(1-\alpha_m)$, respectively. The positive parameter r_m characterizes the size of the mth inclusion, α_m its shape ($0 < \alpha_m < 1$). Introduce the local coordinates (X_m, Y_m) each for a fixed inclusion G_m

$$
\begin{aligned}
X_m &= \frac{1}{r_m}[(x_1 - x'_m)\cos\theta_m + (x_2 - y'_m)\sin\theta_m], \\
Y_m &= \frac{1}{r_m}[(x_1 - x'_m)\sin\theta_m + (x_2 - y'_m)\cos\theta_m],
\end{aligned}
\tag{4.102}
$$

where $a_m = x'_m + \mathrm{i}y'_m$ is the center of the inclusion G_m expressed in terms of complex numbers and θ_m is the angle between the major semi-axis of the ellipse and the x_1-axis. Let G denote the complement of the closures of all domains G_m to the extended complex plane.

Let the domains G and G_m be occupied by materials of the unit and ε_1 permittivity, respectively, where $\varepsilon_1 > 0$. Then, the potentials $u(z)$ and $u_m(z)$ are harmonic in G and G_m $(m = 1, 2, \ldots n)$ and satisfies the perfect contact conditions (2.20). Let the external flux be applied parallel to the x_1-axis. Then, the principal part of $u(z)$ at infinity equals $x_1 = \mathrm{Re}\, z$.

Introduce complex potentials $\varphi(z)$ and $\varphi_m(z)$ analytic in G and G_m, respectively, in such a way that

$$u(z) = \mathrm{Re}\,[\varphi(z)+z], \quad u_m(z) = \frac{2}{1+\varepsilon_1}\mathrm{Re}\,\varphi_m(z). \tag{4.103}$$

Here, the term $x_1 = \mathrm{Re}\, z$ is presented separately; hence, $\varphi(z)$ is analytic at infinity. The considered problem is reduced to the $\mathbb{R}$-linear problem [185]

$$\varphi(t) = \varphi_m(t) - \varrho\,\overline{\varphi_m(t)} - t,\ t \in \partial G_m, \quad m = 1,2,\ldots,n, \tag{4.104}$$

where the scalar real contrast parameter $\varrho = \frac{\varepsilon_1 - 1}{\varepsilon_1 + 1}$ is used.

The $\mathbb{R}$-linear problem (4.104) was reduced in [185], see also Chapter 2, to the integral equations in the space $\mathcal{H}(G^+)$

$$\varphi_k(z) = \varrho\sum_{m=1}^{n}\frac{1}{2\pi\mathrm{i}}\int_{\partial G_m}\frac{\overline{\varphi_m(t)}dt}{t-z} + z, \quad z \in G_k, \quad k = 1,2,\ldots,n. \tag{4.105}$$

The integral operator from (4.105) is bounded in $\mathcal{H}(G^+)$.

Equations (4.105) correspond to Schwarz's method and can be written in the form

$$\varphi_k(z) - \frac{\varrho}{2\pi i}\int_{\partial G_k}\frac{\overline{\varphi_k(t)}}{t-z}dt = \varrho\sum_{m\neq k}\frac{1}{2\pi\mathrm{i}}\int_{\partial G_m}\frac{\overline{\varphi_m(t)}}{t-z}dt + z, \tag{4.106}$$
$$z \in G_k,\ k = 1,2,\ldots,n.$$

At the zeroth approximation, we arrive at the problem for the single inclusion G_k $(k = 1,2,\ldots,n)$

$$\varphi_k^{(0)}(z) - \frac{\varrho}{2\pi\mathrm{i}}\int_{\partial G_k}\frac{\overline{\varphi_k^{(0)}(t)}}{t-z}dt = z, \quad z \in G_k. \tag{4.107}$$

The next approximations are obtained from equations

$$\varphi_k^{(p)}(z) - \frac{\varrho}{2\pi\mathrm{i}}\int_{\partial G_k}\frac{\overline{\varphi_k^{(p)}(t)}}{t-z}dt = \varrho\sum_{m\neq k}\frac{1}{2\pi i}\int_{\partial G_m}\frac{\overline{\varphi_m^{(p-1)}(t)}}{t-z}dt + z,\ z \in G_k \tag{4.108}$$
$$(k = 1,2,\ldots,n), \quad p = 1,2,\ldots.$$

Following Section 2.6.1 equations (4.105) can be solved by successive approximations. The functions $\varphi_k(z)$ are presented in the form of an absolutely convergent series in the convergence disk $|\varrho| < 1$.

4.3.1 REDUCTION TO FUNCTIONAL EQUATIONS

Application of the method of successive approximations in the integral form (4.106) is difficult because it leads to the computation of multiple integrals. In the present section, we reduce the integral equations (4.105) to functional equations, which contain compositions of unknown and known functions instead of integral terms. Hence, it is effective in symbolic and numerical computations.

Consider the Joukowsky conformal mapping

$$z = s_k\left(w + \frac{\alpha_k}{w}\right) + a_k, \tag{4.109}$$

where $s_k = r_k e^{\mathrm{i}\theta_k}$. Let Γ_k denote the cut connecting the foci of the ellipse ∂G_k located at the points $(\pm 2\sqrt{\alpha_k} s_k + a_k)$. Γ_k is considered as a closed curve having two sides. The inverse conformal mapping to (4.109) has the form

$$w = \frac{1}{2}\left[\frac{z-a_k}{s_k} + \sqrt{\left(\frac{z-a_k}{s_k}\right)^2 - 4\alpha_k}\right]. \tag{4.110}$$

The branch of the square root is fixed in such a way that the function $w(z)$ defined by (4.110) is analytic in $\mathbb{C}\backslash\Gamma_k$ and $\lim_{Z\to X\pm i0}\sqrt{Z^2-4\alpha_k} = \pm\mathrm{i}\sqrt{4\alpha_k - X^2}$, where $Z = \frac{z-a_k}{s_k}$ and $X \in \mathbb{R}$. The doubly connected domain $G_k\backslash\Gamma_k$ is mapped onto the annulus $\sqrt{\alpha_k} < |w| < 1$ by the conformal mapping (4.110). The ellipse ∂G_k is transformed onto the unit circle $|w| = 1$ and both sides of the cut Γ_k onto the circle $|w| = \sqrt{\alpha_k}$.

Introduce the function analytic in $\sqrt{\alpha_k} < |w| < 1$ and continuous in $\sqrt{\alpha_k} \le |w| \le 1$

$$\Phi_k(w) = \varphi_k(z) = \varphi_k\left[s_k\left(w + \frac{\alpha_k}{w}\right) + a_k\right] \tag{4.111}$$

Substitute (4.111) in (4.105) and make the analogous change of variables $t = s_k\left(\tau + \frac{\alpha_k}{\tau}\right) + a_k$ in the integrals. Then, (4.105) becomes

$$\Phi_k(w) = \varrho\sum_{m=1}^{n}\frac{1}{2\pi\mathrm{i}}\int_{|\tau|=1}\frac{\overline{\Phi_m(\tau)}(1-\frac{\alpha_m}{\tau^2})d\tau}{\tau + \frac{\alpha_m}{\tau} - \frac{s_k}{s_m}(w + \frac{\alpha_k}{w}) + \frac{a_m - a_k}{s_m}} \tag{4.112}$$

$$+ s_m\left(w + \frac{\alpha_m}{w} + a_m\right), \quad \sqrt{\alpha_k} < |w| < 1, \quad k = 1,2,\ldots,n.$$

Moreover, it follows from the continuity of $\varphi_k(z)$ when z goes across Γ_k that

$$\Phi_k(\tau) = \Phi_k\left(\frac{\alpha_k}{\tau}\right), \quad |\tau| = \sqrt{\alpha_k}. \tag{4.113}$$

Equation (4.113) implies that $\Phi_k(w)$ is represented in the form

$$\Phi_k(w) = \phi_k(w) + \phi_k\left(\frac{\alpha_k}{w}\right), \quad \sqrt{\alpha_k} \le |w| \le 1, \tag{4.114}$$

where $\phi_k(w)$ is analytic in the unit disk $|w| < 1$. Substitution of (4.114) into (4.112) yields

$$\begin{aligned}&\phi_k(w)+\phi_k\left(\tfrac{\alpha_k}{w}\right) = \varrho\textstyle\sum_{m=1}^{n}(R_{km}\phi_m)(w)+g_k(w)+g_k\left(\tfrac{\alpha_k}{w}\right),\\ &\sqrt{\alpha_k} < |w| < 1, \quad k = 1,2,\dots,n,\end{aligned} \tag{4.115}$$

where $(R_{km}\phi_m)(w)$ are integral operators defined by the integrals from (4.112) and $g_k(w) = s_k w + a_k$.

Let P^+ denote the projection operator which transforms a function analytic in $\sqrt{\alpha_m} < |w| < 1$ onto its regular part analytic in the unit disk. This operator can by constructed by the regular part of the Laurent series. The differentiation of (4.115) and application of the operator P^+ yields

$$\psi_k(w) = \varrho\alpha_k e^{-2\mathrm{i}\theta_k}\overline{\psi_k(\alpha_k\overline{w})} - \tag{4.116}$$

$$\varrho\sum_{m\neq k}\frac{\overline{s_m}}{s_k}P^+\left\{\left[\alpha_m\overline{\psi_m(\alpha_m\overline{\beta_{km}(w)})}+\frac{1}{\alpha_m}\overline{\psi_m\left(\frac{\beta_{km}(w)}{\alpha_m}\right)}\right]\beta_{km}'(w)\right\}+1,$$

$$|w| \leq 1,\ k = 1,2,\dots,n,$$

where $\psi_k(w) = s_k^{-1}\phi_k'(w)$ and

$$\beta_{km}(w) = \frac{1}{2}\left\{s_m^{-1}\left[s_k(w+\frac{\alpha_k}{w})+a_k-a_m\right]-\sqrt{\left[\frac{s_k(w+\frac{\alpha_k}{w})+a_k-a_m}{s_m}\right]^2-4\alpha_m}\right\}. \tag{4.117}$$

Here, the exact formulas for $(R_{km}\phi_m)(w)$ from [166] are used.

4.3.2 EFFECTIVE PERMITTIVITY TENSOR

We compute the components ε_{11} and $\varepsilon_{12} = \varepsilon_{21}$. The component ε_{22} can be computed analogously by rotation of the considered structure about $\frac{\pi}{2}$ radian.

First, we calculate the averaged permittivity $\varepsilon_{11}(n)$ and $\varepsilon_{12}(n)$ over a finite cluster containing n inclusions. The number n will be kept in symbolic form that is important for the calculation of conditionally convergent series, as $n \to \infty$, in order to obtain analytical formulas for the effective constants

$$\varepsilon_{11} - \mathrm{i}\varepsilon_{12} = \lim_{n\to\infty}(\varepsilon_{11}(n) - i\varepsilon_{12}(n)). \tag{4.118}$$

The averaged cluster permittivity $\varepsilon_{11}(n)$ and $\varepsilon_{12}(n)$ were calculated in [166] by combination of Green's identity, the conformal mapping (4.109) and Cauchy's residue theorem

$$\varepsilon_{11}(n) - \mathrm{i}\varepsilon_{12}(n) = 1 + 2\varrho\sum_{k=1}^{n}\pi r_k^2(1-\alpha_k^2)\psi_k(0), \tag{4.119}$$

where $\pi r_k^2(1-\alpha_k^2)$ is the area of G_k. Hence, we have to estimate the values $\psi_k(0)$ depending on n where the analytic functions $\psi_k(w)$ satisfy the system of functional equations (4.116).

Introduce the following dimensionless scale parameters

$$\ell_{mk} := \frac{\sqrt{s_m s_k}}{a_k - a_m}, \tag{4.120}$$

where $k, m = 1, 2, \ldots, n$ and $k \neq m$, s_m is determined by (4.109). We formally consider $|\ell_{mk}|$ as small parameters. The system of functional equations (4.116) can be solved by expansion in the powers of ℓ_{mk}. For shortness, the order of a product of L terms $|\ell_{mk}|$ be denoted by $O(\ell^L)$. First, we find from (4.117) by straight-forward computations

$$\beta_{km}(w) = \frac{s_m \alpha_m}{a_k - a_m} - \frac{s_k s_m \alpha_m \left(\alpha_k + w^2\right)}{w\left(a_k - a_m\right)^2} + O(\ell^3), \tag{4.121}$$

$$\beta'_{km}(w) = \frac{s_k s_m \alpha_m \left(\alpha_k - w^2\right)}{w^2 \left(a_k - a_m\right)^2} + \frac{2 s_k^2 s_m \alpha_m \left(w^4 - \alpha_k^2\right)}{w^3 \left(a_k - a_m\right)^3} + O(\ell^3). \tag{4.122}$$

In further computations, we use the symbolic expression for $\beta_{km}(w)$ up to $O(\ell^7)$ extending (4.121) not writing long formulas. Such formulas can be easily obtained analytically for any fixed accuracy $O(\ell^L)$ by Mathematica®. The function $\psi_m(w)$ is analytic in the unit disk, hence, can be expanded in its Taylor series

$$\psi_m(w) = A_{m0} + A_{m1}w + A_{m2}w^2 + A_{m3}w^3 + A_{m4}w^4 + \cdots. \tag{4.123}$$

We see below that the effective permittivity computed to $O(f^4)$ requires only the coefficients A_{mj} for $0 \leq j \leq 3$.

Substitution of the series (4.123) into the right part of (4.116) yields

$$\psi_k(w) - \varrho \alpha_k e^{-2i\theta_k} \overline{\psi_k(\alpha_k \overline{w})} - 1 = -\varrho \sum_{m \neq k} F_{mk}(w), \ |w| \leq 1 \ (k = 1, 2, \ldots, n), \tag{4.124}$$

where $F_{mk}(w)$ up to $O(\ell^7)$ are polynomials of third degree. Symbolic computations can be applied to write them; see the attached codes in [181, Appendix A]. This justifies the cut of the analytic function (4.123) to the polynomial of degree 3. We write below the result of the substitution of these polynomials into the left part of (4.124) and taking the coefficients on the same powers of w. As a result, we obtain an $\mathbb{R}$-linear infinite system of algebraic equations valid up to $O(\ell^7)$

$$A_{k0} - \varrho \alpha_k e^{-2i\theta_k} \overline{A_{k0}} - 1 = \varrho \sum\nolimits_{m \neq k} \left\{ s_m^2 r_m^2 \frac{5\left(\alpha_m^6+1\right)\overline{A_{m2}}\left(2s_k^2\alpha_k+s_m^2\alpha_m\right)+s_m^2\left(\alpha_m^{10}+1\right)\overline{A_{m4}}}{(a_k-a_m)^6} \right.$$
$$+ s_m r_m^2 \frac{2\left(\alpha_m^4+1\right)\overline{A_{m1}}\left(3s_k^2\alpha_k+2s_m^2\alpha_m\right)+s_m^2\left(\alpha_m^8+1\right)\overline{A_{m3}}}{(a_k-a_m)^5}$$
$$\left. + r_m^2 \frac{3\left(\alpha_m^2+1\right)\overline{A_{m0}}\left(s_k^2\alpha_k+s_m^2\alpha_m\right)+s_m^2\left(\alpha_m^6+1\right)\overline{A_{m2}}}{(a_k-a_m)^4} + \frac{r_m^2 s_m\left(\alpha_m^4+1\right)\overline{A_{m1}}}{(a_k-a_m)^3} + \frac{r_m^2\left(\alpha_m^2+1\right)\overline{A_{m0}}}{(a_k-a_m)^2} \right\},$$

$$A_{k1}-\varrho\alpha_k^2 e^{-2i\theta_k}\overline{A_{k1}}=$$
$$-\varrho\sum_{m\neq k}\left\{s_k s_m r_m^2\frac{4(\alpha_m^4+1)\overline{A_{m1}}(3s_k^2\alpha_k+2s_m^2\alpha_m)+5s_m^2(\alpha_m^8+1)\overline{A_{m3}}}{(a_k-a_m)^6}\right.$$
$$\left.+\frac{4s_k s_m^3(\alpha_m^6+1)\overline{s_m}\overline{A_{m2}}}{(a_k-a_m)^5}+\frac{3s_k r_m^2 s_m(\alpha_m^4+1)\overline{A_{m1}}}{(a_k-a_m)^4}+\frac{2s_k r_m^2(\alpha_m^2+1)\overline{A_{m0}}}{(a_k-a_m)^3}\right\},$$

$$A_{k2}-\varrho\alpha_k^3 e^{-2i\theta_k}\overline{A_{k2}}=\varrho\sum_{m\neq k}\left\{\frac{10s_k^2 s_m^2 r_m^2(\alpha_m^6+1)\overline{A_{m2}}}{(a_k-a_m)^6}\right.$$
$$\left.+\frac{6s_k^2 s_m r_m^2(\alpha_m^4+1)\overline{A_{m1}}}{(a_k-a_m)^5}+\frac{3s_k^2 r_m^2(\alpha_m^2+1)\overline{A_{m0}}}{(a_k-a_m)^4}\right\},$$

$$A_{k3}-\varrho\alpha_k^4 e^{-2i\theta_k}\overline{A_{k3}}=-\varrho\sum_{m\neq k}\frac{6s_k^3 s_m r_m^2(\alpha_m^4+1)\overline{A_{m1}}}{(a_k-a_m)^6}, \tag{4.125}$$
$$A_{k4}-\varrho\alpha_k^5 e^{-2i\theta_k}\overline{A_{k4}}=0,\quad k=1,2,\ldots,n. \tag{4.126}$$

The structure of each equation has the form $A-\varrho\alpha e^{-2i\theta_k}\overline{A}=b$. Then,

$$A=\frac{b+\varrho\alpha e^{-2i\theta_k}\overline{b}}{1-\varrho^2\alpha^2}. \tag{4.127}$$

One can see that b and A have the same order in ℓ. Its solution is essentially simplified by neglecting the values of order $O(\ell^7)$ during the solution. Below the designation $A_m^{(j)}$ for A_m within the precision $O(\ell^j)$ is used. Equation (4.126) yields $A_{k4}=0$. It follows from (4.125) that A_{k3} is at least of order $O(\ell^6)$. All the terms with A_{m3} enter in equations with high-order multipliers in ℓ; hence, they vanish. The same concerns A_{k2} having at least the order $O(\ell^4)$. The term A_{k1} is at least of order $O(\ell^3)$. Therefore, the system (4.125)–(4.126) becomes

$$A_{k0}-\varrho\alpha_k e^{-2i\theta_k}\overline{A_{k0}}-1=\varrho\sum_{m\neq k}\left\{r_m^2\frac{3(\alpha_m^2+1)\overline{A_{m0}^{(2)}}(s_k^2\alpha_k+s_m^2\alpha_m)}{(a_k-a_m)^4}\right.$$
$$\left.+\frac{r_m^2 s_m(\alpha_m^4+1)\overline{A_{m1}^{(3)}}}{(a_k-a_m)^3}+\frac{r_m^2(\alpha_m^2+1)\overline{A_{m0}^{(4)}}}{(a_k-a_m)^2}\right\}+O(\ell^7),$$

$$A_{k1}-\varrho\alpha_k^2 e^{-2i\theta_k}\overline{A_{k1}}=-\varrho\sum_{m\neq k}\left\{s_k s_m r_m^2\frac{4(\alpha_m^4+1)\overline{A_{m1}^{(0)}}(3s_k^2\alpha_k+2s_m^2\alpha_m)}{(a_k-a_m)^6}\right.$$
$$\left.+\frac{2s_k r_m^2(\alpha_m^2+1)\overline{A_{m0}^{(3)}}}{(a_k-a_m)^3}\right\}+O(\ell^7),\quad k=1,2,\ldots,n. \tag{4.128}$$

Equation (4.128) gives

$$A_{k0}^{(0)}-\varrho\alpha_k e^{-2i\theta_k}\overline{A_{k0}^{(0)}}=1, \tag{4.129}$$

$$A_{k0}^{(1)}=A_{k0}^{(0)}, \tag{4.130}$$

$$A_{k0}^{(2)} - \varrho\alpha_k e^{-2i\theta_k}\overline{A_{k0}^{(2)}} - 1 = \varrho\sum_{m\neq k}\frac{r_m^2\left(\alpha_m^2+1\right)\overline{A_{m0}^{(0)}}}{\left(a_k-a_m\right)^2}, \tag{4.131}$$

$$A_{k0}^{(3)} - \varrho\alpha_k e^{-2i\theta_k}\overline{A_{k0}^{(3)}} - 1 = \varrho\sum_{m\neq k}\frac{r_m^2\left(\alpha_m^2+1\right)\overline{A_{m0}^{(1)}}}{\left(a_k-a_m\right)^2}. \tag{4.132}$$

Equation (4.128) is reduced for the calculation of the third-order approximation of A_{k1} to the following equation

$$A_{k1}^{(3)} - \varrho\alpha_k^2 e^{-2i\theta_k}\overline{A_{k1}^{(3)}} = -\varrho\sum_{m\neq k}\frac{2s_k r_m^2\left(\alpha_m^2+1\right)\overline{A_{m0}^{(3)}}}{\left(a_k-a_m\right)^3} \tag{4.133}$$

The considered equations are solved by applying (4.127) a few times. First, $A_{k0}^{(0)}$ is found from (4.129)

$$A_{k0}^{(0)} = \frac{1+\varrho\alpha_k e^{2i\theta_k}}{1-\varrho^2\alpha_k^2}. \tag{4.134}$$

The result (4.134) is substituted into (4.131) to determine $A_{k0}^{(1)} = A_{k0}^{(2)}$. The obtain result is substituted into (4.133) to find $A_{k0}^{(3)} = A_{k0}^{(4)}$. Next, $A_{k1}^{(3)}$ is found from (4.133). Ultimately, these results are substituted into (4.128) in order to determine $\psi_k(0) = A_{k0}$ up to $O(\ell^7)$ and further to compute (4.119)

$$\varepsilon_{11}(n) - i\varepsilon_{12}(n) = 1 + 2\varrho\sum_{k=1}^{n}\pi r_k^2(1-\alpha_k^2)\left\{\frac{1+\varrho\alpha_k e^{-2i\theta_k}}{1-\varrho^2\alpha_k^2} + \frac{\varrho}{1-\varrho^2\alpha_k^2}\times\right. \tag{4.135}$$
$$\left.\sum_{m\neq k}\frac{r_m^2(\alpha_m^2+1)}{1-\varrho^2\alpha_m^2}\left[\frac{1+\varrho\alpha_m e^{-2i\theta_m}}{(\overline{a_k-a_m})^2} + \alpha_k\varrho e^{-2i\theta_k}\frac{1+\alpha_m\varrho e^{2i\theta_m}}{(a_k-a_m)^2}\right]\right\} + O(\ell^6).$$

The formula (4.135) and others derived below give the components of the effective permittivity tensor as the limit (4.118). We use the Eisenstein method to take the induced charge equal to zero, see page 62, to get the proper result. Introduce the limits

$$\mathcal{S}_p = \lim_{n\to\infty}\frac{1}{n}\sum_{k=1}^{n}\sum_{m\neq k}\frac{1}{(a_k-a_m)^p},\ \mathcal{S}_{pq} = \lim_{n\to\infty}\frac{1}{n}\sum_{k=1}^{n}\sum_{m\neq k}\sum_{l\neq m}\frac{1}{(a_k-a_m)^p(\overline{a_m-a_l})^q}, \tag{4.136}$$

where $p,q = 2,3,\ldots$.

Consider a part of the doubly periodic composite with $n = MN$ inclusions where N is the number of inclusions per periodicity cell, and M is the number of cells. The location of inclusion in each periodicity cell is arbitrary but the same. Let the number N be fixed, but n and M simultaneously tend to infinity. To precisely describe this limit, we consider the lattice Q, described on page 29. For definiteness, we fix the numbering of a_k in such a way that $a_1, a_2, \ldots a_N$ lie in the zero periodicity cell $Q_{0,0} = \{z \in \mathbb{C}: \ z = t_1\omega_1 + t_2\omega_2, -\frac{1}{2} < t_{1,2} < \frac{1}{2}\}$. All other points are generated by the translations $a_j + m_1\omega_1 + m_2\omega_2$ where $j = 1, 2, \ldots, N$; m_1 and m_2 run over the integer numbers $\mathbb{Z}$. Without loss of generality, the area of Q_{00} is normalized to unity, i.e., $\omega_1 \mathrm{Im}\, \omega_2 = 1$.

The following limit defines the concentration of inclusions

$$f = \lim_{n\to\infty} \frac{1}{M} \sum_{k=1}^{n} \pi r_k^2 (1-\alpha_k^2) = \sum_{k=1}^{N} \pi r_k^2 (1-\alpha_k^2). \tag{4.137}$$

The limit $\mathcal{S}_p$ for the periodic continuation of the set $\{a_1, a_2, \ldots a_N\}$ is calculated by the Eisenstein summation

$$\mathcal{S}_p = \lim_{M\to\infty} \frac{1}{MN} \sum_{k,m=1}^{N}{}^{*} \sum_{m_1,m_2\in\mathbb{Z}}^{e} \frac{1}{(a_k - a_m - m_1\omega_1 - m_2\omega_2)^p} = \frac{1}{N} \sum_{k,m=1}^{N} E_p(a_k - a_m), \tag{4.138}$$

where the term $m_1 = m_2 = 0$ is omitted in the sum $\sum^*$ when $a_s = a_\ell$. It is assumed for shortness that $E_2(a_l - a_s) = S_2$ if $a_l = a_s$ in (4.138).

4.3.3 ANALYTICAL FORMULAS FOR THE EFFECTIVE PERMITTIVITY

The parameters ℓ_{km} defined by (4.120), as $n \to \infty$, are related to the concentration f. The meaning of ℓ_{km} can be easily explained in the case of equal ellipses, i.e., $|s_m| = r$ for all m, as follows. The order of $|\ell_{km}|^2$ is r^2 for fixed α_k and a_k from the unit cell $Q_{(0,0)}$, as $r \to 0$. The concentration (4.137) in this case is of order r^2 for a finite number N, hence $|\ell_{km}|^2 \sim Cf$, as $n \to \infty$, for a positive constant C.

Consider, for instance, the asymptotic equation (4.135). The explicit expression from its right-hand part contains the terms r^2, which will be transformed into the terms including f and f^2 in the limit $n \to \infty$. The term $O(\ell^6)$ becomes of order $O(f^3)$. Therefore, equation (4.135) in the limit $n \to \infty$ becomes an asymptotic equation on $\varepsilon_{11} - i\varepsilon_{12}$ up to $O(f^3)$.

Below, we consider three examples illustrating various distributions of ellipses.

Example 1. Randomly distributed equal unidirectional ellipses

Consider equal randomly distributed unidirectional ellipses when $\alpha_m = \alpha$, $\theta_m = \theta$ and $r_m = r$ are the same for each inclusion. In a more general case, the precise notion of randomly distributed ellipses will be given below. The concentration holds $f = \pi r^2(1-\alpha^2)N$, hence $r^2 = \frac{f}{\pi N(1-\alpha^2)}$. Calculation of the limits in (4.135), as $n \to \infty$, yields the following formula

$$\varepsilon_{11} - \mathrm{i}\varepsilon_{12} = 1 + 2\varrho f\, \frac{1+\varrho\alpha e^{-2\mathrm{i}\theta}}{1-\varrho^2\alpha^2}$$
$$+ \frac{2\varrho^2 f^2 (1+\alpha^2)}{(1-\varrho^2\alpha^2)^2(1-\alpha^2)}\, \frac{1}{\pi} \left[(1+\varrho\alpha e^{-2\mathrm{i}\theta}) e_2 + \varrho\alpha(e^{-2\mathrm{i}\theta} + \varrho\alpha)\overline{e_2} \right] + O(f^3). \tag{4.139}$$

Different distributions of the centers a_k determine different values of e_2 calculated with (4.66). We have $e_2 = \pi$ in the case of the uniform non-overlapping distribution of a_k in the periodicity cell Q_{00}. Then, (4.139) becomes

$$\begin{aligned}\varepsilon_{11} - i\varepsilon_{12} &= 1 + 2\varrho f \, \tfrac{1+\varrho\alpha e^{-2i\theta}}{1-\varrho^2\alpha^2} \\ &+ 2\varrho^2 f^2 \, \tfrac{1+\alpha^2}{(1-\varrho^2\alpha^2)^2(1-\alpha^2)} \left(1 + 2\varrho\alpha e^{-2i\theta} + \varrho^2\alpha^2\right) + O(f^3).\end{aligned} \tag{4.140}$$

The rotation of the structure about $\frac{\pi}{2}$ radian yields

$$\begin{aligned}\varepsilon_{22} + i\varepsilon_{12} &= 1 + 2\varrho f \, \tfrac{1-\varrho\alpha e^{-2i\theta}}{1-\varrho^2\alpha^2} \\ &+ 2\varrho^2 f^2 \, \tfrac{1+\alpha^2}{(1-\varrho^2\alpha^2)^2(1-\alpha^2)} \left(1 - 2\varrho\alpha e^{-2i\theta} + \varrho^2\alpha^2\right) + O(f^3).\end{aligned} \tag{4.141}$$

The formulas (4.140)–(4.141) with $\theta = 0$ correspond to Galeener's formulas (4.90) and Cohen's formulas (4.91). One can see that the corresponding formulas coincide up to $O(f^2)$ and give different results in the second-order approximations. This corresponds to the general result that any self-consistent method gives a correct result up to $O(f^2)$.

Example 2. Randomly distributed equal ellipses

Let $\alpha_m = \alpha$ and $r_m = r$, i.e., equal ellipses are considered. Randomly distributed non-overlapping equal ellipses means that the set of geometrical objects $\{G_1, G_2, \ldots, G_N\}$ is considered as a vector of identically distributed random variables G_k. The location of every domain G_k is characterized by two random parameters a_k and θ_k. The set $\mathcal{A} = \{a_1, a_2, \ldots, a_N\}$ consists of identically distributed random variables a_k. The same concerns the set of angles $\Theta = \{\theta_1, \theta_2, \ldots, \theta_N\}$. The elements of the sets $\mathcal{A}$ and Θ are dependent random variables. The non-overlapping conditions express the dependence between them. These conditions can be written explicitly in the form of inequalities involving α. The identical distribution of the angles θ_k implies that the expected value $\langle e^{i\theta_k} \rangle$ vanishes.

In this case, the precision is increased up to $O(\ell^6)$. The composite discussed is macroscopically isotropic. Hence, it is described by the scalar effective permittivity ε_e. Using the formulas from Section 4.3.2 and applying the relation $\langle e^{i\theta_k} \rangle = 0$, we arrive at the formula

$$\varepsilon_e = 1 + \frac{2\varrho f}{1-\varrho^2\alpha^2} + \frac{2\varrho^2 f^2(1+\alpha^2)}{(1-\varrho^2\alpha^2)^2(1-\alpha^2)} + 2\varrho^3 f^3 \frac{e_{22}}{\pi^2} \frac{(1+\varrho^2\alpha^2)(1+\alpha^2)^2}{(1-\varrho^2\alpha^2)^3(1-\alpha^2)^2} + O(f^4). \tag{4.142}$$

The limit sums from the beginning of Sec.4.3.3 are applied to deduce (4.142).

The formulas (4.142) and (4.92) coincide up to $O(f^2)$ and give different results in the second-order approximations.

Remark 8 *Direct substitution zero instead of* $e^{-2i\theta}$ *in* (4.140) *gives the wrong result for random angles. It is seen from the second-order terms of* (4.140) *and* (4.142).

The formula (4.142) can be improved by consideration of the next terms in f. One can increase the precision $O(\ell^L)$ of the proposed iterative scheme to construct a symbolic algorithm for high-order terms. The algorithm is iterative, which gives a possibility to investigate which terms can participate in the final solution written in the form of an infinite series. We now demonstrate the evolution of the term $(a_k - a_m)^{-2}$ during the iteration applied to (4.128)–(4.128). We take only the part of (4.128)–(4.128) containing this term

$$A_{k0} - \varrho\alpha e^{-2i\theta_k}\overline{A_{k0}} - 1 = \varrho r^2(\alpha^2+1)\sum_{m\neq k}\frac{\overline{A_{m0}}}{(a_k - a_m)^2}, \quad k = 1,2,\ldots n, \quad (4.143)$$

where for simplicity we put equal $r = r_k$ and $\alpha = \alpha_k$.

Introduce for brevity the value

$$\kappa = \frac{\varrho r^2(1+\alpha^2)}{1-\varrho^2\alpha^2}. \quad (4.144)$$

Using (4.127) we apply the iterative scheme to (4.143)

$$\begin{aligned} A_{k0}^{(0)} &= \frac{1+\varrho\alpha e^{-2i\theta_k}}{1-\varrho^2\alpha^2}, \\ A_{k0}^{(p)} &= \frac{1+\varrho\alpha e^{-2i\theta_k}}{1-\varrho^2\alpha^2} + \kappa\sum_{m\neq k}\left[\frac{\overline{A_{m0}^{(p-1)}}}{(a_k-a_m)^2} + \frac{\varrho\alpha e^{-2i\theta_k}A_{m0}^{(p-1)}}{\overline{(a_k-a_m)}^2}\right], \quad p = 1,2,\ldots. \end{aligned} \quad (4.145)$$

For $|\kappa| < 1$ the scheme converges and gives

$$A_{k0} = \frac{1}{1-\varrho^2\alpha^2}\left(1 + \kappa\sum_{k_1\neq k}\frac{1}{(a_k-a_{k_1})^2} + \kappa^2\sum_{k_1\neq k}\sum_{k_2\neq k_1}\frac{1}{(a_k-a_{k_1})^2\overline{(a_{k_2}-a_{k_1})}^2}\right.$$
$$\left. +\kappa^3\sum_{k_1\neq k}\sum_{k_2\neq k_1}\sum_{k_3\neq k_2}\frac{1}{(a_k-a_{k_1})^2\overline{(a_{k_2}-a_{k_1})}^2(a_{k_2}-a_{k_3})^2} + \ldots\right). \quad (4.146)$$

where the relation $\langle e^{-2i\theta_k}\rangle = 0$ is used. A similar method can be applied to investigate other terms.

Substitution the series (4.146) into (4.119) and calculation of the limit $n \to \infty$ yield

$$\varepsilon_e = 1 + \frac{2\varrho f}{1-\varrho^2\alpha^2}\left[1 + \varrho f h + (\varrho f h)^2\frac{e_{22}}{\pi^2} + (\varrho f h)^3\frac{e_{222}}{\pi^3} + \ldots\right] + O(f^5), \quad (4.147)$$

where $h = \frac{1}{1-\varrho^2\alpha^2}\frac{1+\alpha^2}{1-\alpha^2}$. Although there is an infinite series in square brackets, it is important to note that the true precision of (4.147) is limited to $O(f^5)$. This limitation stems from the emergence of complicated higher-order terms in f, which become apparent in (4.69) for circular inclusions. However, it is worth emphasizing that the extended series format holds its advantages due to incorporating these terms and others into the series representation of ε_e. The series expression (4.147) is favored over a third-degree polynomial in f since the higher-order terms are present in the full expansion, though not completely written.

Remark 9 *The precision $O(f^5)$ is shown and justified in the formula (4.147) in order to avoid methodological mistakes. Some specialists in material science have a problematic tendency to present a formula and then validate it with a numerical example using specific data, often for small concentrations or contrast parameters. Such sloppiness develops a misleading approach to the theory of composites. Consider such a trick used for elastic composites [196, Section 2.6] cited more than* 4000 *times. It was assumed [196, formula (2.6.16) on page 57] that the square averaged magnitude of the fluctuation of the micro-strains about the averaged strains should decrease as the size of the micro-constituents decreases. It is true, but it is not said that this estimation holds up to $O(f^2)$. This mistake is unrelated to violating the homogenization principles discussed in Chapter 3. It is just neglecting the asymptotic rules of mathematical analysis.*

This formula (4.147) and others can be considered polynomial approximations in concentration after truncation. At the same time, it is known that rational and other asymptotically equivalent approximations usually give better results than polynomial ones [18,20,93]. The explanation of this observation is based on the general property of Padé approximants to capture singularity in f, which occurs in composites because of the percolation for the critical values f_c. The explicit formulas of the present paper can also confirm this fact. Consider, for instance, formulas (4.142)–(4.147). The inequality $e_{22\ldots2} \geq \pi^L$ [177] becomes equality for square and hexagonal arrays. Then, (4.147) becomes

$$\varepsilon_e = 1 + \frac{2\varrho f}{1-\varrho^2\alpha^2}\left[1+\varrho f h + (\varrho f h)^2 + (\varrho f h)^3 + \ldots\right] + O(f^5) = \tag{4.148}$$

$$1 + \frac{2\left(1-\alpha^2\right) f\varrho}{1 - f\varrho - \alpha^2\left(f\varrho + \varrho^2 + 1\right) + \alpha^4\varrho^2} + O(f^5).$$

This is an example of the correct rational approximation of the polynomial expression within the considered precision. The advanced renormalization methods to establish asymptotically equivalent dependencies can be applied to (4.148) following [91,92].

Example 3. Elliptic and circle inclusions

Let N_1 be the numbers of the elliptic and N_2 circle inclusions per periodicity cell; $N = N_1 + N_2$. Divide the set $\{1,2,\ldots,N\}$ onto $\mathcal{N}_1 = \{1,2,\ldots,N_1\}$ and $\mathcal{N}_2 = \{N_1+1, N_1+2,\ldots,N\}$ corresponding to ellipses and disks, respectively, per the unit periodicity cell. The sets of numbers $\mathcal{L}_1 = \{1,2,\ldots,N_1, N+1, N+2,\ldots,N+N_1,\ldots\}$ correspond to ellipses and $\mathcal{L}_2 = \{N_1+1, N_1+2,\ldots,N, N+N_1+1, N+N_1+2,\ldots,2N,\ldots\}$ to disks on the infinite plane. Let each inclusion have the same area πr_0^2. Then, $\alpha_k = \alpha$, $r_k = \frac{r_0}{\sqrt{1-\alpha^2}}$ for $k = 1,2,\ldots,N_1$ and $\alpha_k = 0$, $r_k = r_0$ for $k = N_1+1, N_1+2,\ldots,N$. The relative concentrations of elliptical and circle inclusions are denoted by $f_1' = \frac{N_1}{N}$ and $f_2' = \frac{N_2}{N}$, respectively, where $f_1' + f_2' = 1$.

In order to apply formula (4.135), consider the following four limits

$$I_{11} = \lim_{M\to\infty} \frac{1}{MN_1^2} \sum_{k\in\mathcal{L}_1} \sum_{m\in\mathcal{L}_1} \frac{1}{(a_k - a_m)^2} = \frac{1}{N_1^2} \sum_{s=1}^{N_1} \sum_{l=1}^{N_1} E_2(a_l - a_s), \tag{4.149}$$

$$I_{22} = \lim_{M\to\infty} \frac{1}{MN_2^2} \sum_{k\in\mathcal{L}_2} \sum_{m\in\mathcal{L}_2} \frac{1}{(a_k - a_m)^2} = \frac{1}{N_2^2} \sum_{s=N_1+1}^{N_2} \sum_{l=N_1+1}^{N_2} E_2(a_l - a_s), \tag{4.150}$$

$$I_{12} = \lim_{M\to\infty} \frac{1}{MN} \sum_{k\in\mathcal{L}_1} \sum_{m\in\mathcal{L}_2} \frac{1}{(a_k - a_m)^2} = \frac{1}{N_1 N_2} \sum_{s=1}^{N_1} \sum_{l=N_1+1}^{N_2} E_2(a_l - a_s), \tag{4.151}$$

$$I_{21} = \lim_{M\to\infty} \frac{1}{MN} \sum_{k\in\mathcal{L}_2} \sum_{m\in\mathcal{L}_1} \frac{1}{(a_k - a_m)^2} = \frac{1}{N_1 N_2} \sum_{s=N_1+1}^{N_2} \sum_{l=1}^{N_1} E_2(a_l - a_s). \tag{4.152}$$

The numerical computations of I_{ij} were performed for the uniform non-overlapping distributions of ellipses with centers a_k for various r_0 and α. The code generates 1500 random samples with $N = 100$ centers following the scheme [62]. The computational results suggest that $I_{ij} = \pi$ for all $i, j = 1, 2$ as it was established for I_{11} and I_{22} in [62].

Taking the limit $n \to \infty$ in (4.135) and using $I_{ij} = \pi$ we arrive at the formula

$$\varepsilon_e = 1 + 2\varrho f \left[\frac{1}{1-\varrho^2\alpha^2} f_1' + f_2' + \varrho (f_1')^2 \frac{1+\alpha^2}{(1-\varrho^2\alpha^2)^2} \frac{1}{1-\alpha^2} \right. \tag{4.153}$$
$$\left. + \varrho f_1' f_2' \frac{1}{1-\varrho^2\alpha^2} \frac{2}{1-\alpha^2} + \varrho (f_2')^2 \right] + O(f^3).$$

The dependence of the effective permittivity on parameters is displayed in Figure 4.12. One can see that the effective permittivity increases with f_1' and with α.

Following **Rule** $\mathbb{R} \to \mathbb{C}$ on page 58, the formulas derived in the present section for ellipses hold for complex permittivity.

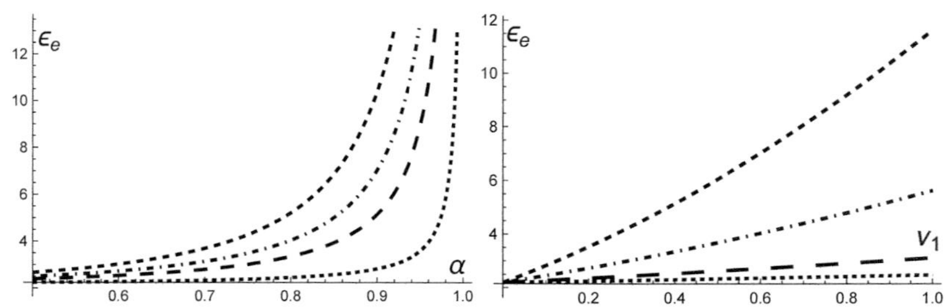

Figure 4.12 Dependence of ε_e on α given by (4.153) for: $\varrho = 0.7$, $f = 0.6$ and $f_1 = 0.1$ (tiny-dashed), $f_1 = 0.4$ (long -dashed), $f_1 = 0.6$ (dot-dashed), $f_1 = 0.9$ (dashed) and dependence of ε_e on $v_1 = f_1$ for: $\varrho = 0.7$, $f = 0.6$ and $\alpha = 0.4$ (tiny-dashed), $\alpha = 0.6$ (long-dashed), $\alpha = 0.8$ (dot-dashed), $\alpha = 0.9$ (dashed).

Part III

Macroscopic Properties of 2D Piezoelectric Composites

5 Fibrous Magneto-electro-elastic Composites

5.1 MATHEMATICAL METHODS FOR MAGNETO-ELECTRO-ELASTIC COMPOSITES

Magneto-electro-elastic (MEE) composites extend the physical notion of piezoelectric composites when the electric and elastic fields are coupled with the magnetic field. MEE composites and multiferroic phases belong to a new class of metamaterials that find wide applications in modern technology. Their peculiar property is the cross-coupling between electric polarization, magnetization, and elastic fields, which creates new possibilities for electronic devices (sensors, actuators, transducers, memories). The introduction of composites offers a practical method for realizing strong interactions of magnetic and electric fields at room temperature.

Piezoelectric materials are locally anisotropic [225]. Therefore, the complex potentials by Kolosov-Muskhelishvili do not fit to model the piezoelectric fields in two-dimensional (2D) composites except in the special cases discussed below. Fortunately, Lekhnitskii extended the method of complex potentials in 1939 to anisotropic 2D elastic problems and summarized the results in his fundamental book [130]. The extension of Lekhnitskii's formalism to the 2D piezoelectric materials was first made perhaps in 1975 by Kosmodamianskii and Lozhkin [121]. At the same time, Kudryavtsev, Parton, and Rakitin [125] investigated a crack in the boundary of a two-phase piezoelectrics. Later, this and other results on the electrostatic waves and the heat conduction in piezoelectrics were summarized in the book [202]. Kosmodamianskii and his coauthors solved the plane problems with holes, cracks, and inclusions in 1975-1984. They considered piezoelectrics with an elliptical hole for a point source in half-plane, etc. [120, 121]. Guz et al. selected the results on the coupled fields up to 1988 in five volumes of the treatise [101]. Unfortunately, these works were published in local, hardly accessible journals and publishing companies in the Russian language.

One may divide the complex potential methods in 2D MEE problems into the analytical method and the method of integral equations. The present section is devoted to an analytical approach. Filshtinsky was included in the group of scientists who worked in piezoelectricity in 1979. His contribution to a method of integral equations for MEE is summarized in [16].

It is worth noting that there is a "parallel Universe" of the same investigations beginning from the paper by Barnett and Lothe [25] published in 1975. This paper [25] extended Stroh's eigenvalue equation (1958) equivalent to Lekhnitskii's

DOI: 10.1201/9781032646633-5

characteristic (1939) equation to dislocated piezoelectric insulators in the whole piezoelectric plane. This corresponds to Kosmodamianskii's result for a point source. Some of the results by Kosmodamianskii, Filshtinsky, and others were repeated in terms of Stroh's formalism later; see [111, 237] and references therein. New results on the mathematical problems of piezoelectricity obtained in the last years can be found in [111, Chapter 11]. Special attention deserves the exact solution in the series of the plane elastic problem for an elliptic anisotropic inclusion embedded in a matrix having another type of anisotropy [111, Chapter 8]. In the previous works, anisotropic media with holes was successfully discussed using conformal mappings and the application of the Riemann-Hilbert type problem. The study of inclusions instead of holes essentially complicates the corresponding boundary value problem for analytic functions since the $\mathbb{R}$-linear condition contains different $\mathbb{R}$-linear shifts on the opposite sides of the contour. Below, we write such a problem in the form (5.67). It is a new boundary value problem in the theory of analytic functions called by $\mathbb{RR}$-linear problem in the present chapter.

In Section 5.2, we develop Lekhnitskii's formalism to 2D MEE problems following Filshtinsky's method [175]. The MEE fields are written in terms of complex potentials. It is worth noting that Filshtinsky's extension of Lekhnitskii's formalism does not reduce to a formal increase of analytic functions from two to four. Section 5.3 is devoted to the statement of boundary value problems for complex potentials in the vector-matrix form. In Section 5.4, we formally present Filshtinsky's method for dielectrics, i.e., for a problem for one complex potential. Section 5.4 does not contain a new result but clearly demonstrates the difference in these approaches and the powerful perspectives of Filshtinsky's modified method.

The considered boundary value problems correspond to two-scale asymptotic homogenization applied to a two-component thermopiezoelectric material in [81]. By applying the Ritz method, local problems were numerically solved in [82]. Gambin [84] gave an in-depth analysis of the influence of microstructure on the properties of elastic, piezoelectric, and thermoplastic composites. The set of problems can be optimized by the reduction of the coupled problem to uncoupled equations when the coupled constitutive matrix is symmetric [151, 219].

The Natanzon-Filshtinsky method was extended in [44] to the square and hexagonal arrays of piezoelectric composites in the anti-plane statement. The problem was reduced to an infinite system of linear algebraic equations solved numerically by the truncation method.[1] A series of exact formulas known for the anti-plane statement of composites was automatically extended to piezoelectric composites by a matrix transformation of the $\mathbb{R}$-problem introduced in [175, 211]. This result completes the constructive theory of the piezoelectric transversely isotropic fibrous composites. More precisely, any formula for the scalar permittivity of a 2D composite can be transformed by a matrix formalism to a formula for the effective tensor of a piezoelectric fibrous composite.

---

[1]It should be noted that the term *closed form solution* for the same problem is used misleadingly in [44], see review [17] and discussion on page 135.

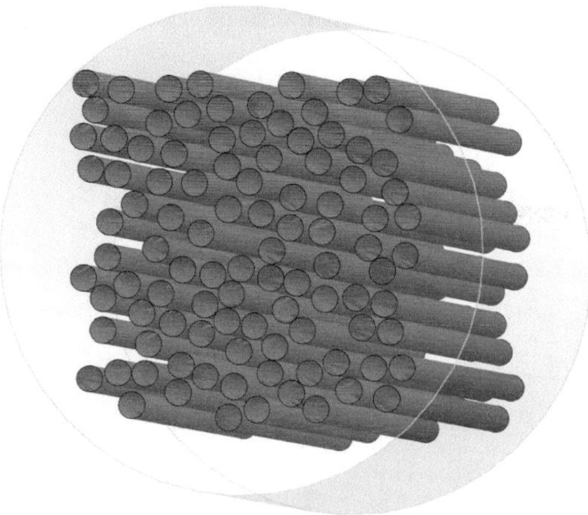

Figure 5.1 Fibrous composite.

5.2 2D EQUATIONS OF MAGNETO-ELECTRO-ELASTICITY

Consider fibrous composites formed by unidirectional infinite cylinders shown in Figure 5.1. It is assumed that the cylinder axis coincides with the x_3-axis and a section parallel to the plane (x_1,x_2) represents the fibrous composite.

Following the previous approach, we use the complex variable $z = x_1 + \mathrm{i}x_2$. Consider a domain $\mathcal{G}$ on the extended complex plane $\widehat{\mathbb{C}} = \mathbb{C} \cup \{\infty\}$ with the Lyapunov boundary $\partial\mathcal{G}$. The domain $\mathcal{G}$ can be unbounded, multiply connected, and even not connected. We will also discuss the continuous boundary $\partial\mathcal{G}$ consisting of a finite number of closed Lyapunov arcs.

In the present section, we introduce complex potentials for the local linear MEE equations in a plane domain $\mathcal{G}$ following Filshtinsky's approach [23, 74, 76].

5.2.1 LINEAR EQUATIONS OF MAGNETO-ELECTRO-ELASTICITY AND POTENTIALS

Let $\partial_j \equiv \frac{\partial}{\partial x_j}$ denote the partial derivative $(i, j = 1,2)$. Introduce the following designations for the local fields. Let σ_{ij} $(i, j = 1,2)$ denote the components of stress tensor, $\epsilon_{ij} = \frac{1}{2}(\partial_i u_j + \partial_j u_i)$ the strains expressed by means of the displacement U_j $(i, j = 1,2)$. The designations for the electric displacement and electric fields D_j and E_j $(j = 1,2)$ and for the magnetic induction and magnetic displacement fields B_j and H_j $(j = 1,2)$ are used following Chapter 1. The equilibrium differential equations of elastic, electric, and magnetic fields have the classic form

$$\partial_1\sigma_{11} + \partial_2\sigma_{12} = 0, \quad \partial_1\sigma_{12} + \partial_2\sigma_{22} = 0, \tag{5.1}$$

$$\partial_1 B_1 + \partial_2 B_2 = 0, \quad \partial_1 D_1 + \partial_2 D_2 = 0, \tag{5.2}$$

$$\partial_2 H_1 - \partial_1 H_2 = 0, \quad \partial_2 E_1 - \partial_1 E_2 = 0. \tag{5.3}$$

The compatibility condition for strains has the form

$$\partial_1^2\epsilon_{22}+\partial_2^2\epsilon_{11}=2\partial_1\partial_2\epsilon_{12}. \tag{5.4}$$

The linear constitutive equations relate the local fields

$$\begin{aligned}
\epsilon_{11} &= s_{11}\sigma_{11}+s_{12}\sigma_{22}+s_{16}\sigma_{12}+g_{11}D_1+g_{21}D_2+p_{11}B_1+p_{21}B_2\\
\epsilon_{22} &= s_{12}\sigma_{11}+s_{22}\sigma_{22}+s_{26}\sigma_{12}+g_{12}D_1+g_{22}D_2+p_{12}B_1+p_{22}B_2\\
2\epsilon_{12} &= s_{16}\sigma_{11}+s_{26}\sigma_{22}+s_{66}\sigma_{12}+g_{16}D_1+g_{26}D_2+p_{16}B_1+p_{26}B_2\\
E_1 &= -g_{11}\sigma_{11}-g_{12}\sigma_{22}-g_{16}\sigma_{12}+\beta_{11}D_1+\beta_{12}D_2+\nu_{11}B_1+\nu_{12}B_2\\
E_2 &= -g_{21}\sigma_{11}-g_{22}\sigma_{22}-g_{26}\sigma_{12}+\beta_{12}D_1+\beta_{22}D_2+\nu_{12}B_1+\nu_{22}B_2\\
H_1 &= -p_{11}\sigma_{11}-p_{12}\sigma_{22}-p_{16}\sigma_{12}+\nu_{11}D_1+\nu_{12}D_2+\chi_{11}B_1+\chi_{12}B_2\\
H_2 &= -p_{21}\sigma_{11}-p_{22}\sigma_{22}-p_{26}\sigma_{12}+\nu_{12}D_1+\nu_{22}D_2+\chi_{12}B_1+\chi_{22}B_2.
\end{aligned}$$

Here, the coefficients s_{ij} describe Hook's elastic law; g_{ij} and p_{ij} piezoelectric and piezomagnetic effects; β_{ij}, ν_{ij} and χ_{ij} dielectric, magnetic and electromagnetic effects.

The constitutive equations express linear interactions between the considered fields. They generalize the constitutive equations (1.43) for pure electromagnetic fields. Consider for definiteness the part of equations when the electric field $\mathbf{E}=(E_1,E_2)^\top$ depends only on the electric displacement $\mathbf{D}=(D_1,D_2)^\top$

$$\begin{cases} E_1=\beta_{11}D_1+\beta_{12}D_2\\ E_2=\beta_{12}D_1+\beta_{22}D_2\end{cases} \iff \mathbf{E}=\mathbf{b}\mathbf{D}, \tag{5.5}$$

where

$$\mathbf{b}=\begin{pmatrix}\beta_{11} & \beta_{12}\\ \beta_{12} & \beta_{22}\end{pmatrix}. \tag{5.6}$$

The second equation (5.5) is equivalent to the first equation (1.43) with $\boldsymbol{\varepsilon}=\mathbf{b}^{-1}$. The study of anisotropic dielectrics will be continued in Section 5.4.

In the case of locally isotropic materials $\beta_{11}=\beta_{22}$ and $\beta_{12}=0$, hence (5.5) becomes

$$E_1=\beta_{11}D_1,\ E_2=\beta_{11}D_2, \iff \mathbf{E}=\beta_{11}\mathbf{D}, \iff \mathbf{D}=\beta_{11}^{-1}\mathbf{E}. \tag{5.7}$$

5.2.2 COMPLEX REPRESENTATION OF FIELD VALUES

Using the elasticity equations (5.1) we introduce the standard Airy function $F_1(x_1,x_2)$ in such a way that

$$\sigma_{11}=\partial_2^2F_1, \quad \sigma_{12}=-\partial_1\partial_2F_1, \quad \sigma_{22}=\partial_1^2F_1. \tag{5.8}$$

The electric and magnetic potentials U and $\varphi^{(H)}$, respectively, satisfy the relations

$$E_1=-\partial_1U, \quad E_2=-\partial_2U, \quad H_1=-\partial_1\varphi^{(H)};\ H_2=-\partial_2\varphi^{(H)}. \tag{5.9}$$

Then, equations (5.3) are fulfilled identically. Along similar lines, the magnetic induction and magnetic displacement are expressed by means of the functions $F_2(x_1,x_2)$ and $F_3(x_1,x_2)$ as follows

$$D_1=\partial_2F_2, \quad D_2=-\partial_1F_2, \quad B_1=\partial_2F_3, \quad B_2=-\partial_1F_3. \tag{5.10}$$

Then, equations (5.2) are fulfilled.

Substitute (5.9) and (5.10) into the right part of the constitutive equations. As a result we obtain an expression of ϵ_{ij}, E_i and H_i in terms of the functions $F_j(x_1,x_2)$ not explicitly written here. Substituting these expressions into three equations (5.3) and (5.4) we arrive at the system of differential equations on $F_j(x_1,x_2)$

$$\sum_{j=1,2,3} L_{ij}(\partial_1,\partial_2)F_j = 0, \quad i = 1,2,3, \tag{5.11}$$

where the differential operators L_{ij} have the form

$$L_{11}(\partial_1,\partial_2) = s_{11}\partial_2^4 - 2s_{16}\partial_1\partial_2^3 + (2s_{12}+s_{66})\partial_1^2\partial_2^2 - 2s_{26}\partial_1^3\partial_2 + s_{22}\partial_1^4,$$

$$L_{12}(\partial_1,\partial_2) = g_{11}\partial_2^3 - g_{22}\partial_1^3 - (g_{21}+g_{16})\partial_1\partial_2^2 + (g_{12}+g_{26})\partial_1^2\partial_2,$$

$$L_{13}(\partial_1,\partial_2) = p_{11}\partial_2^3 - p_{22}\partial_1^3 - (p_{21}+p_{16})\partial_1\partial_2^2 + (p_{12}+p_{26})\partial_1^2\partial_2,$$

$$L_{22}(\partial_1,\partial_2) = -\beta_{11}\partial_2^2 - \beta_{22}\partial_1^2 + 2\beta_{12}\partial_1\partial_2,$$

$$L_{23}(\partial_1,\partial_2) = -\nu_{11}\partial_2^2 - \nu_{22}\partial_1^2 + 2\nu_{12}\partial_1\partial_2,$$

$$L_{33}(\partial_1,\partial_2) = -\chi_{11}\partial_2^2 - \chi_{22}\partial_1^2 + 2\chi_{12}\partial_1\partial_2.$$

We now proceed to derive a representation of $F_j(x_1,x_2)$ in terms of the complex potentials to identically fulfill the differential equations (5.11). Let $A = \{A_{ij}\}$ denote the matrix of cofactors to the 3×3 matrix $L = \{L_{ij}\}$. The components A_{1j} are explicitly written below

$$\begin{aligned}
A_{11}(\partial_1,\partial_2) &= (\beta_{22}\chi_{22} - \nu_{22}^2)\partial_1^4 + (\beta_{11}\chi_{22} - 2\nu_{11}\nu_{12} + \beta_{22}\chi_{11})\partial_1^2\partial_2^2 + \\
&(\beta_{11}\chi_{11} - \nu_{11}^2)\partial_2^4, \\
A_{12}(\partial_1,\partial_2) &= (g_{22}\chi_{22} + p_{22}\nu_{22})\partial_1^5 \\
&+(g_{22}\chi_{11} + p_{22}\nu_{11} + g_{62}\chi_{22} + p_{62}\nu_{22})\partial_1^3\partial_2^2 + (g_{62}\chi_{11} + p_{62}\nu_{11})\partial_1\partial_2^4, \\
A_{13}(\partial_1,\partial_2) &= -(g_{22}\nu_{22} + p_{22}\beta_{22})\partial_1^5 \\
&-(g_{22}\nu_{11} + p_{22}\beta_{11} + g_{62}\nu_{22} + p_{62}\beta_{22})\partial_1^3\partial_2^2 - (g_{62}\nu_{11} + p_{62}\beta_{11})\partial_1\partial_2^4.
\end{aligned} \tag{5.12}$$

We shall need only these components A_{1j}.

Let $\Psi(x_1,x_2)$ be a sufficiently smooth function. Substitute the expressions

$$F_j = A_{1j}\Psi \quad (j = 1,2,3) \tag{5.13}$$

into (5.11). The first equation (5.11) becomes

$$\sum_{j=1,2,3} L_{1j}A_{1j}\Psi \equiv \det(L)\,\Psi = 0, \tag{5.14}$$

where the Laplace expansion formula for the determinant $\det L$ is used. The second and third equations (5.11) are fulfilled identically because of Cramer's rule. One can consider the transformation of the system (5.11) to equation (5.14) as elimination of F_2, F_3 and the introduction of the new unknown function $F_1 = A_{11}\Psi$.

The differential equation (5.14) can be solved by the characteristic equation method [130] based on the representation

$$\Psi(x_1,x_2) = f(x_1+\mu x_2), \tag{5.15}$$

where $f(z)$ is sufficiently smooth function and μ is an undetermined constant. Substitution of (5.15) into (5.14) yields the algebraic equation of 8th power with real coefficients

$$\det L(1,\mu) = 0. \tag{5.16}$$

The positively defined matrices $\{s_{ij}\}$, $\{\beta_{ij}\}$ and $\{\chi_{ij}\}$ imply that the roots of (5.16) are complex [130]. Below, we consider only the case of different roots. Let us take four roots μ_k $(k = 1,2,3,4)$ having the positive imaginary parts and introduce the corresponding complex variables

$$z_k = x_1 + \mu_k x_2 \quad (k = 1,2,3,4). \tag{5.17}$$

Equation (5.17) can be considered as the $\mathbb{R}$-linear transformations of the complex plane

$$z_k = \frac{1-\mathrm{i}\mu_k}{2}z + \frac{1+\mathrm{i}\mu_k}{2}\overline{z} \quad (k = 1,2,3,4). \tag{5.18}$$

Remark 10 *If one of the values μ_k is equal to* i*, say for definiteness $\mu_1 = \mathrm{i}$, equation* (5.18) *becomes $z_1 = z$. This case does not simplify the study since all the other roots are not equal to* i *by the assumption. More precisely, they can be equal to* i*, but then, the representation* (5.13) *must be replaced by another, which holds in the theory of differential equations with multiple roots μ_k. The case $z_k = z$ $(k = 1,2,3,4)$ leads to Kolosov-Muskhelishvili formulas [192].*

Introduce the domain $\mathcal{G}_k = \{z_k = x_1 + \mu_k x_2 \in \mathbb{C} : (x_1,x_2) \in \mathcal{G}\}$. The general solution to the differential equation (5.14) has the form

$$\Psi(x_1,x_2) = 2\mathrm{Re} \sum_{k=1,2,3,4} c_k \varphi_k(z_k), \tag{5.19}$$

where c_k are complex constants, the function $\varphi_k(z_k)$ is analytic in the domain $\mathcal{G}_k$ in the variable z_k.

Let $\Phi_k = \frac{d^6\varphi_k}{dz_k^6}$ denote the sixth-order derivative of the function $\varphi_k(z_k)$. Substitution of (5.19), (5.13) into (5.8), (5.10) yields the complex representation of local fields

$$\begin{pmatrix} \sigma_{11} \\ \sigma_{12} \\ \sigma_{22} \end{pmatrix} = 2\mathrm{Re} \sum_{k=1}^{4} A_{11}(1,\mu_k)\Phi_k(z_k) \begin{pmatrix} \mu_k^2 \\ -\mu_k \\ 1 \end{pmatrix}, \tag{5.20}$$

$$\begin{pmatrix} D_1 \\ D_2 \end{pmatrix} = 2\mathrm{Re} \sum_{k=1}^{4} A_{12}(1,\mu_k)\Phi_k(z_k) \begin{pmatrix} \mu_k \\ -1 \end{pmatrix}, \tag{5.21}$$

$$\begin{pmatrix} B_1 \\ B_2 \end{pmatrix} = 2\mathrm{Re} \sum_{k=1}^{4} A_{13}(1,\mu_k)\Phi_k(z_k) \begin{pmatrix} \mu_k \\ -1 \end{pmatrix}. \tag{5.22}$$

Here, $A_{1j}(1,\mu_k)$ are polynomial expressions in μ_k. The left parts of (5.20)–(5.22) depend on $(x_1,x_2) \in \mathcal{G}$ through the complex variables (5.17) in the right parts.

Substitution of (5.20)–(5.22) into (5.5) yields

$$\begin{pmatrix} E_1 \\ E_2 \end{pmatrix} = 2\mathrm{Re} \sum_{k=1}^{4} \alpha_k^{(E)} \Phi_k(z_k) \begin{pmatrix} 1 \\ \mu_k \end{pmatrix} \tag{5.23}$$

and

$$\begin{pmatrix} H_1 \\ H_2 \end{pmatrix} = 2\mathrm{Re} \sum_{k=1}^{4} \alpha_k^{(H)} \Phi_k(z_k) \begin{pmatrix} 1 \\ \mu_k \end{pmatrix}, \tag{5.24}$$

where

$$\alpha_k^{(E)} = \mu_k[g_{16}A_{11}(1,\mu_k) + \beta_{11}A_{12}(1,\mu_k) + \nu_{11}A_{13}(1,\mu_k)] \tag{5.25}$$

and

$$\alpha_k^{(H)} = \mu_k[p_{16}A_{11}(1,\mu_k) + \nu_{11}A_{12}(1,\mu_k) + \chi_{11}A_{13}(1,\mu_k)]. \tag{5.26}$$

The coefficients on A_{1j} are taken from the constitutive equations (5.5). The strains are expressed by means of the complex potentials as follows

$$\partial_1 U_1 = \epsilon_{11} = 2\mathrm{Re} \sum_{k=1}^{4} r_k \Phi_k(z_k), \quad \partial_2 U_2 = \epsilon_{22} = 2\mathrm{Re} \sum_{k=1}^{4} \mu_k q_k \Phi_k(z_k), \tag{5.27}$$

where

$$r_k = (s_{11}\mu_k^2 + s_{12})A_{11}(1,\mu_k) - g_{21}A_{12}(1,\mu_k) - p_{21}A_{13}(1,\mu_k)$$

and

$$q_k = (s_{12}\mu_k + \frac{s_{22}}{\mu_k})A_{11}(1,\mu_k) - \frac{g_{22}}{\mu_k}A_{12}(1,\mu_k) - \frac{p_{22}}{\mu_k}A_{13}(1,\mu_k).$$

After integration of (5.27), we obtain the elastic displacement up to an additive constant vector

$$\begin{pmatrix} U_1 \\ U_2 \end{pmatrix} = 2\mathrm{Re} \sum_{k=1}^{4} \varphi_k(z_k) \begin{pmatrix} r_k \\ q_k \end{pmatrix}. \tag{5.28}$$

5.3 BOUNDARY VALUE PROBLEM

In the present section, we consider the problem with one hole or inclusion embedded in the MEE host material. It is assumed that the complex plane is divided by a simple closed Lyapunov curve Γ onto the simply connected bounded domain $\mathcal{G}^+$ and its complement $\mathcal{G} \equiv \mathcal{G}^-$ to the extended complex plane $\widehat{\mathbb{C}} = \mathbb{C} \cup \{\infty\}$.

The problem for a hole is stated in the following way. Let a MEE material occupy the domain $\mathcal{G}$. The domain $\mathcal{G}^+$ is considered as a hole. The field is given on the boundary of the hole $\Gamma = \partial\mathcal{G}^+$ oriented in the positive direction, i.e., $\mathcal{G}^+$ is on the left side of Γ. It is necessary to find the field in the domain $\mathcal{G}$. The boundary field on Γ can be given in various forms, as it is in the classic theories of separated electric, magnetic, and elastic fields. We consider one of the possible statements.

Let θ denote the angle between the x_1-axis and the normal outward unit vector $\mathbf{n} = (\cos\theta, \sin\theta)$ to Γ. Then, the arc length vector along Γ has the form $\mathbf{s} = (-\sin\theta, \cos\theta)$. The differential $\mathrm{d}z_k = \mathrm{d}x_1 + \mu_k \mathrm{d}x_2$ and the infinitesimal arc length ds on Γ are related by equation $\mathrm{d}z_k = g_k(\theta)ds$ where $g_k(\theta) = \mu_k\cos\theta - \sin\theta$. The function $g_k(\theta)$ can be written as a function in z and $\overline{z}$ as it is done for a circle in Section 5.5. The normal stresses vector applied to Γ is given by the formula

$$\begin{pmatrix} X_1 \\ X_2 \end{pmatrix} = \begin{pmatrix} \sigma_{11} & \sigma_{12} \\ \sigma_{12} & \sigma_{22} \end{pmatrix} \begin{pmatrix} \cos\theta \\ \sin\theta \end{pmatrix}. \tag{5.29}$$

Using (5.20) we obtain

$$\begin{pmatrix} X_1 \\ X_2 \end{pmatrix} = 2\mathrm{Re}\sum_{k=1}^{4} A_{11}(1,\mu_k) g_k \Phi_k(z_k) \begin{pmatrix} \mu_k \\ -1 \end{pmatrix}. \tag{5.30}$$

The normal components of the vectors (5.21)–(5.22) are determined by formulas

$$D_{\mathbf{n}} = 2\mathrm{Re}\sum_{k=1}^{4} A_{12}(1,\mu_k) g_k \Phi_k(z_k) \tag{5.31}$$

and

$$B_{\mathbf{n}} = 2\mathrm{Re}\sum_{k=1}^{4} A_{13}(1,\mu_k) g_k \Phi_k(z_k). \tag{5.32}$$

The boundary values of the stress vector and the scalars $D_{\mathbf{n}}$ and $B_{\mathbf{n}}$ on Γ can be given in the form of the vector-function $\mathbf{X} = (X_1, X_2, D_{\mathbf{n}}, B_{\mathbf{n}})^T$ depending on $(x_1, x_2) \in \Gamma$. Introduce the matrices

$$\mathbf{A} = \begin{pmatrix} \mu_1 A_{11}(1,\mu_1) & \mu_2 A_{11}(1,\mu_2) & \mu_3 A_{11}(1,\mu_3) & \mu_4 A_{11}(1,\mu_4) \\ -A_{11}(1,\mu_1) & -A_{11}(1,\mu_2) & -A_{11}(1,\mu_3) & -A_{11}(1,\mu_4) \\ A_{12}(1,\mu_1) & A_{12}(1,\mu_2) & A_{12}(1,\mu_3) & A_{12}(1,\mu_4) \\ A_{13}(1,\mu_1) & A_{13}(1,\mu_2) & A_{13}(1,\mu_3) & A_{13}(1,\mu_4) \end{pmatrix} \tag{5.33}$$

and

$$\mathbf{g} = \begin{pmatrix} g_1 & 0 & 0 & 0 \\ 0 & g_2 & 0 & 0 \\ 0 & 0 & g_3 & 0 \\ 0 & 0 & 0 & g_4 \end{pmatrix}. \tag{5.34}$$

Then, the boundary conditions on Γ can be written in the form of the vector-matrix Riemann-Hilbert problem with the shifts (5.18)

$$2\mathrm{Re}\,\mathbf{A}\mathbf{g}\,\boldsymbol{\Phi} \circ \boldsymbol{\gamma}(z) = \mathbf{X}(z), \quad z \in \Gamma, \tag{5.35}$$

where the matrix-functions $\mathbf{A}$ and $\mathbf{g}$ are defined on Γ. The shift $\boldsymbol{\gamma}(z)$ of the curve Γ is introduced in every component of $\boldsymbol{\Phi}$

$$\boldsymbol{\Phi} \circ \boldsymbol{\gamma}(z) := (\Phi_1(z_1), \Phi_2(z_2), \Phi_3(z_3), \Phi_4(z_4))^T. \tag{5.36}$$

Remark 11 *The Riemann-Hilbert problem* (5.35) *with shifts can be considered as a Riemann-Hilbert problem on a Riemann surface obtained from the complex plane by gluing together the corresponding curves through shifts. Such boundary value problems on Riemann surfaces can be solved by constructive methods developed in [9, 88, 89] and works cited therein.*

Assume that the stresses σ_{ij}^{∞}, the electric displacement D_i^{∞} and the magnetic induction B_i^{∞} are given at infinity. Substituting the infinite point into (5.20)–(5.22) we obtain seven real equations on four complex constants $C_k = \Phi_k(\infty)$ $(k = 1, 2, 3, 4)$

$$\begin{pmatrix} \sigma_{11}^{\infty} \\ \sigma_{12}^{\infty} \\ \sigma_{22}^{\infty} \end{pmatrix} = 2\text{Re} \sum_{k=1}^{4} A_{11}(1, \mu_k) C_k \begin{pmatrix} \mu_k^2 \\ -\mu_k \\ 1 \end{pmatrix},$$

$$\begin{pmatrix} D_1^{\infty} \\ D_2^{\infty} \end{pmatrix} = 2\text{Re} \sum_{k=1}^{4} A_{12}(1, \mu_k) C_k \begin{pmatrix} \mu_k \\ -1 \end{pmatrix}, \tag{5.37}$$

$$\begin{pmatrix} B_1^{\infty} \\ B_2^{\infty} \end{pmatrix} = 2\text{Re} \sum_{k=1}^{4} A_{13}(1, \mu_k) C_k \begin{pmatrix} \mu_k \\ -1 \end{pmatrix}.$$

One of the complex constant, say C_4, can be taken real, i.e., $\text{Im}\, C_4 = 0$. Therefore, the number of real unknowns in the system (5.37) holds seven. We do not investigate general equations on C_k in the present section and assume that a set of constants $\{C_k,\ k = 1, 2, 3, 4\}$ is determined. Anyway, the existence and uniqueness of the solution for the given field at infinity [175] implies the unique existence of the constants C_k satisfying the system (5.37).

5.4 DIELECTRICS

In the present section, we consider a particular case when all the effects vanish besides the electric displacement $\mathbf{D} = (D_1, D_2)^{\top}$ and electric field $\mathbf{E} = (E_1, E_2)^{\top}$. The following equations are considered in a simply connected domain $\mathcal{G}$

$$\partial_1 D_1 + \partial_2 D_2 = 0, \quad \partial_2 E_1 - \partial_1 E_2 = 0, \tag{5.38}$$

and

$$E_1 = \beta_{11} D_1 + \beta_{12} D_2, \quad E_2 = \beta_{12} D_1 + \beta_{22} D_2. \tag{5.39}$$

It is assumed for definiteness that infinity belongs to $\mathcal{G}$. The complement of the closure $\mathcal{G} \cup \Gamma$ to the extended complex plane is considered as a hole. We formally follow the general scheme described in the previous section in order to explicitly demonstrate Filshtinsky's method and to compare it to the classic method of complex potential in the considered simple case.

The Airy function $F \equiv F_2$ from equations (5.10) is introduced in such a way that

$$D_1 = \partial_2 F, \quad D_2 = -\partial_1 F. \tag{5.40}$$

Substitute (5.40) into (5.39)

$$E_1 = \beta_{11}\partial_2 F - \beta_{12}\partial_1 F, \quad E_2 = \beta_{12}\partial_2 F - \beta_{22}\partial_1 F. \tag{5.41}$$

Substituting these expressions into the second equation (5.38) and using the symmetry of the dielectric tensor we arrive at the differential equation

$$(\beta_{11}\partial_2^2 - 2\beta_{12}\partial_1\partial_2 + \beta_{22}\partial_1^2)F = 0. \tag{5.42}$$

The matrix operator (5.12) is introduced in order to reduce the system of differential equations (5.11) to the differential equation (5.14). This step is skipped for the differential equation (5.42). The function $F(x_1,x_2)$ is found from the representation

$$F(x_1,x_2) = f(x_1 + \mu x_2) \tag{5.43}$$

which yields the second order algebraic equation [130]

$$\beta_{11}\mu^2 - 2\beta_{12}\mu + \beta_{22} = 0. \tag{5.44}$$

The quadratic equation (5.44) has two complex conjugated roots since its discriminant $4(\beta_{12}^2 - \beta_{11}\beta_{22})$ is negative due to the positiveness of the dielectric tensor [130]. Therefore, the function (5.43) can be expressed through a function $\varphi(z_1)$ analytic in the variable $z_1 = x_1 + \mu x_2$

$$F(x_1,x_2) = 2\mathrm{Re}\,\varphi(z_1). \tag{5.45}$$

Here, for definiteness the root $\mu = \frac{1}{\beta_{11}}(\beta_{12} + \mathrm{i}\Delta_\beta)$ with the positive imaginary part is taken. The designation $\Delta_\beta = \sqrt{\beta_{11}\beta_{22} - \beta_{12}^2}$ is introduced for shortness.

The local fields in $\mathcal{G}$ are expressed in terms of the complex potential $\varphi(z_1)$. Let $z_1 = x + \mathrm{i}y$, where

$$x = x_1 + x_2\frac{\beta_{12}}{\beta_{11}},\ y = x_2\frac{\Delta_\beta}{\beta_{11}} \quad \Leftrightarrow \quad x_1 = x - \frac{\beta_{12}}{\Delta_\beta}y,\ x_2 = \frac{\beta_{11}}{\Delta_\beta}y. \tag{5.46}$$

Calculate the derivative of the analytic function $\varphi(z_1)$ through the partial derivatives of its real part $U(x,y) = \frac{1}{2}F(x_1,x_2)$

$$\varphi'(z_1) = \partial_x U - \mathrm{i}\partial_y U = \frac{1}{2}\partial_1 F - \frac{\mathrm{i}}{2\Delta_\beta}(-\beta_{12}\partial_1 F + \beta_{11}\partial_2 F). \tag{5.47}$$

This formula can be written in the form

$$\mathrm{Re}\,\varphi'(z_1) = \frac{1}{2}\partial_1 F, \quad \mathrm{Im}\,\varphi'(z_1) = \frac{1}{2\Delta_\beta}(\beta_{12}\partial_1 F - \beta_{11}\partial_2 F). \tag{5.48}$$

Express the derivatives $\partial_1 F$ and $\partial_2 F$ in terms of Re $\varphi'(z_1)$ and Im $\varphi'(z_1)$

$$\partial_1 F = 2\text{Re}\,\varphi'(z_1), \quad \partial_2 F = 2\left(\frac{\beta_{12}}{\beta_{11}}\text{Re}\,\varphi'(z_1) - \frac{\Delta_\beta}{\beta_{11}}\text{Im}\,\varphi'(z_1)\right) \tag{5.49}$$

The second equation (5.49) is equivalent to the following one

$$\partial_2 F = 2\text{Re}[\mu\varphi'(z_1)]. \tag{5.50}$$

Equations (5.49) and (5.40) imply that

$$D_1 = 2\left(\frac{\beta_{12}}{\beta_{11}}\text{Re}\,\varphi'(z_1) - \frac{\Delta_\beta}{\beta_{11}}\text{Im}\,\varphi'(z_1)\right), \; D_2 = -2\text{Re}\,\varphi'(z_1). \tag{5.51}$$

Equations (5.39) yield

$$E_1 = -2\Delta_\beta \text{Im}\,\varphi'(z_1), \quad E_2 = -\frac{2\Delta_\beta}{\beta_{11}}\left(\beta_{12}\text{Im}\,\varphi'(z_1) + \Delta_\beta \text{Re}\,\varphi'(z_1)\right). \tag{5.52}$$

One can check that the second equation (5.52) can be written in the form $E_2 = 2\Delta_\beta \text{Im}\,[\mu\varphi'(z_1)]$. Equations (5.51)–(5.52) express the components of dielectric field in terms of the complex potential $\varphi'(z_1)$.

Remark. *A potential ϕ can be used instead of the function F in the following way. Using* (5.39) *express the displacement* $(D_1, D_2)^\top$ *in terms of* $(E_1, E_2)^\top$

$$D_1 = \frac{1}{\Delta_\beta^2}(\beta_{22}E_1 - \beta_{12}E_2), \quad D_2 = \frac{1}{\Delta_\beta^2}(-\beta_{12}E_1 + \beta_{11}E_2). \tag{5.53}$$

Substitute the relations

$$E_1 = -\partial_1\phi, \quad E_2 = -\partial_2\phi \tag{5.54}$$

into (5.53). *Using the first equation* (5.38), *we arrive at the same differential equation* (5.42) *for ϕ. It follows from the theory of differential equations that a complex potential $\varphi^{(E)}$ real part of which coincides with ϕ is a linear combination of analytic functions in z_1 and $\overline{z_1}$, i.e., $\varphi^{(E)}(z_1) = C_1\varphi_1(z_1) + C_2\varphi_2(\overline{z_1})$.*

The domain $\mathcal{G}$ is mapped to the domain $\mathcal{G}_1$ by the transformation $z_1 = x_1 + \mu x_2$. Therefore, a boundary value problem of one complex variable z_1 arises. This essentially simplifies the constructive study of the problems. For instance, the boundary value problem for a hole with the prescribed normal component of the displacement can be written as the classic boundary value problem of complex analysis. Let Γ be oriented in the counterclockwise sense and $\mathbf{n} = (n_1, n_2)$ denote the outward unit vector to Γ and $\mathbf{s} = (n_2, -n_1)$ the tangent vector to Γ. We have

$$D_n = -\mathbf{D}\cdot\mathbf{n} = \mathbf{s}\cdot\nabla F \quad \text{on } \Gamma. \tag{5.55}$$

Integrating (5.55) along Γ and using (5.45), we arive at the boundary value problem

$$2\text{Re}\,\varphi(z_1) = \int D_n\, ds, \quad z_1 \in \Gamma', \tag{5.56}$$

where Γ' is obtained from Γ by the shift $z_1 = x_1 + \mu x_2$. The indefinite integral in (5.56) contains an arbitrary additive constant.

Consider the problem with $\mathbf{D_n} = 0$ and the normalized electric field prescribed at infinity $\mathbf{E}_0 = (\cos\theta_0, \sin\theta_0)^\top \equiv \exp(\mathrm{i}\theta_0)$. The value $\varphi'(\infty)$ is determined by means of (5.52)

$$\varphi'(\infty) = \frac{1}{2\Delta_\beta^2}[(\beta_{12}+\mathrm{i})\cos\theta_0 - \beta_{11}\sin\theta_0]. \tag{5.57}$$

The condition (5.57) is equivalent to the Laurent series at infinity

$$\varphi(z_1) = \frac{z_1}{2\Delta_\beta^2}[(\beta_{12}+\mathrm{i})\cos\theta_0 - \beta_{11}\sin\theta_0] + \alpha_0 + \frac{\alpha_1}{z_1} + \ldots. \tag{5.58}$$

The boundary condition (5.56) becomes

$$\operatorname{Re}\varphi(z_1) = 0, \quad z_1 \in \Gamma'. \tag{5.59}$$

The arbitrary constant is fixed as zero without loss of generality since a complex potential is defined up to an additive constant. Thus, we arrive at the classic boundary value problem of the analytic function theory [79]. Let a simply connected domain $\mathcal{G}'$ contain infinity and be bounded by a smooth curve Γ'. It is necessary to find a function $\varphi(z_1)$ analytic in $\mathcal{G}'$ and continuously differentiable in its closure except at infinity where $\varphi(z_1)$ has a simple pole described by (5.58) with the boundary condition (5.59). The problem (5.58)–(5.59) can be solved in terms of the conformal mapping of $\mathcal{G}'$ onto a disk. Section 5.5 contains an example of the general MEE medium with a circular hole. The general vector-matrix problem for a circular hole is decomposed into scalar problems such as (5.58)–(5.59).

Consider now another boundary condition when instead of $\mathbf{D_n}$, the tangent electric field E_s is given on the boundary of the hole

$$\mathbf{E_s} = \mathbf{E}\cdot\mathbf{s} \quad \text{on } \Gamma. \tag{5.60}$$

The method of solution can be based on the potential ϕ defined by (5.54) instead of the Airy function F. Such an approach repeats the above method.

The situation becomes complicated when both fields $\mathbf{E}$ and $\mathbf{D}$ have to be involved in the boundary condition. Such a problem arises when, instead of the hole, we consider an inclusion occupied by another medium. The corresponding boundary conditions (1.51) are written in Section 1.4 and will be written now in a slightly changed designation for anisotropic phases.

Let $\mathcal{G}^+$ denote the inclusion domain and $\mathcal{G}^- \equiv \mathcal{G}$ the host domain. We use the above presentations for the local fields and potentials, adding the superscripts $^\pm$ to indicate the domain under consideration. Let the contact between $\mathcal{G}^+$ and $\mathcal{G}^-$ be perfect. Then, the conditions (1.51) hold. According to the designations used in the present Chapter, we write

$$\mathbf{E_s^+} = \mathbf{E_s^-}, \quad \mathbf{D_n^+} = \mathbf{D_n^-} \quad \text{on } \Gamma. \tag{5.61}$$

Following (5.55) we write the second equality (5.61) in the form

$$\mathbf{s}\cdot\nabla F^{+} = \mathbf{s}\cdot\nabla F^{-} \quad \text{on } \Gamma. \tag{5.62}$$

The Airy functions $F^{\pm}$ in $\mathcal{G}^{\pm}$ are expressed by means of complex potentials. Using (5.49) we write (5.62) in terms of the derivatives of complex potentials

$$\left(s_1 + s_2\frac{\beta_{12}^{+}}{\beta_{11}^{+}}\right)\text{Re}\,\varphi'^{+}(z_1^{+}) - s_2\frac{\Delta_{\beta^{+}}}{\beta_{11}^{+}}\text{Im}\,\varphi'^{+}(z_1^{+}) = \tag{5.63}$$
$$\left(s_1 + s_2\frac{\beta_{12}^{-}}{\beta_{11}^{-}}\right)\text{Re}\,\varphi'^{-}(z_1^{-}) - s_2\frac{\Delta_{\beta^{-}}}{\beta_{11}^{-}}\text{Im}\,\varphi'^{-}(z_1^{-}) \text{ on } \Gamma.$$

One can integrate equation (5.62) along Γ and write it in terms of the complex potentials

$$\text{Re}\,\varphi^{+}(z_1^{+}) = \text{Re}\,\varphi^{-}(z_1^{-}) \quad \text{on } \Gamma. \tag{5.64}$$

Here, $(x_1,x_2)\in\Gamma$, the transformations $z_1^{+} = x_1 + \mu^{+}x_2$ and $z_1^{-} = x_1 + \mu^{-}x_2$ map the curves Γ onto Γ^{+} and Γ^{-}, respectively. The curves Γ^{+} and $(-\Gamma^{-})$ are oriented in the counterclockwise sense and bound the domains $\mathcal{G}'^{+}$ and $\mathcal{G}'^{-}$. One can consider the transformation of the equation (5.64) into (5.63) as the differentiation along the curve Γ.

Consider now the first relation (5.61) in the form $\mathbf{s}\cdot\mathbf{E}^{+} = \mathbf{s}\cdot\mathbf{E}^{-}$ on Γ. Using (5.52) we can write it as follows

$$\Delta_{\beta^{+}}\left[\left(s_1 + s_2\frac{\beta_{12}^{+}}{\beta_{11}^{+}}\right)\text{Im}\,\varphi'^{+}(z_1^{+}) + s_2\frac{\Delta_{\beta^{+}}}{\beta_{11}^{+}}\text{Re}\,\varphi'^{+}(z_1^{+})\right] = \tag{5.65}$$
$$\Delta_{\beta^{-}}\left[\left(s_1 + s_2\frac{\beta_{12}^{-}}{\beta_{11}^{-}}\right)\text{Im}\,\varphi'^{-}(z_1^{-}) + s_2\frac{\Delta_{\beta^{-}}}{\beta_{11}^{-}}\text{Re}\,\varphi'^{-}(z_1^{-})\right] \text{ on } \Gamma.$$

Integration of (5.65) along Γ yields

$$\text{Im}\,\varphi^{+}(z_1^{+}) = \gamma\,\text{Im}\,\varphi^{-}(z_1^{-}) \quad \text{on } \Gamma, \tag{5.66}$$

where $\gamma = \frac{\Delta_{\beta^{-}}}{\Delta_{\beta^{+}}}$. The transformation of (5.66) into (5.65) can be considered as the differentiation along the curve Γ.

Multiply (5.66) by i and add (5.64). As a result we obtain one complex equation equivalent to two real equations (5.64) and (5.66)

$$\varphi^{+}[z_1^{+}(t)] = \frac{\gamma+1}{2}\varphi^{-}[z_1^{-}(t)] - \frac{\gamma-1}{2}\overline{\varphi^{-}[z_1^{-}(t)]}, \quad t\in\Gamma. \tag{5.67}$$

Following [185] and [93, Chapter 2] the condition (5.67) in the particular case $z_1^{+}(t)\equiv z_1^{-}(t)\equiv t$ is called by the $\mathbb{R}$-linear conjugation problem. In the case with the shifts defined by (5.18), i.e. by the $\mathbb{R}$-linear shifts

$$z_1^{\pm}(t) = \frac{1-\mathrm{i}\mu^{\pm}}{2}t + \frac{1+\mathrm{i}\mu^{\pm}}{2}\bar{t} \tag{5.68}$$

we call it by the $\mathbb{RR}$-linear problem. The exact solution of this problem in the case of ellipse Γ follows from [111, Chapter 8].

We have $\mu^{\pm} = \mathrm{i}$ in the case of locally isotropic constitutes. Then, the shifts become identical transformations, and we arrive at the classic $\mathbb{R}$-linear problem [185] and [93, Chapter 2]. The method of functional equations and the generalized alternating method of Schwarz were applied in the above books, and works cited therein to solve the problem for locally isotropic media. In all other cases, including $\mu^+ = \mu^-$, the method of integral equations due to Filshtinsky was developed in [200].

5.5 CIRCULAR HOLE

We now consider the case when Γ is the unit circle, a particular case of the elliptic inclusions considered by Hwu [111, Chapter 8]. It is useful to understand the structure of the boundary value problem in this simple case. We now fix the first coordinate of (5.36) and write it in the extended form

$$2\mathrm{Re}[a_{11}g_1\Phi_1(z_1)+a_{12}g_2\Phi_2(z_2)+a_{13}g_3\Phi_3(z_3)+a_{14}g_4\Phi_4(z_4)] = X_1(z),\ z\in\Gamma, \tag{5.69}$$

where a_{1j} are the elements of the matrix (5.33). Using the parametric equation $z = \exp(\mathrm{i}\theta)$ of the unit circle we write the function $g_k(\theta)$ as the function of z and $\overline{z}$

$$g_k(z) = \frac{\mu_k+\mathrm{i}}{2}z+\frac{\mu_k-\mathrm{i}}{2}\overline{z} \quad (k=1,2,3,4). \tag{5.70}$$

Using the relations $\overline{z} = z^{-1}$ on the unit circle, (5.70) and (5.17) we can write the boundary condition (5.69) in the form

$$\mathrm{Re}\sum_{k=1}^{4} a_{1k}\left((\mu_k+\mathrm{i})z+\frac{\mu_k-\mathrm{i}}{z}\right)\Phi_k\left[\gamma_k(z)\right] = X_1(z),\ |z|=1, \tag{5.71}$$

where

$$z_k = \gamma_k(z) \equiv \frac{1-\mathrm{i}\mu_k}{2}z+\frac{1+\mathrm{i}\mu_k}{2z}. \tag{5.72}$$

The shift (5.72) conformally transforms the domain $|z|>1$ onto the domain $\mathcal{G}_k$, the exterior of the ellipse defined by the parametric equations on the plane $z_k = x_k + \mathrm{i}y_k$

$$x_k = \cos\theta - \mathrm{Im}\,\mu_k\sin\theta,\quad y_k = \mathrm{Re}\,\mu_k\sin\theta \quad (0\le\theta<2\pi). \tag{5.73}$$

The infinite point $z=\infty$ is mapped onto infinity. The ellipse degenerates to the slit $(-1,1)$ on the real axis, when $\mathrm{Re}\,\mu_k = 0$.

Introduce the function analytic in $|z|>1$ and continuously differentiable in $|z|\ge 1$

$$\Omega_1(z) = \sum_{k=1}^{4} a_{1k}\left((\mu_k+\mathrm{i})z+\frac{\mu_k-\mathrm{i}}{z}\right)\Phi_k\left[\gamma_k(z)\right]-w_1z, \tag{5.74}$$

where $w_1 = \sum_{k=1}^{4} a_{1k}(\mu_k + \mathrm{i})C_k$, the constants C_k are found from the system (5.37). The boundary condition (5.71) can be written in the form

$$\mathrm{Re}\left[\overline{\Omega_1\left(\frac{1}{\overline{z}}\right)} + w_1 z\right] = X_1(z),\ |z| = 1, \tag{5.75}$$

where the function in brackets is analytic in the unit disk. The Schwarz operator for the unit disk [79], [185, Section 2.7.1] solves the problem (5.75). After application of the transformation $z \longmapsto \frac{1}{\overline{z}}$ and the complex conjugation we get

$$\Omega_1(z) = -\frac{\overline{w_1}}{z} - \frac{1}{2\pi \mathrm{i}} \int_{|\zeta|=1} X_1(\zeta) \frac{\zeta + z}{\zeta - z} \frac{d\zeta}{\zeta},\ |z| > 1. \tag{5.76}$$

The same arguments can be applied to other coordinates of the boundary value problem (5.36). We obtain

$$\Omega_m(z) = -\frac{\overline{w_m}}{z} - \frac{1}{2\pi \mathrm{i}} \int_{|\zeta|=1} X_m(\zeta) \frac{\zeta + z}{\zeta - z} \frac{d\zeta}{\zeta},\ |z| > 1, \tag{5.77}$$

where

$$w_m = \sum_{k=1}^{4} a_{mk}(\mu_k + \mathrm{i})C_k \quad (m = 1,2,3,4). \tag{5.78}$$

The complex potentials $\Phi_k(z_k)$ satisfy the system of linear algebraic equations

$$\sum_{k=1}^{4} a_{mk}\left((\mu_k + \mathrm{i})z + \frac{\mu_k - \mathrm{i}}{z}\right)\Phi_k(z_k) = \Omega_m(z) + w_m z\ (m = 1,2,3,4). \tag{5.79}$$

After its solution, the local fields are calculated by the formulas (5.20)–(5.22).

Consider the case when the forces on the boundary $|z| = 1$ are absent and $B_{\mathbf{n}} = D_{\mathbf{n}} = 0$. This implies that $X_m(\zeta) = 0$ $(m = 1,2,3,4)$, hence,

$$\Omega_m(z) = -\frac{\overline{w_m}}{z}. \tag{5.80}$$

We must find the constants C_k from the system of equations (5.37). Let these constants be found. Calculate w_m by (5.78) and substitute (5.80) in (5.79)

$$\sum_{k=1}^{4} a_{mk}\left((\mu_k + \mathrm{i})z + \frac{\mu_k - \mathrm{i}}{z}\right)\Phi_k(z_k) = w_m z - \frac{\overline{w_m}}{z}\ (m = 1,2,3,4). \tag{5.81}$$

It is convenient to write (5.81) the result in the vector-matrix form

$$\mathbf{A}\mathbf{x} = \mathbf{w}, \tag{5.82}$$

where $\mathbf{A} = \{a_{mk}\}$, the vector $\mathbf{x}$ has the coordinates $\left((\mu_k + \mathrm{i})z + \frac{\mu_k - \mathrm{i}}{z}\right)\Phi_k(z_k)$ and the vector $\mathbf{w}$ the coordinates $w_m z - \frac{\overline{w_m}}{z}$. Let $\mathbf{x} = \mathbf{A}^{-1}\mathbf{w}$ be the solution of the system (5.82) consisting of four linear algebraic equations. Then, we obtain the expressions

$$\left((\mu_k + \mathrm{i})z + \frac{\mu_k - \mathrm{i}}{z}\right)\Phi_k(z_k) = (\mathbf{A}^{-1}\mathbf{w})_k(z) \quad k = 1,2,3,4, \tag{5.83}$$

where $(\mathbf{A}^{-1}\mathbf{w})_k(z)$ denotes the kth coordinate of the vector $\mathbf{A}^{-1}\mathbf{w}(z)$. Ultimately, the formula (5.83) yields

$$\Phi_k(z_k) = \frac{(\mathbf{A}^{-1}\mathbf{w})_k(z)}{(\mu_k + \mathrm{i})z + \frac{\mu_k - \mathrm{i}}{z}}, \quad k = 1,2,3,4, \tag{5.84}$$

where $z \in \mathcal{G}$ is expressed in terms of $z_k \in \mathcal{G}_k$ by the function inverse to (5.72)

$$z = \frac{z_k - \sqrt{z_k^2 - \mu_k^2 - 1}}{1 - \mathrm{i}\mu_k} \quad (k = 1,2,3,4). \tag{5.85}$$

The branch of the square root in (5.85) is selected in the complex plane with the excluded cut connecting the points $z_k = \pm\sqrt{\mu_k^2 + 1}$. The local fields are calculated by (5.20)–(5.24).

5.6 TRANSVERSELY ISOTROPIC PIEZOELECTRIC FIBROUS COMPOSITES

The complex potentials are constructed above for 2D locally anisotropic MEE media. The locally isotropic media correspond to the characteristic equation (5.16) with the same roots $\mu_k = \mathrm{i}$. The transformations (5.18) become identical $z_k = z$. The representations (5.21)–(5.22) are simplified, but (5.20) becomes more complicated. More precisely, the representation (5.20) has to be replaced with the Kolosov-Muskhelishvii formulas [192]. Such a change is similar to the different representations of fundamental solutions of linear differential equations with constant coefficients for distinct and coinciding eigenvalues, as noted in Remark 10.

From now, only locally isotropic fibrous composites will be considered in this chapter. More precisely, we restrict ourselves by piezoelectric fibrous composites in a stationary electromagnetic field when the electric charge induces elastic stresses and deformations and vice versa. The geometry of composites can be described by two-dimensional geometry in a section perpendicular to the unidirectional fibers parallel to the x_3–direction as displayed in Figure 5.1. It is additionally assumed that all the components of the considered fiber composite are locally isotropic, and the composite is transversely isotropic on the macroscale.

Different methods have been applied to estimate the effective constants of such piezoelectric fibrous composites. The theory of homogenization of piezoelectric composites and exact formulas for the layered piezoelectric composites were described in the series of the works [29, 81–83, 85, 234]. Milgrom and Shtrikman [151] studied the effective properties of polycrystals made of uniaxial crystals. Schulgasser [216] established exact relations for overall moduli of piezoelectric composites consisting of two and many transversely isotropic phases.

Duality transformations were systematically applied in [8] to vector–matrix problems. The authors imposed conditions on the statistical properties of the local quantities to deduce exact explicit and implicit formulas when continuous local laws describe media. These conditions and formulas are analogous to Keller-Matheron-Dykhne [69, 116, 137] ones deduced for scalar problems. It is worth noting that formulas obtained in the framework of the two–dimensional theory of duality transformations [8, 69, 137] are exact, and formulas deduced are approximate and hold only for dilute and weakly inhomogeneous composites. The theory of the self-consistent methods was systematically discussed by Kananun and Levin [115]; its limitations were analyzed in [169, 179] and in the previous chapter. A method of integral equations for doubly periodic problems, including anisotropic piezoelectric composites, was developed in [24], [97], [75]. A rigorous mathematical study of MEE boundary value problems based on the theory of singular integral equations was presented in [46, 47, 189].

Bravo-Castillero et al. [43] applied the method of Filshtinsky developed in 1964-1970 to the piezoelectric square array of circular inclusions. This paper and others are presented as the benchmark of the series of results concerning the "closed-form analytical formulas" for the effective properties. By *closed-form analytical formulas*, these authors mean the effective constants presented in a form that explicitly contains some values. These values can be numerically computed for fixed, regular arrays by the truncation of infinite systems of linear algebraic equations following the method of Filshtinsky.

Here, following the lines of the footnote on page 20, we stress that the terms *exact formulas* and *approximate analytical formulas* are used in the present book in a commonly accepted meaning [17]. In particular, the effective piezoelectric constants will given in such a form that only the effective constants are in the left part of the formula, and the right part includes all the given geometrical and physical parameters explicitly in symbolic form, for instance, centers and radii of inclusions. Contrary to [43], we have no parameter in the right parts of our formulas that a hidden numerical procedure should be separately used.

The paper [43] outlines the previous results for transversely isotropic fibrous composites. Namely, it is sufficient to solve three basic problems: two 2D pure elastic problems and one coupled piezoelectric anti-plane problem. We refer to analytical approximate formulas for the elastic constants in [67]. The piezoelectric anti-plane problem has been solved in [175, 211].

The piezoelectric anti-plane problem was first written as the vector-matrix $\mathbb{R}$–linear problem in a class of doubly periodic functions. Applying the zero-order approximation yields a vector–matrix extension of the famous Clausius–Mossotti approximation (5.147). This result agrees with [8] and [115] obtained in another form. The vector–matrix problem is decomposed onto two scalar $\mathbb{R}$–linear problem. This reduction allows us to directly apply all the exact and approximate analytical formulas obtained before for scalar problems. In particular, coupled anti-plane problems presented in [100] are reduced to scalar problems. This implies that truncated systems of order N^2 for the numerical solution used in [100] are reduced to systems of order N by a simple linear transformation.

5.7 LOCAL EQUATIONS AND COMPLEX POTENTIALS

Anti-plane problem is stated for fibers parallel to the x_3–axis when the elastic and electric forces do not depend on x_3; the stress tensor has only the non-zero components σ_{13} and σ_{23}, the displacement has the form $\mathbf{u} = (0,0,u_3)$.

Let a smooth oriented curve L locally divide the domains $\mathcal{G}^+$ and $\mathcal{G}^-$ on the plane (x_1,x_2) occupied by different materials. The electric field strength $\mathbf{E}(x_1,x_2)$ in $\mathcal{G}^+$ and $\mathcal{G}^-$ is expressed through the electrostatic potential by equations (5.54). The electric displacement vector and the stresses satisfy

$$\frac{\partial D_1}{\partial x_1} + \frac{\partial D_2}{\partial x_2} = 0, \quad \frac{\partial \sigma_{13}}{\partial x_1} + \frac{\partial \sigma_{23}}{\partial x_2} = 0. \tag{5.86}$$

The combined effect of the anti-plane deformation and the electric fields in each section of the fibrous composite follows from the coupled equations (5.5)

$$\begin{aligned} \sigma_{i3} &= c_{44}\frac{\partial u_3}{\partial x_i} - d_{12}E_i, \\ D_i &= d_{12}\frac{\partial u_3}{\partial x_i} + \epsilon_{11}E_i \quad (i = 1,2). \end{aligned} \tag{5.87}$$

where d_{12} stands for the piezoelectric modulus, ϵ_{11} the permittivity and c_{44} the shear modulus. Equations (5.87) correspond to the inverse general equations (5.5) with some vanishing coefficients.

These equations imply that the functions u_3 and ϕ from (5.54) are harmonic in the domains $\mathcal{G}^+$ and $\mathcal{G}^-$. Therefore, they can be expressed in terms of the complex potentials

$$u_3(z) = \text{Re}\,\varphi_1(z), \; \phi(z) = \text{Re}\,\varphi_2(z), \tag{5.88}$$

The functions $\varphi_1(z)$ and $\varphi_2(z)$ are analytic in $\mathcal{G}^\pm$ and continuously differentiable in the closures of the considered domains. The following representations take place in the closures $\mathcal{G}^\pm \cup L$ of the considered domains

$$\sigma_{13} - i\sigma_{23} = \sum_{j=1,2} d_{1j}\varphi_j'(z), \; E_1 - iE_2 = -\varphi_2'(z)\;, D_1 - iD_2 = \sum_{j=1,2} d_{2j}\varphi_j'(z), \tag{5.89}$$

where $d_{11} = c_{44}$, $d_{22} = -\epsilon_{11}$.

The main force vector $P_1^{(AB)}$ and the electric flux $P_2^{(AB)}$ along an arc AB have the form [24]

$$P_1^{(AB)} = \int_{AB} \sigma_{\mathbf{n}3}ds = -\text{Im} \sum_{j=1,2} d_{1j}\varphi_j(z)|_A^B, \tag{5.90}$$

$$P_2^{(AB)} = \int_{AB} D_{\mathbf{n}}ds = -\text{Im} \sum_{j=1,2} d_{2j}\varphi_j(z))|_A^B,$$

where $\mathbf{n} = (n_1,n_2)$ is the unit vector to AB, $\sigma_{\mathbf{n}3} = \sigma_{13}n_1 + \sigma_{23}n_2$ and $D_{\mathbf{n}} = D_1n_1 + D_2n_2$.

5.8 PERFECT CONTACT AND $\mathbb{R}$–LINEAR PROBLEM

Let the mechanical and electric contact between media that occupy the domains $\mathcal{G}^+$ and $\mathcal{G}^-$ be perfect. The conditions (5.61) give the ideal electric contact. The perfect mechanical contact means that the normal limit stresses and the displacements from both sides of L coincide:

$$\sigma_{13}^+ n_1 + \sigma_{23}^+ n_2 = \sigma_{13}^- n_1 + \sigma_{23}^- n_2, \quad u_3^+ = u_3^- \text{ on } L. \tag{5.91}$$

Here, $\mathbf{n} = (n_1, n_2)$ is the outward normal vector to L.

Using the complex potentials we write (5.91) and (5.61) in the form

$$\text{Im} \sum_{j=1,2} d_{1j}^+ \varphi_j^+(t) = \text{Im} \sum_{j=1,2} d_{1j}^- \varphi_j^-(t), \tag{5.92}$$

$$\text{Re } \varphi_1^+(t) = \text{Re } \varphi_1^-(t), \ t \in L \tag{5.93}$$

and

$$\text{Im} \sum_{j=1,2} d_{2j}^+ \varphi_j^+(t) = \text{Im} \sum_{j=1,2} d_{2j}^- \varphi_j^-(t), \tag{5.94}$$

$$\text{Re } \varphi_2^+(t) = \text{Re } \varphi_2^-(t), \ t \in L, \tag{5.95}$$

where

$$\varphi_j^+(t) = \lim_{t \to z \in G^+} \varphi_j^+(z), \quad \varphi_j^-(t) = \lim_{t \to z \in \mathcal{G}^-} \varphi_j^-(z). \tag{5.96}$$

Equations (5.92), (5.94) and (5.95) are obtained from the first equation (5.91) and equations (5.61) by integration along L [185].

Introduce the vector–functions

$$\Phi^\pm(z) = \begin{pmatrix} \varphi_1^\pm(z) \\ \varphi_2^\pm(z) \end{pmatrix}. \tag{5.97}$$

Then two real relations (5.93) and (5.95) become

$$\text{Re } \Phi^+(t) = \text{Re } \Phi^-(t), \ t \in L. \tag{5.98}$$

The conditions (5.92) and (5.94) can be written in the form

$$\text{Im } \mathbf{d}_+ \Phi^+(t) = \text{Im } \mathbf{d}_- \Phi^-(t), \ t \in L, \tag{5.99}$$

where

$$\mathbf{d}_\pm \equiv \begin{pmatrix} d_{11}^\pm & d_{12}^\pm \\ d_{12}^\pm & d_{22}^\pm \end{pmatrix} = \begin{pmatrix} c_{44}^\pm & d_{12}^\pm \\ d_{12}^\pm & -\epsilon_{11}^\pm \end{pmatrix}. \tag{5.100}$$

Two real vector–matrix conditions (5.98) and (5.99) can be cast in the complex vector form

$$\Phi^+(t) = \frac{1}{2}\mathbf{d}_+^{-1}(\mathbf{d}_- + \mathbf{d}_+)\Phi^-(t) - \frac{1}{2}\mathbf{d}_+^{-1}(\mathbf{d}_- - \mathbf{d}_+)\overline{\Phi^-(t)}, \ t \in L. \tag{5.101}$$

The matrix $\mathbf{d}_+$ is invertible since its determinant $-(c_{44}\epsilon_{11}+d_{12}^2)$ is negative.

It is convenient to introduce the vector–function

$$\Omega^-(z) = \frac{1}{2}\mathbf{d}_+^{-1}(\mathbf{d}_- + \mathbf{d}_+)\Phi^-(z), \quad z \in \mathcal{G}^- \tag{5.102}$$

and the normalized matrix

$$\mathbf{R} = \mathbf{d}_+^{-1}(\mathbf{d}_- - \mathbf{d}_+)(\mathbf{d}_- + \mathbf{d}_+)^{-1}\mathbf{d}_+ = -(\mathbf{I} - \mathbf{D}_+^{-1}\mathbf{d}_-)(\mathbf{I} + \mathbf{D}_+^{-1}\mathbf{d}_-)^{-1}. \tag{5.103}$$

Then, (5.101) takes the form of the following $\mathbb{R}$–linear conjugation condition

$$\Phi^+(t) = \Omega^-(t) - \mathbf{R}\,\overline{\Omega^-(t)},\ t \in L. \tag{5.104}$$

Introduce the vector–functions

$$\Psi^+(z) = \frac{d}{dz}\Phi^+(z),\ z \in G^+, \quad \Psi^-(z) = \frac{d}{dz}\Omega^-(z),\ z \in \mathcal{G}^-. \tag{5.105}$$

Differentiate the conditions (5.104) along the curve L on the tangent vector $\mathbf{s}$. Then (5.104) becomes (see analogous scalar manipulations)

$$\Psi^+(t) = \Psi^-(t) + \overline{n(t)}^2\,\mathbf{R}\,\overline{\Psi^-(t)},\ t \in L. \tag{5.106}$$

Here, the unit outward normal vector $\mathbf{n} = (n_1, n_2)$ is presented as the complex value $n(t) = n_1 + in_2$ when t belongs to L.

Equations (5.90) can be written in the vector–matrix form

$$\mathbf{P}^{(AB)} = \begin{pmatrix} P_1^{(AB)} \\ P_2^{(AB)} \end{pmatrix} = -\mathrm{Im}\,\mathbf{d}_+\Phi^+(z)|_A^B, \tag{5.107}$$

where for definiteness, it is assumed that the arc AB lies in $\mathcal{G}^+$.

5.9 DOUBLE PERIODIC STATEMENT

We consider a two–dimensional lattice Q defined by the zero-th periodicity cell $Q \equiv Q_{(0,0)}$ described on page 29. Let mutually disjoint domains G_k $(k = 1,2,\ldots,N)$ are located in $Q_{(0,0)}$. The union of all the inclusions forms the non-connected domain $G^- = \cup_{k=1}^N G_k$. The multiply connected domain G^+ is defined as the complement of $G^- \cup \partial G^-$ to the open parallelogram $Q_{(0,0)}$, i.e., $G^+ = Q_{(0,0)} \setminus \cup_{k=1}^N (G^- \cup \partial G^-)$, where ∂G^- denotes the boundary of G^-.

We study the electrical and elastic fields in the doubly periodic composites when the domains G^+ and G^- are occupied by piezoelectric materials whose properties are described by the matrices (5.100). It is convenient to introduce the complex potentials $\Omega_k(z) = \Omega(z)$ and $\Psi_k(z) = \Psi(z)$ in G_k $(k = 1,2,\ldots,N)$. The complex potential $\Psi_0(z) \equiv \Psi^+(z)$ is analytic in $\mathcal{G}^+$ and doubly periodic, i.e., $\Psi_0(z) = \Psi_0(z + m_1\omega_1 + m_2\omega_2)$ for all $m_1, m_2 \in \mathbb{Z}$. The complex potential $\Phi_0(z) \equiv \Phi^+(z)$ is quasiperiodic, i.e.,

$$\Phi_0(z+\omega_j) - \Phi_0(z) = C_j \quad (j = 1,2), \tag{5.108}$$

where $C_j = (C_{1j}, C_{2j})^T$ are constant vectors which model the external piezoelectric field applied to the composite in the following way. Let A and B from (5.90) have the complex coordinates z and $z+\omega_j$, respectively. Then, (5.108) yields

$$P_{\ell j} := P_\ell^{(z,z+\omega_j)} = -\text{Im} \sum_{m=1,2} d_{\ell m}^+ C_{mj} \ (\ell, j = 1,2). \tag{5.109}$$

It can also be written in the form

$$\mathbf{P}_j^{(AB)} = -\text{Im}\, \mathbf{d}_+ C_j \quad (j = 1,2), \tag{5.110}$$

where (5.107) is used. The constants $P_{\ell j}$ can be expressed in terms of the average stresses and induction. First, introduce the average value of a doubly periodic function $f(x_1, x_2)$ over the periodicity cell $Q_{(0,0)}$ as the double integral

$$\langle f \rangle = \int_{Q_{(0,0)}} f(x_1, x_2) \mathrm{d}x_1 \mathrm{d}x_2, \tag{5.111}$$

where the area $|Q_{(0,0)}| = \omega_1 \text{Im}\, \omega_2 = 1$. Periodicity of $f(x_1, x_2)$ implies that the integral (5.111) can be replaced by the integral over a parallelogram $Q_{(0,0)}$ with the vertices $\pm\frac{1}{2}\omega_1 \pm \frac{i}{2}\text{Im}\, \omega_2$,

$$\langle f \rangle = \int_{Q_{(0,0)}} f(x_1, x_2) \mathrm{d}x_1 \mathrm{d}x_2 = \int_{-\frac{\omega_1}{2}}^{\frac{\omega_1}{2}} \mathrm{d}x_1 \int_{-\frac{\text{Im}\,\omega_2}{2}}^{\frac{\text{Im}\,\omega_2}{2}} f(x_1, x_2) \mathrm{d}x_2. \tag{5.112}$$

Using (5.90) consider, for instance, the constant

$$P_{11} = -\int_{-\frac{\omega_1}{2}}^{\frac{\omega_1}{2}} \sigma_{23}(x_1, x_2) \mathrm{d}x_1. \tag{5.113}$$

Integrate (5.113) by x_2 from $-\frac{\text{Im}\,\omega_2}{2}$ to $\frac{\text{Im}\,\omega_2}{2}$ and divide the result by $\text{Im}\,\omega_2$. In accordance with (5.111) we obtain

$$P_{11} = -\omega_1 \langle \sigma_{23} \rangle. \tag{5.114}$$

Similar arguments yield

$$P_{12} = -\text{Im}\,[\omega_2(\langle \sigma_{13} \rangle - i\langle \sigma_{23} \rangle)],\ P_{21} = -\omega_1 \langle D_2 \rangle,\ P_{22} = -\text{Im}\,[\omega_2(\langle D_1 \rangle - i\langle D_2 \rangle)]. \tag{5.115}$$

Let the average values $\langle \sigma_{\ell 3} \rangle$ and $\langle D_\ell \rangle$ $(\ell = 1,2)$ are known. Then, the constants $P_{\ell m}$ $(\ell, m = 1,2)$ are calculated by (5.114)–(5.115). Thus, we arrive at the following

Boundary value problem: Given constants $P_{\ell j}$ $(\ell, j = 1,2)$. To find the vector–function $\Phi^+(z) \equiv \Phi_0(z)$ analytic in G^+ and continuously differentiable in the closure of G^+; $\Phi_0(z)$ fulfills the quasiperiodicity conditions (5.108), where the constants $C_{\ell j}$ satisfy (5.109). To find the vector–functions $\Phi^-(z) \equiv \Omega_k(z)$ analytic in G_k

$(k = 1,2,\dots,N)$ and continuously differentiable in $G_k \cup \partial G_k$. The boundary values of these vector-functions satisfy the $\mathbb{R}$-linear condition

$$\Phi_0(t) = \Omega_k(t) - \mathbf{R}\,\overline{\Omega_k(t)},\ t \in \partial G_k\ (k = 1,2,\dots,N). \tag{5.116}$$

This boundary value problem can be considered as a vector-matrix generalization of the scalar problem discussed in Section 2.5 of Chapter 2. This problem in the theory of composites provides the following standard method to compute the effective piezoelectric constants [24]. First, the averaged values $\langle\sigma_{\ell 3}\rangle$ and $\langle D_\ell\rangle$ $(\ell = 1,2)$ are fixed. After, the constants $P_{\ell j}$ are calculated with (5.114)–(5.115) and the problem (5.116) is solved in a class of quasi-periodic functions. Here, the jumps C_j from (5.108) satisfy the conditions (5.109). Then, the averaged deformations $\langle\frac{\partial u_3}{\partial x_\ell}\rangle$ and electric fields $\langle E_\ell\rangle$ $(\ell = 1,2)$ are computed as linear combinations of $\langle\sigma_{\ell 3}\rangle$ and $\langle D_\ell\rangle$. This linear dependence determines the effective piezoelectric tensor $\mathbf{P}$. More precisely, a few independent problems should be solved to determine all the components of $\mathbf{P}$. For instance, two problems should be solved for macroscopically isotropic composites.

We propose another method to determine $\mathbf{P}$, which is convenient in symbolic computations. Let the host medium and inclusions be made from the same material. This implies that $\mathbf{D}_+ = \mathbf{D}_-$ and $\mathbf{R} = \mathbf{0}$ by (5.103). Then, the $\mathbb{R}$–linear condition (5.116) becomes the condition of analytic continuation of $\Phi_0(z)$ into G_k $(k = 1,2,\dots,N)$. Therefore, this degenerate case corresponds to the problem when a quasi-periodic vector–function $\Phi_0(z)$ analytic in $Q_{(0,0)}$ has to be found. The components of $\Phi_0(z)$ have to be linear functions $az+b$. It is convenient to take two linearly independent solutions to the degenerate problem

$$\Phi_0^{(1)}(z) = (z,0)^T \text{ and } \Phi_0^{(2)}(z) = (0,z)^T. \tag{5.117}$$

Then, (5.108) yields

$$C_j^{(1)} = \omega_j \begin{pmatrix} 1 \\ 0 \end{pmatrix} \text{ and } C_j^{(2)} = \omega_j \begin{pmatrix} 0 \\ 1 \end{pmatrix}. \tag{5.118}$$

In the next section, we come back to the general condition (5.116) and solve the problem (5.116), (5.108) with the prescribed jumps (5.118). After, we calculate the averaged fields and symbolically compute $\mathbf{P}$ by the averaged piezoelectric law.

5.10 EFFECTIVE ELASTIC CONSTANTS FOR PIEZOELECTRIC PROBLEMS

This section is devoted to calculating the effective piezoelectric tensor $\mathbf{P}$ by the method outlined at the end of the previous section. The inclusions G_k are assumed to be distributed so that the composite is transversely isotropic in macroscale in the plane perpendicular to the fibers. The average equations (5.87) for transversely isotropic composites take the form

$$\begin{aligned}\langle\sigma_{i3}\rangle &= c_{44}^*\langle\frac{\partial u_3}{\partial x_i}\rangle - d_{12}^*\langle E_i\rangle,\\ \langle D_i\rangle &= d_{12}^*\langle\frac{\partial u_3}{\partial x_i}\rangle + \epsilon_{11}^*\langle E_i\rangle \quad (i = 1,2).\end{aligned} \tag{5.119}$$

Equations (5.119) determine the effective piezoelectric tensor $\mathbf{D}^{\mathrm{h}}$. For transversely isotropic composites, $\mathbf{D}^{\mathrm{h}}$ can be presented by the matrix

$$\mathbf{D}^{\mathrm{h}} \equiv \begin{pmatrix} d_{11}^* & d_{12}^* \\ d_{12}^* & d_{11}^* \end{pmatrix} = \begin{pmatrix} c_{44}^* & d_{12}^* \\ d_{12}^* & -\epsilon_{11}^* \end{pmatrix}. \tag{5.120}$$

The signs in the second column correspond to equations (5.119). They are chosen for the convenience of further calculations. Moreover, it makes the tensor $\mathbf{D}^{\mathrm{h}}$ symmetric. Equations (5.119) can be written in the vector-matrix form

$$\begin{pmatrix} \langle\sigma_{i3}\rangle \\ \langle D_i\rangle \end{pmatrix} = \mathbf{D}^{\mathrm{h}} \begin{pmatrix} \langle\frac{\partial u_3}{\partial x_i}\rangle \\ -\langle E_i\rangle \end{pmatrix} \quad (i = 1,2). \tag{5.121}$$

To find the components of $\mathbf{D}^{\mathrm{h}}$, we calculate two sets of average values from (5.119) corresponding to two different external fields for $i = 1$. First, using the representations (5.88) and the definition (5.97) of the vector–functions $\Phi_k(z)$ we calculate the integral

$$\begin{pmatrix} \langle\sigma_{13}\rangle \\ \langle D_1\rangle \end{pmatrix} = \mathbf{D}_+ \int_{G^+} \mathrm{Re}\ \Phi_0'(z)\mathrm{d}x_1\mathrm{d}x_2 + \mathbf{D}_- \sum_{k=1}^{N} \int_{G_k} \mathrm{Re}\ \Phi_k'(z)\mathrm{d}x_1\mathrm{d}x_2. \tag{5.122}$$

The vector–functions $\Phi_k(z)$ and $\Omega_k(z)$ are related by equation (see (5.102))

$$\Omega_k(z) = \frac{1}{2}\mathbf{d}_+^{-1}(\mathbf{d}_- + \mathbf{d}_+)\Phi_k(z), \quad z \in G_k. \tag{5.123}$$

In order to transform (5.122) we use Green's formula (2.179) in the complex form

$$\int_Q \mathrm{Re}\ w'(z)\mathrm{d}x_1\mathrm{d}x_2 = \int_{\partial Q} \mathrm{Re}\ w(z)\mathrm{d}x_2, \tag{5.124}$$

where $z = x_1 + ix_2$. The double integrals from (5.122) can be reduced to the linear integrals

$$\int_{G^+} \mathrm{Re}\ \Phi_0'(z)\mathrm{d}x_1\mathrm{d}x_2 = \int_{\partial Q_{(0,0)}} \mathrm{Re}\ \Phi_0(t)\mathrm{d}x_2 - \sum_{k=1}^{N} \int_{\partial G_k} \mathrm{Re}\ \Phi_0(t)\mathrm{d}x_2 \tag{5.125}$$

and

$$\int_{G_k} \mathrm{Re}\ \Phi_k'(z)\mathrm{d}x_1\mathrm{d}x_2 = \int_{\partial G_k} \mathrm{Re}\ \Phi_k(t)\mathrm{d}t, \tag{5.126}$$

where $t = x_1 + ix_2 \in \partial G_k$. The integral on $\partial Q_{(0,0)}$ from (5.125) can be calculated through the increments (5.108)

$$\int_{\partial Q_{(0,0)}} \mathrm{Re}\ \Phi_0(t)\mathrm{d}x_2 = \mathrm{Re}\ C_1\ \mathrm{Im}\ \omega_2. \tag{5.127}$$

Formula (5.122) becomes

$$\begin{pmatrix} \langle\sigma_{23}\rangle \\ \langle D_2\rangle \end{pmatrix} = \mathbf{D}_+\mathrm{Re}\ C_1\ \mathrm{Im}\ \omega_2 + \sum_{k=1}^{N} \int_{\partial G_k} (\mathbf{D}_-\mathrm{Re}\ \Phi_k(t) - \mathbf{D}_+\mathrm{Re}\ \Phi_0(t))\mathrm{d}t. \tag{5.128}$$

Application of (5.98), (5.123) and (5.103) to the latter integrand yields

$$\mathbf{D}_{-}\text{Re }\Phi_k(t) - \mathbf{D}_{+}\text{Re }\Phi_0(t) = 2\mathbf{R}\text{ Re }\Omega_k(t). \tag{5.129}$$

Then, (5.128) becomes

$$\left(\begin{array}{c} \langle \sigma_{13} \rangle \\ \langle D_1 \rangle \end{array} \right) = \mathbf{D}_{+} \left[\text{Re } C_1 \text{ Im } \omega_2 + 2\mathbf{R} \sum_{k=1}^{N} J_k \right], \tag{5.130}$$

where

$$J_k = \int_{\partial G_k} \text{Re }\Omega_k(t) \text{dx}_2. \tag{5.131}$$

Application of similar arguments yields

$$\left(\begin{array}{c} \langle \frac{\partial u_3}{\partial x_1} \rangle \\ -\langle E_1 \rangle \end{array} \right) = \text{Re } C_1 \text{ Im } \omega_2. \tag{5.132}$$

Equations (5.121), (5.130) and (5.132) yield formulas for the effective tensor $\mathbf{D}^{\text{h}}$ in terms of the integrals (5.131) in the following way. Substitute two vectors (5.118) into (5.130) and (5.132) instead of C_1 and C_2. Then, we obtain two vectors

$$\text{Re } C_1^{(1)} \text{ Im } \omega_2 = \left(\begin{array}{c} 1 \\ 0 \end{array} \right) \tag{5.133}$$

and

$$\text{Re } C_1^{(2)} \text{ Im } \omega_2 = \left(\begin{array}{c} 0 \\ 1 \end{array} \right). \tag{5.134}$$

The corresponding local fields can be averaged, and equation (5.121) applied to these fields yields

$$\left(\begin{array}{cc} \langle \sigma_{13}^{(1)} \rangle & \langle \sigma_{13}^{(2)} \rangle \\ \langle D_1^{(1)} \rangle & \langle D_1^{(2)} \rangle \end{array} \right) = \mathbf{D}^{\text{h}}. \tag{5.135}$$

The first column of the left part of (5.135) can be determined by (5.130) with the constant vectors C_1 and C_2 given by (5.118). The second column can be determined analogously, as described below.

Consider two $\mathbb{R}$-linear problems (5.116) stated at the end of Sec. 5.9

$$\Phi_0^{(\ell)}(t) = \Omega_k^{(\ell)}(t) - \mathbf{R}\,\overline{\Omega_k^{(\ell)}(t)},\ t \in \partial G_k\ (k = 1,2,\ldots,N;\ \ell = 1,2) \tag{5.136}$$

with the quasi-periodicity conditions (5.108) when the jumps of $\Phi_0^{(\ell)}(z)$ are given by (5.118)

$$\Phi_0^{(1)}(z+\omega_j) - \Phi_0^{(1)}(z) = \omega_j \left(\begin{array}{c} 1 \\ 0 \end{array} \right),\ \Phi_0^{(2)}(z+\omega_j) - \Phi_0^{(2)}(z) = \omega_j \left(\begin{array}{c} 0 \\ 1 \end{array} \right)\ (j = 1,2). \tag{5.137}$$

Here, ℓ is the number of the $\mathbb{R}$–linear problem ($\ell = 1,2$); k is the number of the inclusion and j is the number of the jump condition in each ℓth problem.

Introduce the vector–functions

$$\Psi_0^{(\ell)}(z) = \frac{d}{dz}\Phi_0^{(\ell)}(z), \quad \Psi_k^{(\ell)} z) = \frac{d}{dz}\Omega_k^{(\ell)}(z) \tag{5.138}$$

and the matrices

$$\mathbf{\Psi}_k(z) = \left(\Psi_k^{(1)}(z), \Psi_k^{(2)} z)\right), \quad k = 1,2,\ldots,N. \tag{5.139}$$

Differentiation of (5.136) on a tangent parameter of ∂G_k yields [185]

$$\Psi_0^{(\ell)}(t) = \Psi_k^{(\ell)}(t) + \mathbf{R}\,\overline{n^2(t)\Psi_k^{(\ell)}(t)},\ t \in \partial G_k\ (k = 1,2,\ldots,N), \tag{5.140}$$

where $n(t)$ is the normal outward vector to ∂G_k expressed in terms of complex values. Differentiation of (5.137) implies double periodicity of $\Psi_0^{(\ell)}(z)$. Using (5.131) and (5.124) we introduce the matrix

$$\mathbf{J}_k = -\frac{1}{2i}\int_{\partial G_k} (\Omega_k^{(1)}(t), \Omega_k^{(2)}(t))d\bar{t} = \int_{G_k} \mathbf{\Psi}_k(z)\ \mathrm{d}x_1\mathrm{d}x_2. \tag{5.141}$$

Then, (5.135) can be written in the extended form

$$\mathbf{D}^{\mathrm{h}} = \mathbf{D}_+\left(\mathbf{I} + 2\mathbf{R}\sum_{k=1}^{N}\mathbf{J}_k\right). \tag{5.142}$$

Let mutually disjoint disks $G_k = \{z \in \mathbb{C} : |z - a_k| < r\}$ $(k = 1,2,\ldots,N)$ are located in $Q_{(0,0)}$. The normal vector to the circle $|t - a_k| = r$ has the form $n(t) = \frac{1}{r}(t - a_k)$. Then (5.140) becomes

$$\Psi_0^{(\ell)}(t) = \Psi_k^{(\ell)}(t) + \left(\frac{r}{t - a_k}\right)^2 \mathbf{R}\,\overline{\Psi_k^{(\ell)}(t)},\ |t - a_k| = r\ (k = 1,2,\ldots,N). \tag{5.143}$$

The integral (5.141) can be calculated by the mean value theorem for harmonic functions

$$\mathbf{J}_k = \pi r^2 \mathbf{\Psi}_k(a_k). \tag{5.144}$$

Substitution of (5.144) into (5.142) yields

$$\mathbf{D}^{\mathrm{h}} = \mathbf{D}_+\left(\mathbf{I} + 2f\mathbf{R}\,\frac{1}{N}\sum_{k=1}^{N}\mathbf{\Psi}_k(a_k)\right). \tag{5.145}$$

Example 6 *For dilute and/or weakly inhomogeneous composites, the vector–functions* $\Psi_k^{(\ell)}(z)$ *satisfying* (5.143) *are approximated by constant vectors obtained by differentiation of* (5.117). *Hence, the matrices* $\mathbf{\Psi}_k(z)$ *are approximated by the unit matrix:*

$$\mathbf{\Psi}_k(z) \approx \mathbf{I}. \tag{5.146}$$

Substitution of these approximations into (5.145) *and application of the simple rational approximation in* f *yield*

$$\mathbf{D}^{\mathrm{h}} \approx \mathbf{D}_{+} (\mathbf{I} + f\mathbf{R}) (\mathbf{I} - f\mathbf{R})^{-1}. \tag{5.147}$$

The latter formula extends the famous Clausius–Mossotti approximation to piezoelectric fiber composites. Direct computations show that the matrix in the right part of (5.147) *is symmetric.*

5.11 HIGH ORDER APPROXIMATE ANALYTICAL FORMULAS

The $\mathbb{R}$–linear vector–matrix problem (5.116) is the key to the high-order approximate analytical formulas for the effective piezoelectric tensor $\mathbf{D}^{\mathrm{h}}$ determined by (5.142). The scalar $\mathbb{R}$–linear problems and their application to the effective permittivity tensor were discussed in previous chapters. In order to use these constructive results for the scalar problem, one can try to decompose the vector-matrix problem (5.116) onto two scalar ones.

The matrix $\mathbf{R}$ is defined by (5.103) through two positively determined matrices $\mathbf{D}_{+}$ and $\mathbf{D}_{-}$. Let it be presented in the form

$$\mathbf{R} = \mathbf{T}\boldsymbol{\varrho}'\mathbf{T}^{-1}. \tag{5.148}$$

where $\mathbf{T}$ is a non–singular matrix and $\boldsymbol{\varrho}'$ is diagonal

$$\boldsymbol{\varrho}' = \begin{pmatrix} \varrho_1' & 0 \\ 0 & \varrho_2' \end{pmatrix}. \tag{5.149}$$

Introduce the contrast parameters

$$\varrho_1 = \frac{d_{11}^- - d_{11}^+}{d_{11}^- + d_{11}^+}, \ \varrho_2 = \frac{d_{22}^- - d_{22}^+}{d_{22}^- + d_{22}^+}, \ \varrho_3 = \frac{d_{12}^- - d_{12}^+}{d_{12}^- + d_{12}^+}, \tag{5.150}$$

and the constants

$$c = \frac{d_{11}^- + d_{11}^+}{d_{12}^- + d_{12}^+}, \ e = -\frac{d_{22}^- + d_{22}^+}{d_{12}^- + d_{12}^+}, \ h = ce. \tag{5.151}$$

Then, the matrix $\mathbf{R}$ can be written in the form

$$\mathbf{R} = \frac{1}{1+h} \begin{pmatrix} \varrho_3 + h\varrho_1 & -e(\varrho_2 - \varrho_3) \\ c(\varrho_1 - \varrho_3) & \varrho_3 + h\varrho_2 \end{pmatrix}. \tag{5.152}$$

It is assumed that the matrix $\mathbf{R}$ is not singular, hence

$$\det \mathbf{R} = \frac{1}{1+h}(h\varrho_1\varrho_2 + \varrho_3^2) \neq 0. \tag{5.153}$$

Introduce the value

$$\Delta = h[h(\varrho_1 - \varrho_2)^2 - 4(\varrho_1 - \varrho_3)(\varrho_2 - \varrho_3)]. \tag{5.154}$$

Let the constant be positive. In this case, the matrix $\mathbf{T} = \{t_{m\ell}\}$ is real and has the form

$$\mathbf{T} = \begin{pmatrix} \frac{h(\varrho_1-\varrho_2)-\sqrt{\Delta}}{2c(\varrho_1-\varrho_3)} & \frac{h(\varrho_1-\varrho_2)+\sqrt{\Delta}}{2c(\varrho_1-\varrho_3)} \\ 1 & 1 \end{pmatrix}. \tag{5.155}$$

The inverse matrix $\mathbf{T}^{-1} = \{t^*_{m\ell}\}$ is given by

$$\mathbf{T}^{-1} = \frac{1}{2\sqrt{\Delta}} \begin{pmatrix} -c(\varrho_1-\varrho_3) & h(\varrho_1-\varrho_2)+\sqrt{\Delta} \\ c(\varrho_1-\varrho_3) & -h(\varrho_1-\varrho_2)+\sqrt{\Delta} \end{pmatrix}. \tag{5.156}$$

The real elements of the matrix (5.149) reads as follows

$$\begin{aligned} \varrho'_1 &= \tfrac{1}{2(1+h)}[h(\varrho_1+\varrho_2)+2\varrho_3-\sqrt{\Delta}], \\ \varrho'_2 &= \tfrac{1}{2(1+h)}[h(\varrho_1+\varrho_2)+2\varrho_3+\sqrt{\Delta}]. \end{aligned} \tag{5.157}$$

5.11.1 THE CASE $\Delta \geq 0$

The similarity relation (5.148) between the matrices $\mathbf{R}$ and $\boldsymbol{\varrho}'$ are used. The eigenvalues (5.157) are real in the considered case $\Delta \geq 0$ and

$$|\varrho'_j| \leq 1 \quad (j = 1,2). \tag{5.158}$$

To prove this, we consider the equivalent relation

$$|h(\varrho_1+\varrho_2)+2\varrho_3 \pm \sqrt{\Delta}| \leq 2(h+1). \tag{5.159}$$

Further, consider the case $h(\varrho_1+\varrho_2)+2\varrho_3 \geq 0$. In the opposite case, the proof is similar. It is sufficiently to demonstrate that $h(\varrho_1+\varrho_2)+2\varrho_3+\sqrt{\Delta} \leq 2(h+1)$. The latter inequality for positive Δ given by (5.154) is equivalent to

$$h^2(\varrho_1-\varrho_2)^2 - 4h(\varrho_1-\varrho_3)(\varrho_2-\varrho_3) \leq [2(h+1-\varrho_3)-h(\varrho_1+\varrho_2)]^2. \tag{5.160}$$

The use of simple transformations yields

$$h(\varrho_1+\varrho_2-\varrho_1\varrho_2) \leq h+(1-\varrho_3)^2. \tag{5.161}$$

This inequality follows the form

$$\varrho_1+\varrho_2-\varrho_1\varrho_2 = 1-(1-\varrho_1)(1-\varrho_2) \leq 1.$$

Therefore, (5.158) is proved.

Substitute the representation (5.148) into (5.140) and multiply the result by $\mathbf{T}^{-1}$

$$\mathbf{T}^{-1}\Psi_0^{(\ell)}(t) = \mathbf{T}^{-1}\Psi_k^{(\ell)}(t) + \boldsymbol{\varrho}'\mathbf{T}^{-1}\,\overline{n^2(t)\Psi_k^{(\ell)}(t)},\ t \in \partial G_k \\ (k = 1,2,\ldots,N;\ \ell = 1,2). \tag{5.162}$$

Introduce the vector–functions

$$\omega_k^{(\ell)}(z) = \mathbf{T}^{-1}\Psi_k^{(\ell)}(z) \tag{5.163}$$

and the matrix

$$\mathbf{w}_k(z) = (\omega_k^{(1)}(z), \omega_k^{(2)}(z)). \tag{5.164}$$

Then, (5.162) becomes

$$\omega_0^{(\ell)}(t) = \omega_k^{(\ell)}(t) + \boldsymbol{\varrho}' \, \overline{n^2(t)\omega_k^{(\ell)}(t)}, \; t \in \partial G_k \; (k = 1,2,\dots,N; \; \ell = 1,2) \tag{5.165}$$

since the matrix $\mathbf{T}$ is real. The vector–matrix problem (5.165) can be decoupled onto the scalar problems

$$\begin{array}{r} \omega_0^{(m,\ell)}(t) = \omega_k^{(m,\ell)}(t) + \varrho'_m \overline{n^2(t)\omega_k^{(m,\ell)}(t)}, \; t \in \partial G_k \\ (k = 1,2,\dots,N; \; \ell, m = 1,2) \end{array} \tag{5.166}$$

where $\omega_k^{(m,\ell)}(t)$ is the mth coordinate of the vector $\omega_k^{(\ell)}(t)$ (or the element (m,ℓ) of the matrix $\mathbf{w}_k(t)$).

The method of functional equations discussed in Chapter 4 can be applied to the scalar problems (5.166) due to (5.158) holds. Consider non-overlapping circular inclusions. Then (5.166) becomes

$$\begin{array}{r} \omega_0^{(m,\ell)}(t) = \omega_k^{(m,\ell)}(t) + \varrho'_m \left(\frac{r}{t-a_k}\right)^2 \overline{\omega_k^{(m,\ell)}(t)}, \; |t - a_k| = r \\ (k = 1,2,\dots,N; \; \ell, m = 1,2) \end{array} \tag{5.167}$$

Introduce the matrix

$$\boldsymbol{\psi} := \begin{pmatrix} \psi_{11} & \psi_{12} \\ \psi_{21} & \psi_{22} \end{pmatrix} = \frac{1}{N} \sum_{k=1}^{N} \boldsymbol{\Psi}_k(a_k). \tag{5.168}$$

The effective tensor $\mathbf{D}^{\mathrm{h}}$ has the form (5.145) where by (5.163)–(5.164)

$$\boldsymbol{\psi} = \mathbf{T} \frac{1}{N} \sum_{k=1}^{N} \mathbf{w}_k(a_k). \tag{5.169}$$

In the zero-th approximation, $\boldsymbol{\Psi}_k(z) \approx \mathbf{I}$ (see (5.146)). Then, the zero-th approximation for $\mathbf{w}_k(z)$ is determined by (5.163)

$$\mathbf{w}_k(z) \approx \mathbf{T}^{-1}. \tag{5.170}$$

Therefore, we arrive at four scalar problems (5.166) ($\ell, m = 1,2$) with the zero–th approximations

$$\omega_k^{(m,\ell)}(z) \approx t^*_{m\ell}, \tag{5.171}$$

where $t^*_{m\ell}$ denote the elements of the matrix $\mathbf{T}^{-1}$. The scalar problem (5.166), (5.171) formally coincided with the permittivity problem with the contrast parameter ϱ'_m and the external flux expressed by the vector $t^*_{m\ell}(1,0)^T$. The problems (5.167), (5.171) were solved in Sections ... for arbitrary locations of non-overlapping inclusions $|z - a_k| < r$ ($k = 1,2,\dots,N$) in the periodicity cell.

For macroscopically isotropic composites and real ϱ'_m $(m = 1,2)$, the ratio σ_e of the effective permittivity to the permittivity of the host material can be written in the form

$$\sigma_e(\varrho'_m) = 1 + 2\varrho'_m f P(\varrho'_m), \tag{5.172}$$

where the solution of the problem (5.167) and $P(\varrho'_m)$ are related by equation

$$\frac{1}{N}\sum_{k=1}^{N} \omega_k^{(m,\ell)}(a_k) = t^*_{m\ell} P(\varrho'_m). \tag{5.173}$$

The function

$$P(\varrho'_m) = \frac{\sigma_e(\varrho'_m) - 1}{2\varrho'_m f} = 1 + A_1 f + A_2 f^2 + \ldots \tag{5.174}$$

coincides with the scalar series $1 + A_1 f + A_2 f^2 + \ldots$ from (4.68) written for real ϱ'_m, see exactly the scalar series in [93].

Substitution of (5.173) into (5.169) yields

$$\boldsymbol{\psi} = \begin{pmatrix} t_{11}t^*_{11}P(\varrho'_1) + t_{12}t^*_{21}P(\varrho'_2) & t_{11}t^*_{12}P(\varrho'_1) + t_{12}t^*_{22}P(\varrho'_2) \\ t_{21}t^*_{11}P(\varrho'_1) + t_{21}t^*_{12}P(\varrho'_2) & t_{21}t^*_{12}P(\varrho'_1) + t_{22}t^*_{22}P(\varrho'_2) \end{pmatrix}. \tag{5.175}$$

Application of the explicit formulas (5.155)–(5.156) to the elements $\psi_{m\ell}$ of the matrix (5.168) yields

$$\begin{aligned}
\psi_{11} &= \tfrac{1}{2\sqrt{\Delta}}\left\{[h(\varrho_2 - \varrho_1) + \sqrt{\Delta}]P(\varrho'_1) - [h(\varrho_2 - \varrho_1) - \sqrt{\Delta}]P(\varrho'_2)\right\},\\
\psi_{12} &= \tfrac{e}{\sqrt{\Delta}}(\varrho_2 - \varrho_3)[P(\varrho'_1) - P(\varrho'_2)],\\
\psi_{21} &= \tfrac{c}{\sqrt{\Delta}}(\varrho_3 - \varrho_1)[P(\varrho'_1) - P(\varrho'_2)],\\
\psi_{22} &= \tfrac{1}{2\sqrt{\Delta}}\left\{[h(\varrho_2 - \varrho_1) + \sqrt{\Delta}]P(\varrho'_2) - [h(\varrho_2 - \varrho_1) - \sqrt{\Delta}]P(\varrho'_1)\right\}
\end{aligned} \tag{5.176}$$

The formula (5.145) can be written in the form

$$\mathbf{D}^{\mathrm{h}} = \mathbf{D}_+(\mathbf{I} + 2f\mathbf{R}\boldsymbol{\psi}), \tag{5.177}$$

Symbolic computations of $P(\varrho')$ for various locations of the disks [93] and substitution of the results into (5.176)–(5.177) yield analytical formulas for the effective tensor $\mathbf{D}^{\mathrm{h}}$. Using (5.174), one can obtain the asymptotic formula

$$P(\varrho') = 1 + \varrho' f - \varrho'^3 f^3 \tag{5.178}$$

$$+\left(\frac{\varrho' f}{\pi}\right)^2\left[(1+2\varrho' f)e_{22} - \frac{2f}{\pi}(1+2\varrho' f)e_{33} + \frac{3f^2}{\pi^2}e_{44} + \left(\frac{\varrho' f}{\pi}\right)^2 e_{2222}\right] + O(f^5),$$

where

$$e_{pp} = \frac{(-1)^p}{N^{p+1}}\sum_{m=1}^{N}\sum_{\substack{k=1\\k\neq m}}^{N}|E_p(a_k - a_m)|^2 \quad (p = 2,3,4), \tag{5.179}$$

$$e_{2222} = \frac{1}{N^5} \sum_{k=1}^{N} \left| \sum_{\substack{m=1 \\ m\neq k}}^{N} \sum_{\substack{\ell=1 \\ \ell\neq m}}^{N} E_2(a_k - a_m)\overline{E_2(a_m - a_\ell)} \right|^2 .$$

Here, $E_p(z)$ denotes the Eisenstein function of order p. An exact formula for the effective permittivity of regular arrays of disks, hence for $P(\varrho')$, is written in Section 4.1.3.

Example 7 *Consider a random composite when the disks obey the uniform–non-overlapping distribution. Then, $P(\varrho')$ can be approximated up to $O(f^6)$ by the following expression [62]*

$$\begin{aligned} P(\varrho') \approx\ & 1 + f\varrho' + f^2\varrho'^2(2.4922 - 0.0344\varrho' - 0.07315\varrho'^2 - 0.3998\varrho'^3) \\ & + f^3\varrho'^2(-3.4145 + 3.6826\varrho' + 3.15395\varrho'^2 + 2.2585\varrho'^3) \\ & + f^4\varrho'^2(2.107 - 6.2109\varrho' - 5.22995\varrho'^2 - 3.9301\varrho'^3) \\ & + f^5\varrho'^2(-0.1731 + 3.5434\varrho' + 3.3554\varrho'^2 + 2.9949\varrho'^3). \end{aligned} \tag{5.180}$$

It is shown in [62] that the precision $O(f^5)$ in (5.180) *is sufficient to give excellent results for $|\varrho'| < 1$. For $|\varrho'|$ close to unity, it is better to use Padé approximations [62].*

Example 8 *Consider non-overlapping equal ellipses of the semi-axis $r(1 \pm \alpha)$ in the unit square lattice. Let the angle between the major axis and the axis OX_1 be a random uniformly distributed variable. Then, $P(\varrho')$ can be found from* (5.174) *and* (4.142)

$$P(\varrho') = \frac{1}{1-\varrho^2\alpha^2} + \frac{\varrho f(1+\alpha^2)}{(1-\varrho^2\alpha^2)^2(1-\alpha^2)} + \varrho^2 f^2 \frac{e_{22}}{\pi^2} \frac{(1+\varrho^2\alpha^2)(1+\alpha^2)^2}{(1-\varrho^2\alpha^2)^3(1-\alpha^2)^2} + O(f^3). \tag{5.181}$$

Here, e_{22} denotes the structural sum. If the centers of ellipses $\mathcal{A} = \{a_1, a_2, \ldots, a_N\}$ are fixed, i.e., a_k are deterministic values, e_{22} is directly calculated by (5.179) *with* $p = 2$.

Let $\mathcal{A}$ be considered as a random set of centers, i.e., a probabilistic distribution of the centers given in the form of simulation rules. It should be noted that a formal analytical formula for a probabilistic distribution of $\mathcal{A}$ can be written in terms of multi-point correlation functions [30, 123, 226]. However, immense computational restrictions make this method of correlation functions virtually impossible to compute the correlation functions numerically. Straightforward Monte Carlo simulations of e_{22} over $\mathcal{A}$ obviate the need for correlation functions and yield the estimation of the mathematical expectation $\mathbb{E}[e_{22}]$. After its substitution into (5.181) *instead of e_{22} we get $\mathbb{E}[P(\varrho')]$, next the mathematical expectation of the effective tensor $\mathbb{E}[\mathbf{D}^h]$ by* (5.175), (5.176), (5.177).

Example 9 *Checkerboard two–two-dimensional composites can be considered a limit case of the square inclusions of permittivities ε_1 and $\varepsilon_2 = 1$. Application of Dykhne's formula $\varepsilon_e = \sqrt{\varepsilon_1\varepsilon_2}$ [69] to* (5.172) *with $f = \frac{1}{2}$ yields*

$$P(\varrho') = \frac{1}{2\varrho' f}\left(\sqrt{\frac{1+\varrho'}{1-\varrho'}} - 1\right) = \frac{1}{2f}(1 - \sqrt{\varepsilon_1}), \tag{5.182}$$

where $\varrho' = \frac{\varepsilon_1 - 1}{\varepsilon_1 + 1}$. *Next, the effective tensor* $\mathbf{D}^h$ *is given by exact formulas* (5.175), (5.176) *and* (5.177).

If the permittivities ε_1 *and* ε_2 *are used without normalization, equation* (5.182) *becomes*

$$P(\varrho') = \frac{1}{2\varrho' f}\left(\sqrt{\frac{1+\varrho'}{1-\varrho'}} - 1\right) = \frac{1}{2f}\left(1 - \sqrt{\frac{\varepsilon_1}{\varepsilon_2}}\right), \tag{5.183}$$

where the value $P(\varrho')$ *is again dimensionless. The dimensional units are contained in the matrix* $\mathbf{D}_+$ *of* (5.177).

General exact formulas for vector-matrix problems obtained in the framework of the theory of duality transformation can be found in [8].

Remark 12 *The case* $\Delta = 0$ *equivalent to* $h(\varrho_1 - \varrho_2)^2 = 4(\varrho_1 - \varrho_3)(\varrho_2 - \varrho_3)$ *can be investigated by the same method as* $\Delta > 0$. *The difference involves replacing the diagonal matrix* ϱ' *with the Jordan normal form.*

5.11.2 THE CASE $\Delta < 0$

The present section extends the previous results to the case $\Delta < 0$. The direct application of the decomposition (5.148) with the complex matrix $\mathbf{T}$ does not work, as shown in the following example. Let $\varrho_1 = \varrho_2 > \varrho_3$. Then, $\Delta = -4h(\varrho_1 - \varrho_3)^2 < 0$, the following eigenvalues are complex in general,

$$\varrho_1' = \frac{h\varrho_1 + \varrho_3 - \mathrm{i}\sqrt{h}(\varrho_1 - \varrho_3)}{h+1}, \quad \varrho_2' = \overline{\varrho_1'}. \tag{5.184}$$

The decomposition (5.148) with the complex matrix $\mathbf{T}$ given by (5.155) yields the special $\mathbb{R}$-linear vector-matrix problem

$$\omega_0^{(\ell)}(t) = \omega_k^{(\ell)}(t) - \begin{pmatrix} 0 & \varrho_1' \\ \overline{\varrho_1'} & 0 \end{pmatrix} \overline{\omega_k^{(\ell)}(t)}, \; t \in \partial G_k \; (k = 1, 2, \ldots, N; \; \ell = 1, 2). \tag{5.185}$$

Here, the following relation is used

$$\varrho' \mathbf{T}^{-1} \overline{\mathbf{T}} = \begin{pmatrix} 0 & \varrho_1' \\ \overline{\varrho_1'} & 0 \end{pmatrix}. \tag{5.186}$$

It is worth noting that (5.185) cannot be decomposed onto two scalar problems contrary to (5.165).

The complex values ϱ_j' $(j = 1, 2)$ always satisfy (5.158) because (5.148) and (5.153) imply that $|\varrho_1'|^2 = \det \mathbf{R} \leq 1$. Rewrite equations (5.176) in the slightly another form

$$\psi_{11} = \frac{1}{2}[P(\varrho_1')+P(\varrho_2')] - \frac{h(\varrho_1-\varrho_2)}{2\sqrt{\Delta}}[P(\varrho_1')-P(\varrho_2')], \tag{5.187}$$

$$\psi_{12} = \frac{e}{\sqrt{\Delta}}(\varrho_2-\varrho_3)[P(\varrho_1')-P(\varrho_2')],$$

$$\psi_{21} = \frac{c}{\sqrt{\Delta}}(\varrho_3-\varrho_1)[P(\varrho_1')-P(\varrho_2')],$$

$$\psi_{22} = \frac{1}{2}[P(\varrho_1')+P(\varrho_2')] + \frac{h(\varrho_1-\varrho_2)}{2\sqrt{\Delta}}[P(\varrho_1')-P(\varrho_2')].$$

In the considered case $\Delta < 0$ the values ϱ_1' and $\varrho_2' = \overline{\varrho_1'}$ satisfy the quadratic equation with real coefficients

$$\varrho'^2 - \frac{h(\varrho_1+\varrho_2)+2\varrho_3}{1+h}\varrho' + \frac{h\varrho_1\varrho_2+\varrho_3^2}{1+h} = 0. \tag{5.188}$$

It follows from Vieta's formula and inequalities $|\varrho_m| < 1$, $h > 0$ that $|\varrho_1'|^2 = |\varrho_2'|^2 = \left|\frac{h\varrho_1'\varrho_2'+\varrho_3^2}{1+h}\right| < 1$. Therefore, the complex roots ϱ_1' and $\overline{\varrho_1'}$ lie in the unit disk, the convergence domain of the contrast expansion. The function $P(\varrho')$ satisfies the relations

$$P(\varrho') + P(\overline{\varrho'}) = 2\mathrm{Re}\,[P(\varrho')]\,,\ P(\varrho') - P(\overline{\varrho'}) = 2\mathrm{i}\ \mathrm{Im}\,[P(\varrho')]\,, \tag{5.189}$$

since its coefficients A_k from (5.174) are real. Substitute $\varrho_2' = \overline{\varrho_1'}$ into (5.187) and use the relations (5.189)

$$\psi_{11} = \mathrm{Re}\left[P(\varrho_1')\right] - \frac{h(\varrho_1-\varrho_2)}{\sqrt{|\Delta|}}\mathrm{Im}\left[P(\varrho_1')\right], \tag{5.190}$$

$$\psi_{12} = \frac{2e}{\sqrt{|\Delta|}}(\varrho_2-\varrho_3)\mathrm{Re}\left[P(\varrho_1')\right],\ \psi_{21} = \frac{2c}{\sqrt{|\Delta|}}(\varrho_3-\varrho_1)\mathrm{Im}\left[P(\varrho_1')\right],$$

$$\psi_{22} = \mathrm{Re}\left[P(\varrho_1')\right] + \frac{h(\varrho_1-\varrho_2)}{|\Delta|}\mathrm{Im}\left[P(\varrho_1')\right].$$

One can see that the components of the matrix $\boldsymbol{\psi}$ are real.

We now apply the arguments of analytic continuation. The formulas (5.177) and (5.187) hold for $\Delta > 0$ with the corresponding values (5.157) substituted in the argument of the function (5.174). Consider the ball $\mathbb{B} = \{(\varrho_1', \varrho_2') \in \mathbb{C}^2 :\ |\varrho_1'| \le q,\ |\varrho_2'| \le q\}$ in the two-dimensional complex space $\mathbb{C}^2$ of the arbitrarily fixed radius $q < 1$. One can suppose that the formulas (5.177) and (5.187) hold in the subset $\mathbb{B}_+ = \{(\varrho_1', \varrho_2') \in \mathbb{B} :\ \Delta > 0\}$. The absolute convergence of the series (5.174) in the disk $|\varrho'| \le q$ on the complex plane implies its uniform convergence on the parameters ϱ_1' and ϱ_2'; hence, the analytic continuation on these parameters into the whole ball $\mathbb{B}$. Therefore, the formula (5.177) for positive Δ with (5.187) is analytically continued on the parameters to the set $\Delta < 0$ of $\mathbb{B}$. As a result of the continuation, we arrive at the same formula (5.177) but with ψ given by (5.190).

The justification of (5.177) and (5.190) will be completed by the statement that the effective tensor $\mathbf{D}^{\mathrm{h}}$ is an analytic function of the parameters in the analyticity domain $\mathbb{B}$. It was established in [129, 136] for a scalar problem. It is suggested that the same result is true for the considered piezoelectric problem.

5.11.3 EXAMPLE

The described method allows the writing of exact formulas for the effective piezoelectric tensor $\mathbf{D}^{\mathrm{h}}$ through the known exact formulas for the scalar effective permittivity. Before giving an example, we have to say that our formulas are *exact* in a sense explained in [17] contrary to the misleading terminology used in [43].

As an example, consider the square array of circular cylinders. The function $P(\varrho')$ can be written explicitly by the series (5.174)

$$P(\varrho) = 1 + \varrho f \tag{5.191}$$

$$+\frac{\varrho f}{\pi}\sum_{k=1}^{\infty}\varrho^k \sum_{m_1=1}^{\infty}\cdots\sum_{m_k=1}^{\infty} s_{m_1}^{(1)} s_{m_2}^{(m_1)} \ldots s_{m_k}^{(m_{k-1})} s_1^{(m_k)} \left(\frac{f}{\pi}\right)^{2(m_1+\ldots+m_k)-k}.$$

The effective piezoelectric tensor $\mathbf{D}^{\mathrm{h}}$ is given by the exact formulas (5.177)–(5.187). The exact formula (5.191) can be transformed to the following asymptotic formula, including the percolation regimes $(|\varrho|f) \sim \frac{\pi}{4}$ [93, Chapter 6]

$$2\varrho f P(\varrho) \simeq \sqrt{\frac{\frac{\pi}{4}+\varrho f}{\frac{\pi}{4}-\varrho f}}\sqrt[4]{\frac{1+1.454\varrho f+0.388(\varrho f)^2}{1-1.454\varrho f+0.388(\varrho f)^2}} - 1. \tag{5.192}$$

Ultimately, the effective piezoelectric tensor of the square array is estimated by the formula (5.177), where ψ is given by (5.187) or by (5.190), the function $P(\varrho')$ is calculated exactly by (5.191) or approximately by (5.192).

5.12 BOUNDS ON THE EFFECTIVE PERMITTIVITY

5.12.1 CONTRAST EXPANSION APPROACH

The present section is devoted to analytical contrast bounds for the effective constants of random composites first derived in general form in [32, 33, 94, 137, 147, 152–154]. The reasons for favoring bounds on the effective properties of composites were discussed in [155, Chapter 21]. It was stressed that bounds serve as a benchmark to test approximation formulas. One can find a comprehensive treatment of the theory of bounds and their applications in [51, 155] and references therein. The famous constructive Wiener [233], and Hashin-Shtrikman [103] bounds, and other similar bounds, including the bounds for piezoelectric composites [39] can be considered as 1- and 2-point correlation bounds fulfilled for all composites. Hence, they are universal and do not take into account the geometrical diversity of dispersed random composites. On one side, this is an advantage of the 2-point correlation bounds since they fit any composite. Conversely, dispersed composites with various structural properties, e.g., disorder, clusterization, and so forth [67, 195], could be estimated by employing higher-order bounds. The local field bounds for random composites useful in technological applications were derived in [12, 63, 133, 134].

The theory of higher-order contrast bounds for random composites was developed in [226, Ch. 21]. However, a virtual impossibility of computing the correlation functions of higher orders restricts their applications. A much more computationally effective method of structural sums was proposed in [37, 162] and developed in [62, 67, 93, 164, 174, 180].

This section focuses on analytical bounds for the effective permittivity tensor $\boldsymbol{\varepsilon}_{\perp}$ of two-phase 2D dispersed composites with isotropic phases. We consider only the case of real-valued permittivity of phases. The obtained results are extended to piezoelectric fibrous composites by a decomposition matrix method developed in [175, 211] and discussed above in this chapter. The theoretical basis of contrast bounds was developed in [30, 32, 147, 153] and summarized in [51, 155, 226].

The contrast expansion approach is based on two dimensionless contrast parameters, ϱ and z. The first dimensionless contrast parameter ϱ is introduced through the permittivity of inclusions ε_1 and matrix ε_2

$$\varrho = \frac{\varepsilon_1 - \varepsilon_2}{\varepsilon_1 + \varepsilon_2}. \tag{5.193}$$

It is related to the ratio of the permittivities as follows

$$\frac{\varepsilon_1}{\varepsilon_2} = \frac{1+\varrho}{1-\varrho}. \tag{5.194}$$

The tensor $\boldsymbol{\varepsilon}_{\perp}$ can be expanded in the series [93, 155, 226]

$$\boldsymbol{\varepsilon}_{\perp} = \varepsilon_2 \sum_{k=0}^{\infty} \mathbf{b}_k \varrho^k, \tag{5.195}$$

the matrix coefficients $\mathbf{b}_k$ depend on microstructure. The second contrast parameter

$$z = \frac{\varepsilon_1}{\varepsilon_2} - 1 \tag{5.196}$$

is related to the first one by the relations

$$z = \frac{2\varrho}{1-\varrho} \iff \varrho = \frac{z}{z+2}. \tag{5.197}$$

The tensor $\boldsymbol{\varepsilon}_{\perp}$ can be expanded in the series [93, 155, 226]

$$\boldsymbol{\varepsilon}_{\perp} = \varepsilon_2 \sum_{m=0}^{\infty} \mathbf{c}_m z^m. \tag{5.198}$$

If the coefficients $\mathbf{b}_k$ are known, $\mathbf{c}_k$ are found by the substitution of

$$\varrho = z \sum_{l=0}^{\infty} (-1)^l \frac{z^l}{2^{l+1}} \tag{5.199}$$

into (5.195) and selection of the terms on the same powers of z. The series (5.195) converges absolutely in the unit disk $|\varrho| < 1$ of the complex plane [93, 180, 184]. The series (5.198) converges absolutely in the unit disk $|z| < 1$ due to the conformal mapping (5.197) of the disk $|\varrho| < 1$ onto the half-plane Re $z > -1$.

5.12.2 CONTRAST EXPANSIONS FOR MACROSCOPICALLY ISOTROPIC COMPOSITE

We consider N non-overlapping inclusions G_k $(k = 1,2,\ldots,N)$ of permittivity ε_1 following the description on page 29. Let G denote the host domain of permittivity ε_2. It is assumed that the inclusions $G_k = \{z \in \mathbb{C} : |z - a_k| < r\}$ are equal disks of radius r with the centers a_k.

We will use the formula (4.68) in the form

$$\frac{1}{\varepsilon_2}\boldsymbol{\varepsilon}_\perp = (1+2\varrho f)I + 2\varrho f \sum_{k=1}^{\infty} \mathbf{A}_k f^k, \tag{5.200}$$

where a few first matrices $\mathbf{A}_k$ are given by (4.69).

Substituting the coefficients $\mathbf{A}_k$ into (4.68) and selecting the term on the same powers of ϱ, we arrive at the contrast expansion (5.195). For simplicity of presentation, we consider macroscopically isotropic composites when $\boldsymbol{\varepsilon}_\perp = \varepsilon_e \mathbf{I}$ with a scalar ε_e. It is also for simplicity assumed that ε_1 and ε_2 are real.

First, we derive a few first coefficients in the series corresponding to (5.195)

$$\frac{\varepsilon_e}{\varepsilon_2} = \sum_{k=0}^{\infty} b_k \varrho^k \tag{5.201}$$

and to (5.198)

$$\frac{\varepsilon_e}{\varepsilon_2} = \sum_{k=0}^{\infty} c_k z^k =: F(z). \tag{5.202}$$

Here, ε_e is considered as a function of ϱ in (5.201) and as a function of z in (5.202). It follows from [32] that

$$b_0 = c_0 = 1, \quad b_1 = 2f, \quad b_2 = 2f^2, \quad c_1 = f,\ c_2 = \frac{f(1-f)}{2}. \tag{5.203}$$

The coefficients c_j $(j = 0,1,2)$ are obtained from the low-order expansion (5.195) and (5.197)

$$\frac{\varepsilon_e}{\varepsilon_2} = 1 + \frac{2fz}{z+2} + 2\left(\frac{fz}{z+2}\right)^2 + O(|z|^3) = 1 + fz + \frac{1}{2}f(1-f)z^2 + O(|z|^3). \tag{5.204}$$

The second-order coefficient agrees with the coefficient $\mathbf{A}_1$ from (4.69) since $e_2 = \pi$ for macroscopically isotropic composites [93].

The third-order contrast coefficients can be calculated by the following formula [93, p.90]

$$b_3 = 2f^3 \sum_{n=2}^{\infty} \frac{(-1)^n(n-1)}{\pi^n} e_{nn} f^{n-2}, \quad c_3 = \frac{1}{8}(b_3 + 2f - 4f^2). \tag{5.205}$$

Swap the constants ε_1 and ε_2 in (5.201). Then, the effective permittivity considered as a function $\varepsilon_e(\varrho)$ in ϱ satisfies the relation

$$\frac{\varepsilon_e(-\varrho)}{\varepsilon_1} = \sum_{k=0}^{\infty} b_k(-\varrho)^k. \tag{5.206}$$

Keller's identity can be written in the form [154, Section VI]

$$\varepsilon_e(\varrho)\varepsilon_e(-\varrho) = \varepsilon_1\varepsilon_2. \tag{5.207}$$

Substitute the series (5.201) and (5.206) into Keller's identity and expand the result into series in ϱ. The coefficients on ϱ^{2k} yield algebraic relations expressing b_{2k} through the previous coefficients. In the case $k = 2$, we arrive at the exact formula

$$b_4 = 2f(b_3 - f^3). \tag{5.208}$$

5.12.3 CONTRAST BOUNDS

In the present section, we apply the contrast expansion scheme presented in Section 5.12.2 to the high-order contrast bounds. The theoretical justification of the contrast bounds is based on the properties of the Stieltjes integral and Padé approximations [32]. Below, we outline the computational formulas for the bounds. Consider the case $\varepsilon_1 > \varepsilon_2$ for definiteness.

Consider the function $F(z)$ defined by (5.202) and its Padé approximant of order (m,n)

$$\varepsilon_{m,n}[F](z) := \frac{P_m(z)}{Q_n(z)} = \frac{p_0 + p_1 z + \ldots + p_m z^m}{q_0 + q_1 z + \ldots + q_n z^n}. \tag{5.209}$$

The Padé approximant (5.209) is defined by the asymptotic relation

$$F(z)Q_n(z) - P_m(z) = O(|z|^{m+n+1}). \tag{5.210}$$

The Padé approximant (5.209) is uniquely determined by an additional condition on its coefficients, for instance, $q_0 = 1$. It follows from (5.210) that if the coefficients c_k $(k = 0, 1, \ldots, K)$ of $F(z)$ are known, the Padé approximant of order (m,n) can be easily constructed for $m + n + 1 \leq K$.

Lower and upper bounds on ε_e are obtained by the Padé approximants [155, Section 27.4]

$$\varepsilon_{2M+1}^U = \varepsilon_2\, \varepsilon_{M+1,M}[F], \quad \varepsilon_{2M}^L = \varepsilon_2\, \varepsilon_{M,M}[F]. \tag{5.211}$$

Other lower and upper bounds can be constructed by the Padé approximant $\varepsilon_{m,n}[G](z)$ for the function

$$G(z) := \left[\frac{\varepsilon_e}{\varepsilon_1}\right]^{-1} = \frac{z+1}{F(z)}. \tag{5.212}$$

We have

$$\varepsilon_{2M+1}^L = \varepsilon_1 [\varepsilon_{M+1,M}[G]]^{-1}, \quad \varepsilon_{2M}^U = \varepsilon_1 [\varepsilon_{M,M}[G]]^{-1}. \tag{5.213}$$

The bounds ε_K^L and ε_K^U given by (5.211) and (5.213) ($K = 2M, 2M+1$) are called the K-point bounds. One needs at least K first coefficients in the series (5.202) to construct the K-point bounds. Symbolic and numerical operators to expand into the Taylor series and find Padé approximants are implemented in various packages; for instance, Mathematica®.

We have $\varepsilon_{1,0}[F](z) = 1 + fz$. Hence,

$$\varepsilon_1^U = \varepsilon_2(1+fz) = \varepsilon_1 f + \varepsilon_2(1-f). \tag{5.214}$$

Along similar lines $[\varepsilon_{1,0}[G](z)]^{-1} = \frac{1}{1+(1-f)z}$ and

$$\varepsilon_1^L = \frac{\varepsilon_1}{1+(1-f)z} = \left(\frac{f}{\varepsilon_1} + \frac{1-f}{\varepsilon_2}\right)^{-1}. \tag{5.215}$$

The 1-point bounds (5.214)–(5.215) were first obtained by Wiener [233].

The Padé approximants $\varepsilon_{1,1}[F](z)$ and $[\varepsilon_{1,1}[G](z)]^{-1}$ yield the 2-point bounds equivalent to the Hashin-Shtrikman bounds [32, 71, 153]

$$\varepsilon_2^L = \varepsilon_2 \frac{\varepsilon_1(1+f)+\varepsilon_2(1-f)}{\varepsilon_1(1-f)+\varepsilon_2(1+f)}, \quad \varepsilon_2^U = \varepsilon_1 \frac{\varepsilon_1 f + \varepsilon_2(2-f)}{\varepsilon_1(2-f)+\varepsilon_2 f}. \tag{5.216}$$

The 3-point bounds were written in terms of the Torquato-Milton parameter [226, formulas (20.59) and (20.66)]

$$\zeta_1 = \frac{b_3 - 2f^3}{2(f(1-f)}, \tag{5.217}$$

where b_3 is given by (5.205). The above exact formula (5.217) for ζ_1 holds for an arbitrary set of non-overlapping disks. Milton [155, Chapter 15] and Torquato [226, Section 20.1.6] calculated ζ_1 in the form of a triple integral including n-point correlation functions ($n = 1, 2, 3$) [226, formula (20.66)] and derived the corresponding 3-point bounds. The method of correlation functions was constructively applied in [155, 226] to some special composites, e.g., phase-inversion-symmetric structures.

The high-order coefficients and bounds for square and hexagonal arrays were numerically calculated in [224]. After the derivation of the exact formulas for the tensor $\boldsymbol{\varepsilon}_\perp$ (4.60), the bounds for a regular array of cylinders lost their significance. Here, we address exact formulas in the proper sense [17]. A series of papers beginning from [43] have been published in 2001-2023. They contain numerically derived results for a regular array called the "exact formulas". It will be demonstrated in Figure 5.14 that the "exact formula" [43] is even out of bounds for the effective constants.

5.12.4 NUMERICAL EXAMPLES

It is helpful to explain the implementation details to derive high-order contrast bounds in analytic form for arbitrarily located non-overlapping N disks per periodicity cell and for large K. The effective permittivity tensor of non-overlapping disks is given by

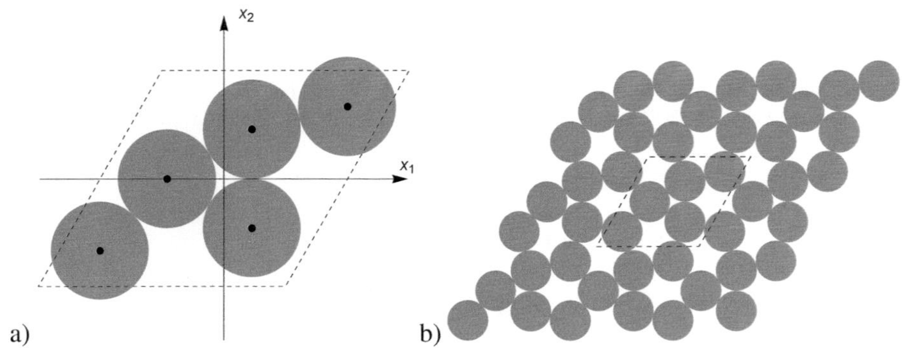

Figure 5.2 Optimal packing of five disks: a) five disks on Q_{hex}, b) double periodic packing.

(4.68)–(4.69) up to $O(f^7)$. The precision in f can be increased automatically. A high-order analytical formula can require thousands of pages [170, Supplement]; the main difficulty is its visualization. However, one shouldn't look at a long formula introducing it in the interface. It is sufficient to derive it as an iteration process and locate the result in the computer kernel. Next, one can use the function $\varepsilon_e(\varrho, f, e_{m_1,\ldots,m_q}(\mathbf{a}))$ for the computation without caring about the visualization of its complete expression. In the second step, the calculation of the structural sums $e_{m_1,\ldots,m_q}(\mathbf{a})$ for a given set $\mathbf{a}$ is computed by a fast matrix algorithm described in [67, Chapter 2], The algorithm takes seconds to calculate the required structural sums for $\sum_{l=1}^{q} m_l \le 32$.

Numerical examples are discussed below. The same method is applied to two essentially different structures. The first example concerns five packing disks in the hexagonal periodicity cell displayed in Figure 5.2. This structure is macroscopically isotropic [59] and contains percolation chains. Therefore, its effective permittivity becomes infinite for the perfectly conducting touching inclusions. The second example concerns 100 non-overlapping disks randomly selected in the square periodicity cell; see Figure 5.6. The centers of disks are independent and identically distributed random variables. 100 centers are simulated once and fixed in further computations. The second structure does not reach the percolation regime since the touching disks of maximally possible radius do not form a percolation chain.

Example 1. Five packing disks in the hexagonal periodicity cell

Let Q_{hex} denote the hexagonal periodicity cell defined by the fundamental translation vectors $\boldsymbol{\omega}_1 = \frac{\sqrt{2}}{\sqrt[4]{3}}$ and $\boldsymbol{\omega}_2 = \frac{\sqrt{2}}{\sqrt[4]{3}} \exp(\frac{i\pi}{6})$. Consider the double periodic optimal packing of five disks on the flat torus represented by Q_{hex} [59]. The radius of optimal packing is equal to $r_c = \frac{\sqrt{11}-\sqrt{3}}{4\sqrt{2}\sqrt[4]{3}} \approx 0.2128$. The exact values of the disk center $\mathcal{A}$ are written in [170]. Numerical approximate values of $\mathcal{A}$ are given below

$$\mathcal{A} \approx \{0.123+0.213i, -0.246, 0.123-0.213i, 0.537+0.31i, -0.537-0.31i\}. \quad (5.218)$$

The discussed optimal packing is displayed in Figure 5.2.

The set $\mathcal{A}$ forms a macroscopically isotropic structure. Its effective permittivity $\varepsilon_{\mathcal{A}}(f,\varrho)$ can be represented in the form of double powers series in f and ϱ, and

$$\begin{aligned}\frac{\varepsilon_{\mathcal{A}}(f,\varrho)}{\varepsilon_2} &= 1+2f\varrho+2f^2\varrho^2+2.0429f^3\varrho^3+0.522075f^4\varrho^3+0.421356f^5\varrho^3\\
&+0.00630323f^6\varrho^3+0.0891653f^7\varrho^3+0.0553012f^8\varrho^3+0.00141888f^9\varrho^3\\
&+0.00590768f^{10}\varrho^3+0.000050285f^{11}\varrho^3+0.00152799f^{12}\varrho^3+2.08579f^4\varrho^4\\
&+1.04415f^5\varrho^4+0.842713f^6\varrho^4+0.0126065f^7\varrho^4+0.178331f^8\varrho^4+0.110602f^9\varrho^4\\
&+0.00283777f^{10}\varrho^4+0.0118154f^{11}\varrho^4+0.0001f^{12}\varrho^4+2.16631f^5\varrho^5+1.50999f^6\varrho^5\\
&+1.55812f^7\varrho^5+0.154633f^8\varrho^5+0.562255f^9\varrho^5+0.408484f^{10}\varrho^5+0.0180911f^{11}\varrho^5\\
&+1.12273f^{12}\varrho^5+2.24775f^6\varrho^6+1.99822f^7\varrho^6+2.42788f^8\varrho^6+0.516909f^9\varrho^6\\
&+1.04207f^{10}\varrho^6+0.757946f^{11}\varrho^6+0.0998668f^{12}\varrho^6+2.3631f^7\varrho^7+2.48993f^8\varrho^7\\
&+3.41721f^9\varrho^7+1.27768f^{10}\varrho^7+1.59983f^{11}\varrho^7+1.30881f^{12}\varrho^7+2.48099f^8\varrho^8\\
&+3.0213f^9\varrho^8+4.56f^{10}\varrho^8+2.3946f^{11}\varrho^8+2.4639f^{12}\varrho^8+2.6304f^9\varrho^9+3.603f^{10}\varrho^9\\
&+5.82016f^{11}\varrho^9+3.91998f^{12}\varrho^9+2.78445f^{10}\varrho^{10}+4.24013f^{11}\varrho^{10}+7.26766f^{12}\varrho^{10}\\
&+2.96868f^{11}\varrho^{11}+4.96673f^{12}\varrho^{11}+3.16024f^{12}\varrho^{12}+O(f^{13}|\varrho|^{13}). \qquad (5.219)\end{aligned}$$

The function $\varepsilon_{\mathcal{A}}(f,\varrho)$ is analytic in the unit disk $|\varrho|<1$ on the complex plane. It is represented by the series (5.201) absolutely convergent in $|\varrho|<1$. Approximation of the series (5.201) up to $O(\varrho^{13})$ and truncation of every coefficient b_k ($k=1,2,\ldots,12$) up to $O(f^{13})$ yields the asymptotic formula (5.219). Substitute $\varrho=\frac{z}{z+2}$ into (5.219) and expand the result on the powers of z. We arrive at the asymptotic formula valid up to $O(f^{13}|\varrho|^{13})$. Next, we find the Padé approximants and the contrast bounds (5.211) and (5.213). The long formulas for the bounds with f in the symbolic form are given in [170, Supplement]. Below the bounds (5.211) with $m=2$ for $f=0.7$ are written

$$\varepsilon_5^U=\frac{1+1.406\varrho+0.197\varrho^2-0.209\varrho^3}{1+0.00624\varrho-0.791\varrho^2-0.0155\varrho^3},\ \varepsilon_4^L=\frac{1+0.7\varrho-0.16\varrho^2}{1-0.7\varrho-0.16\varrho^2}. \qquad (5.220)$$

The series $\varepsilon_{\mathcal{A}}(f,\pm 1)$ converge for $0\le f<f_c$ where the critical concentration $f_c=5\pi r_c^2\approx 0.711597$ corresponds to the touching disks. The percolation regime $\varepsilon_{\mathcal{A}}(f,1)\to+\infty$, as $f\to f_c$, can be investigated by the renormalization approach [93]. Below, we write the simple asymptotic formula

$$\frac{\varepsilon_{\mathcal{A}}(f,1)}{\varepsilon_2}=\frac{0.843+1.09f+0.289f^2-0.033f^3+0.147f^4+0.41f^5-1.4f^6}{\sqrt{f_c-f}} +O(f^7). \qquad (5.221)$$

The K-point bounds are found and displayed in Figure 5.3 with the fixed ratio $\frac{\varepsilon_1}{\varepsilon_2}=20 \iff \varrho=0.9$, in Figure 5.4 with the critical concentration $f=f_c$, and in Figure 5.5 with the fixed ratio $\frac{\varepsilon_1}{\varepsilon_2}=100 \iff \varrho=0.98$. The numerical values of points displayed in Figure 5.5 are given in Tables 5.1.

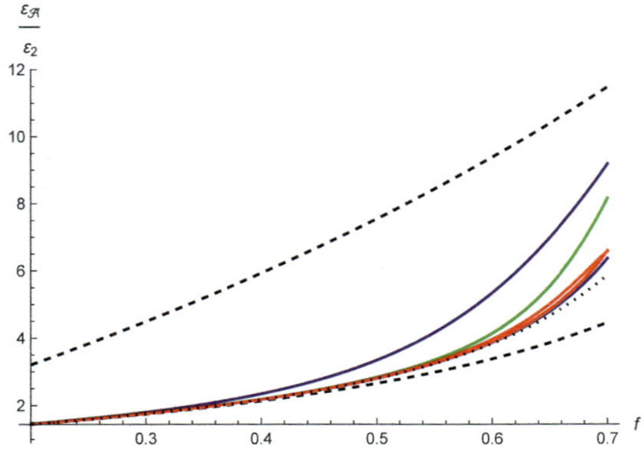

Figure 5.3 The K-point bounds for $\frac{1}{\varepsilon_2}\varepsilon_{\mathcal{A}}(f, 0.9)$. Data are for black-dashed ($K = 2$, i.e., Hashin-Shtrikman bounds), blue ($K = 4$), green ($K = 5$), and red ($K = 6$). The polynomial approximation (5.223) is shown by black-dotted. In the black-and-white version, the dashed, blue, green, and red lines go from the top.

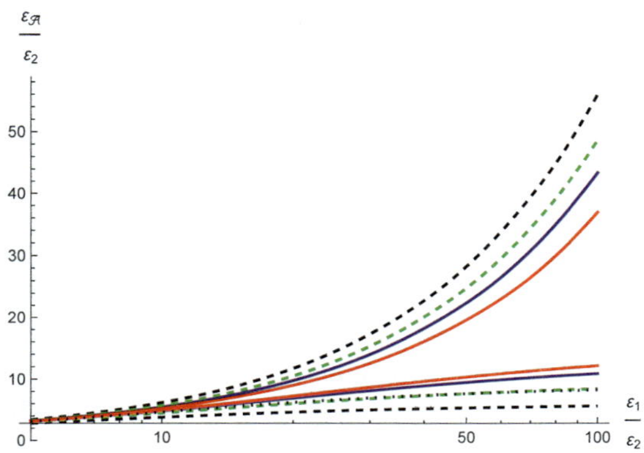

Figure 5.4 The log-plot of the K-point bounds for $\frac{1}{\varepsilon_2}\varepsilon_{\mathcal{A}}\left(f_c, \frac{\varepsilon_1-\varepsilon_2}{\varepsilon_1+\varepsilon_2}\right)$. Data are for black-dashed ($K = 2$), green-dashed ($K = 3$), second line from the top, blue ($K = 4$), and fourth line from the top, red ($K = 5$). The polynomial approximation (5.223) is shown by black-dotted. In the black-white version, the dashed, green-dashed, blue, and red lines go from the top and from the bottom.

One can observe in Figure 5.3 the almost coinciding lower and upper 6-point bounds (fourth line from the top, red in the online version) for $\frac{\varepsilon_1}{\varepsilon_2} < 20$. The polynomial approximation (5.223) (black-dotted) lies below the lower 6-point bound in Figures 5.3 and 5.4. This means that a polynomial approximation is inaccurate for the percolation regime near $f \sim f_c$ as noted in [93].

Further simulations demonstrate that the higher-order Padé approximants are not stable when calculated with the concentration f in symbolic form. The results become excellent when a Padé approximant in ϱ is found numerically for a fixed f. The 11-point bounds (points) and the polynomial approximation (5.223) (black line)

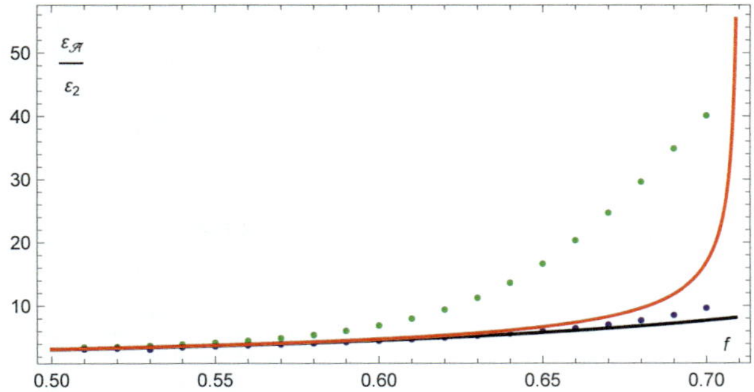

Figure 5.5 The 11-point bounds for $\frac{1}{\varepsilon_2}\varepsilon_{\mathcal{A}}(f, 0.98)$ (blue and green points), the polynomial approximation (5.223) (black line), and $\frac{1}{\varepsilon_2}\varepsilon_{\mathcal{A}}(f, 1)$ calculated by (5.221) red line).

are displayed in Figure 5.5. The value $\frac{1}{\varepsilon_2}\varepsilon_{\mathcal{A}}(f, 1)$ (red line) estimated using the renormalization method is added for completeness.

Example 2. Random 100 disks in the periodic unit square cell

Consider an example of $N = 100$ non-overlapping disks in the unit square displayed in Figure 5.6. The disks are simulated by sequence location described in [62] with the radius threshold $r_0 = 0.035$. This means the centers a_k are uniformly distributed in the unit square with the conditions $dist(a_k, a_m) > 2r_0$ $(m \neq k)$, where the non-overlapping condition holds in the plane torus topology. More precisely, the distance is introduced as follows

$$dist(z_1, z_2) := \min_{s_1, s_2 = 0, \pm 1} |z_1 - z_2 + s_1 + \mathrm{i}s_2|. \tag{5.222}$$

In the limit case $r_0 = 0$, the set of centers $\mathcal{A} = \{a_1, a_2, \ldots, a_N\}$ satisfies the Poisson process.

We stress that the centers a_k are uniformly distributed in the unit square, not the disks $|z - a_k| < r_0$ $(r_0 > 0)$. These distributions of $\mathcal{A}$ determine different stochastic processes [122]. The replacement of the distance (5.222) by the Euclidean metric $dist_E(z_1, z_2) = |z_1 - z_2|$ also perturbs the required uniform i.i.d. (independent and

Table 5.1

The numerical values of the 11-point bounds displayed in Figure 5.5 vs concentration f: ε_{11}^L (blue points), ε_{11}^U (green points).

f	0.51	0.52	0.53	0.54	0.55	0.56	0.57	0.58	0.59	0.6
ε_{11}^L	3.15	3.27	3.10	3.46	3.60	3.75	3.91	4.08	4.27	4.48
ε_{11}^U	3.47	3.54	3.67	3.91	4.17	4.49	4.89	5.40	6.05	6.89

f	0.61	0.62	0.63	0.64	0.65	0.66	0.67	0.68	0.69	0.7
ε_{11}^L	4.71	4.97	5.27	5.61	6.00	6.45	7.00	7.68	8.52	9.61
ε_{11}^U	7.98	9.40	11.3	13.6	16.6	20.3	24.7	29.6	34.9	40.1

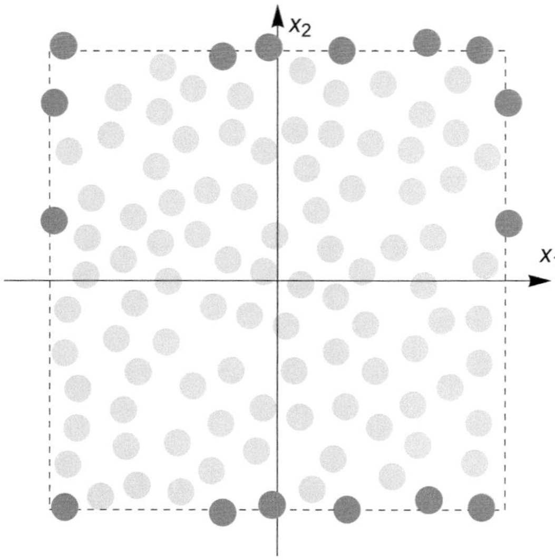

Figure 5.6 100 disks in the unit square periodicity cell. The dark gray disks on the opposite sides of the square are identified by pairs according to the torus topology.

identically distributed) non-overlapping disks. Doubly periodic structure (plane torus) is the key assumption of the homogenization theory, which eliminates the influence of boundaries that tacitly exist and disturbs the results in the framework of SCM, EMA, and other similar empirical approximations discussed in Chapter 3.

The effective permittivity $\varepsilon_{100}(f,\varrho)$ can be represented in the form of double series in the powers f and ϱ, and

$$\begin{aligned}\frac{\varepsilon_{100}(f,\varrho)}{\varepsilon_2} &= 1+2f\varrho+2f^2\varrho^2+2.30974f^3\varrho^3+0.358151f^4\varrho^3+0.440133f^5\varrho^3\\ &+0.283968f^6\varrho^3+0.17321f^7\varrho^3+0.0763706f^8\varrho^3+0.0349588f^9\varrho^3+0.0192229f^{10}\varrho^3\\ &+0.0100547f^{11}\varrho^3+0.00455705f^{12}\varrho^3+2.63336f^4\varrho^4+0.764796f^5\varrho^4+0.828513f^6\varrho^4\\ &+0.591477f^7\varrho^4+0.357103f^8\varrho^4+0.281438f^9\varrho^4+0.255881f^{10}\varrho^4+0.175048f^{11}\varrho^4\\ &+0.0607134f^{12}\varrho^4+3.20538f^5\varrho^5+1.42227f^6\varrho^5+1.71443f^7\varrho^5+1.69485f^8\varrho^5\\ &+1.62367f^9\varrho^5+2.09162f^{10}\varrho^5+2.7258f^{11}\varrho^5+3.72375f^{12}\varrho^5+3.87034f^6\varrho^6\\ &+2.25248f^7\varrho^6+2.86381f^8\varrho^6+3.14345f^9\varrho^6+3.1104f^{10}\varrho^6+4.40338f^{11}\varrho^6\\ &+5.4525f^{12}\varrho^6+4.87402f^7\varrho^7+3.5852f^8\varrho^7+4.91752f^9\varrho^7+5.87811f^{10}\varrho^7\\ &+6.99559f^{11}\varrho^7+10.3469f^{12}\varrho^7+6.07698f^8\varrho^8+5.34872f^9\varrho^8+7.69997f^{10}\varrho^8\\ &+9.75023f^{11}\varrho^8+12.5134f^{12}\varrho^8+7.80964f^9\varrho^9+8.07344f^{10}\varrho^9+12.3949f^{11}\varrho^9\\ &+16.5787f^{12}\varrho^9+9.93356f^{10}\varrho^{10}+11.7505f^{11}\varrho^{10}+18.8899f^{12}\varrho^{10}+12.9369f^{11}\varrho^{11}\\ &+17.2854f^{12}\varrho^{11}+16.6825f^{12}\varrho^{12}++O(f^{13}|\varrho|^{13}). \qquad (5.223)\end{aligned}$$

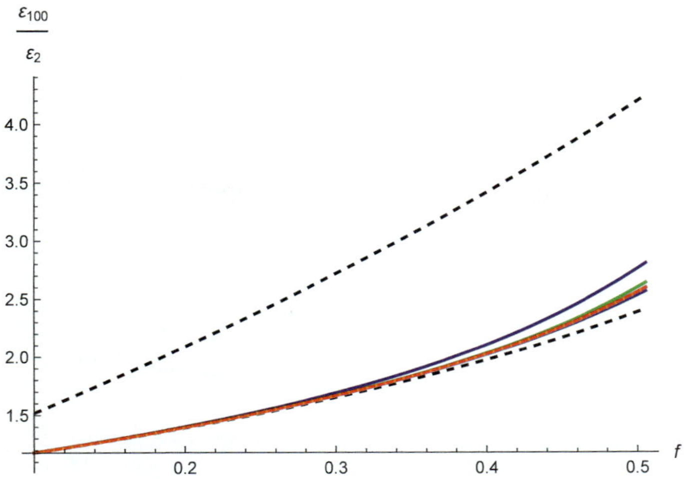

Figure 5.7 The K-point bounds for $\frac{\varepsilon_{100}(f,0.818)}{\varepsilon_2}$. Data are for black-dashed (K = 2, i.e., Hashin-Shtrikman bounds), blue (K = 4), green (K = 6), and red (K = 7). The polynomial approximation (5.223) coincides with the lower and upper 7-point bounds (red). The dashed, blue, green, red, and black lines go from the top in the black-white version.

The series $\varepsilon_{100}(f,\varrho)$ converge for $0 \leq f \leq f_c$ where $f_c = 0.505482$. The first touching disks arise for the radius $r_c = 0.0401123$. This structure does not form a percolation chain. Hence, the series $\varepsilon_{100}(f,\varrho)$ converges in the closed disk $|\varrho| \leq 1$ on the complex plane [185].

The K-point bounds are found by (5.211) and (5.213) and displayed in Figure 5.7 with the fixed ratio $\frac{\varepsilon_1}{\varepsilon_2} = 10 \iff \varrho = 0.818$ and in Figure 5.8 with the critical concentration $f = f_c$. One can note that the lower 6-point bound, the lower and upper 7-point bounds, and the polynomial approximation (5.223) almost coincide as functions of f in Figure 5.7. The lower 4-point bound, the lower and upper 7-point bounds, and the polynomial approximation (5.223) almost coincide for $1 \leq \frac{\varepsilon_1}{\varepsilon_2} \leq 100$ in Figure 5.8 and close the lower Hashin-Shtrikman bound. The part $1 \leq \frac{\varepsilon_1}{\varepsilon_2} \leq 10$ from Figure 5.8 is displayed in Figure 5.9, where the lower and upper 7-point bounds, and the polynomial approximation coincide. Therefore, the 7-point upper bound for $f = f_c$

$$\varepsilon_7^U = \varepsilon_1 \frac{3.63466\varepsilon_1^3 + 13.7384\varepsilon_1^2\varepsilon_2 + 7.83067\varepsilon_1\varepsilon_2^2 + 0.924842\varepsilon_2^3}{\varepsilon_1^4 + 8.22711\varepsilon_1^3\varepsilon_2 + 13.5477\varepsilon_1^2\varepsilon_2^2 + 3.35568\varepsilon_1\varepsilon_2^3 - 0.0019548\varepsilon_2^4} \tag{5.224}$$

can be used for an accurate approximation of $\varepsilon_{100}\left(f_c, \frac{\varepsilon_1 - \varepsilon_2}{\varepsilon_1 + \varepsilon_2}\right)$.

5.13 BOUNDS FOR PIEZOELECTRIC FIBER COMPOSITES

The effective tensor of the considered piezoelectric composite has the form (5.142), where the elements of the matrix $\boldsymbol{\psi}$ in the case $\Delta > 0$ are given by the expressions

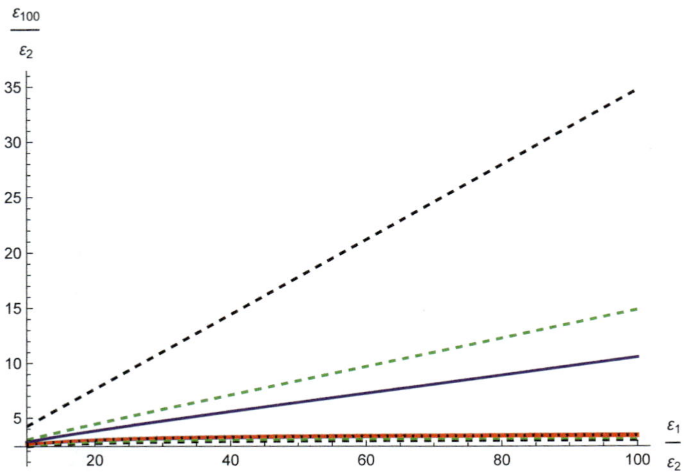

Figure 5.8 The K-point bounds for $\frac{1}{\varepsilon_2}\varepsilon_{100}\left(f_c, \frac{\varepsilon_1-\varepsilon_2}{\varepsilon_1+\varepsilon_2}\right)$. Data are for black-dashed ($K = 2$), green-dashed ($K = 3$), blue ($K = 4$), and red ($K = 7$). The polynomial approximation (5.223) shown by dots coincides with the lower and upper 7-point bounds (red). The dashed, dashed green, blue, and red lines go from the top in the black-white version.

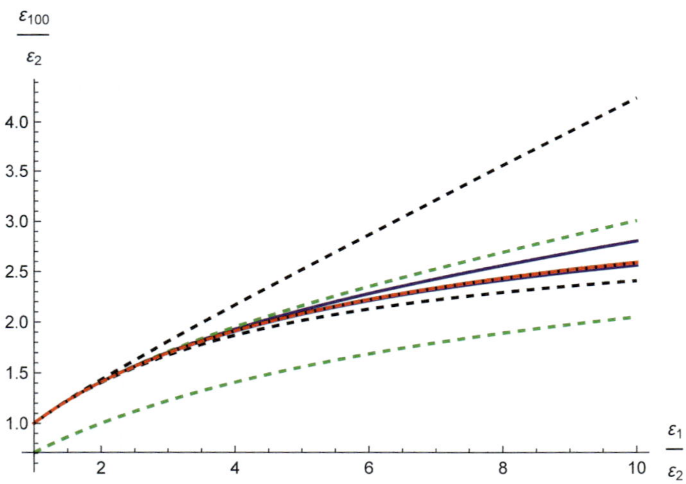

Figure 5.9 The same designations as in Figure 5.8 for the ratio $1 \leq \frac{\varepsilon_1}{\varepsilon_2} \leq 10$.

(5.176) written here in the slightly modified form

$$\psi_{11} = \frac{1}{2}[P(\varrho_1') + P(\varrho_2')] - \frac{h(\varrho_1 - \varrho_2)}{2\sqrt{\Delta}}[P(\varrho_1') - P(\varrho_2')],$$
$$\psi_{12} = \frac{e(\varrho_2 - \varrho_3)}{\sqrt{\Delta}}[P(\varrho_1') - P(\varrho_2')],\ \psi_{21} = \frac{c(\varrho_3 - \varrho_1)}{\sqrt{\Delta}}[P(\varrho_1') - P(\varrho_2')], \quad (5.225)$$
$$\psi_{22} = \frac{1}{2}[P(\varrho_1') + P(\varrho_2')] + \frac{h(\varrho_1 - \varrho_2)}{2\sqrt{\Delta}}[P(\varrho_1') - P(\varrho_2')],$$

and in the case $\Delta < 0$ by the expressions (5.190).

It follows from (5.225) and (5.190) that

$$\psi_{11} = \frac{1}{2}[P(\varrho_1') + P(\varrho_2')] - \frac{c(\varrho_1 - \varrho_2)}{2(\varrho_2 - \varrho_3)}\psi_{12},$$
$$\psi_{12} = \frac{e(\varrho_2 - \varrho_3)}{\sqrt{\Delta}}[P(\varrho_1') - P(\varrho_2')],\ \psi_{21} = \frac{c(\varrho_3 - \varrho_1)}{e(\varrho_2 - \varrho_3)}\psi_{12}, \quad (5.226)$$
$$\psi_{22} = \frac{1}{2}[P(\varrho_1') + P(\varrho_2')] + \frac{c(\varrho_1 - \varrho_2)}{2(\varrho_2 - \varrho_3)}\psi_{12}.$$

Hence, ψ_{11}, ψ_{12} and ψ_{22} are estimated by $P(\varrho_1')$ and $P(\varrho_2')$. The component ψ_{21} can be determined through ψ_{12} by a simple formula.

If some contrast bounds ϱ^L and ϱ^U are constructed for the effective conductivity ϱ'^{h}, we obtain from (5.173)

$$P^L(\varrho') \equiv \frac{\varrho^L(\varrho') - 1}{2\varrho' f} \leq P(\varrho') \leq \frac{\varrho^U(\varrho') - 1}{2\varrho' f} \equiv P^U(\varrho'). \quad (5.227)$$

Substitution of (5.227) into (5.225) and next the result into (5.142) yields the contrast bounds for the components of the effective piezoelectric tensor (5.142).

We illustrate the bounds (5.227) using the asymptotic formula

$$P(\varrho) = 1 + f\varrho + \frac{f^2\varrho^2}{\pi^2}(1 + 2f\varrho)e_{22} - \frac{2f^3\varrho^2}{\pi^3}e_{33} - f^3\varrho^3 + O(f^4). \quad (5.228)$$

This formula for a macroscopically isotropic composite is based on (4.68)–(4.69) and (5.207), for details, see [177]. The 4-point contrast bounds for the approximation up to $O(f^5)$ take the form

$$\varrho^U(\varrho') = \frac{P_4^U(\varrho')}{Q_4^U(\varrho')}, \quad \varrho^L(\varrho') = \frac{P_4^L(\varrho')}{Q_L^U(\varrho')}, \quad (5.229)$$

where

$$P_4^U(\varrho') = \pi e_{22} f^2 \varrho'((2f-1)\varrho' + 1) + 2e_{33} f^3 \varrho'(-2f\varrho' + \varrho' - 1)$$
$$-\pi^3\left(2f^3\varrho'^2 - 2f^2(\varrho'-1)\varrho' + f(\varrho'-1)^2 + \varrho' - 1\right),$$
$$Q_4^U(\varrho') = (\varrho'-1)\left(-\pi e_{22} f^2 \varrho' + 2e_{33} f^3 \varrho' + \pi^3(f\varrho' + f - 1)\right),$$
$$P_4^L(\varrho') = 2\pi f e_{33}\left[\pi^2\left(4f^2\varrho^2 + 3f\varrho + 1\right) - 2e_{22} f^2 \varrho^2\right] + 4e_{33}^2 f^4 \varrho^2$$
$$+\pi^2\left(\pi^2 - e_{22}\right)\left(\pi^2\left(f^2\varrho^2 + f\varrho + 1\right) - e_{22} f^2 \varrho^2\right),$$
$$Q_L^U(\varrho) = 2\pi f e_{33}\left[\pi^2(f\varrho + 1) - 2e_{22} f^3 \varrho'^2\right] + 4e_{33}^2 f^4 \varrho'^2$$
$$+\pi^2\left(\pi^2 - e_{22}\right)\left(\pi^2\left(f^2\varrho'^2 - f\varrho' + 1\right) - e_{22} f^2 \varrho'^2\right).$$

Substitution of (5.229)–(5.230) into (5.180) yields the pare of bounds

$$p^L(\varrho_j) \leq p(\varrho_j) \leq p^U(\varrho_j), \quad j = 1,2. \tag{5.230}$$

Find the matrix $\boldsymbol{\psi}$ from (5.225)

$$\boldsymbol{\psi}(\mathbf{D}^{\mathrm{h}}) = \frac{1}{2f}\mathbf{R}^{-1}(\mathbf{D}_+^{-1}\mathbf{D}^{\mathrm{h}} - \mathbf{I}). \tag{5.231}$$

The dependence of the matrix $\boldsymbol{\psi}$ on $\mathbf{D}^{\mathrm{h}}$, i.e., on three effective constants d_{11}^{h}, $d_{12}^{\mathrm{h}} = d_{21}^{\mathrm{h}}$ and d_{22}^{h} is stressed here. Consider the component $\psi_{11}(\mathbf{D}^{\mathrm{h}})$. Equations (5.225) implies that it can be written in the form

$$\psi_{11}(\mathbf{D}^{\mathrm{h}}) = \frac{1}{2}p(\varrho_1')\left[1 - \frac{h(\varrho_1 - \varrho_2)}{2\sqrt{\Delta}}\right] + \frac{1}{2}p(\varrho_2')\left[1 + \frac{h(\varrho_1 - \varrho_2)}{2\sqrt{\Delta}}\right]. \tag{5.232}$$

Consider for definiteness the case when the parameters of components satisfy the inequalities

$$1 - \frac{h(\varrho_1 - \varrho_2)}{2\sqrt{\Delta}} > 0,\ 1 + \frac{h(\varrho_1 - \varrho_2)}{2\sqrt{\Delta}} > 0,\ \frac{e(\varrho_2 - \varrho_3)}{\sqrt{\Delta}} < 0. \tag{5.233}$$

Using (5.230) and (5.233) we obtain from (5.232)

$$\begin{aligned}\frac{1}{2}p^L(\varrho_1')\left[1 - \frac{h(\varrho_1 - \varrho_2)}{2\sqrt{\Delta}}\right] + \frac{1}{2}p^L(\varrho_2')\left[1 + \frac{h(\varrho_1 - \varrho_2)}{2\sqrt{\Delta}}\right] \leq \psi_{11}(\mathbf{D}^{\mathrm{h}}) \leq \\ \leq \frac{1}{2}p^U(\varrho_1')\left[1 - \frac{h(\varrho_1 - \varrho_2)}{2\sqrt{\Delta}}\right] + \frac{1}{2}p^U(\varrho_2')\left[1 + \frac{h(\varrho_1 - \varrho_2)}{2\sqrt{\Delta}}\right].\end{aligned} \tag{5.234}$$

The inequality (5.234) can be considered as the lower and upper bounds on the linear combination of the effective constants d_{11}^{h}, d_{12}^{h} and d_{22}^{h} including in $\psi_{11}(\mathbf{D}^{\mathrm{h}})$. Of course, the bounds (5.234) will change for other signs of brackets. Analogous bounds hold for other $\boldsymbol{\psi}$ components. Then (5.226),(5.233) and (5.230) yield

$$\frac{e(\varrho_2 - \varrho_3)}{\sqrt{\Delta}}[p^U(\varrho_1') - p^L(\varrho_2')] \leq \psi_{12}(\mathbf{D}^{\mathrm{h}}) \leq \frac{e(\varrho_2 - \varrho_3)}{\sqrt{\Delta}}[p^L(\varrho_1') - p^U(\varrho_2')] \tag{5.235}$$

and

$$\begin{aligned}\frac{1}{2}p^L(\varrho_1')\left[1 + \frac{h(\varrho_1 - \varrho_2)}{2\sqrt{\Delta}}\right] + \frac{1}{2}p^L(\varrho_2')\left[1 - \frac{h(\varrho_1 - \varrho_2)}{2\sqrt{\Delta}}\right] \leq \psi_{22}(\mathbf{D}^{\mathrm{h}}) \leq \\ \leq \frac{1}{2}p^U(\varrho_1')\left[1 + \frac{h(\varrho_1 - \varrho_2)}{2\sqrt{\Delta}}\right] + \frac{1}{2}p^U(\varrho_2')\left[1 - \frac{h(\varrho_1 - \varrho_2)}{2\sqrt{\Delta}}\right].\end{aligned} \tag{5.236}$$

Example.

At the end of this section, we consider a numerical example illustrating the strategy of the applications of the derived theoretical bounds. The selected data and the obtained intermediate numerical results demonstrate various options that arise during the implementation.

Let the host and inclusions of the piezoelectric components be described by equations (5.100) with the dimensionless matrices

$$\mathbf{D}_+ = \begin{pmatrix} 3 & 0 \\ 0 & -1 \end{pmatrix}, \quad \mathbf{D}_- = \begin{pmatrix} 1 & 0.7 \\ 0.7 & -7 \end{pmatrix}. \tag{5.237}$$

Consider the location of 100 disks from Section 5.12.4 for which $e_{22} = 11.3981$ and $e_{33} = -2.7762$. Calculate $\varrho_1' = -0.473$ and $\varrho_2' = 0.749$ by (5.159). The bounds $p^L(\varrho_j)$ and $p^U(\varrho_j)$ from (5.230) take the numerical form

$$P^L(\varrho_1') = \frac{1 - f - 0.04972f^2 - 0.05749f^3}{1 - 0.5272f - 0.5461f^2 - 0.08467f^3}, \tag{5.238}$$

$$P^U(\varrho_1') = \frac{1 + 1.1563f - 1.0935f^2}{1 + 1.6291f - 0.5814f^2 - 0.5971f^3 - 0.04629f^4},$$

$$P^L(\varrho_2') = \frac{1 - f + 0.4627f^2 + 0.535f^3}{1 - 1.7492f + 0.8653f^2 + 0.1342f^3}, \tag{5.239}$$

$$P^U(\varrho_2') = \frac{1 + 1.1563f + 1.73267f^2}{1 + 0.4071f + 0.7794f^2 - 1.4992f^3 - 0.1162f^4}.$$

The bounds (5.230) for $P(\varrho_j')$ $(j = 1, 2)$ in the considered example are displayed in Figure 5.10. In the considered numerical example, one can take $p(\varrho_1') \approx P^L(\varrho_1') \approx P^U(\varrho_1')$ because the graphs for the bounds almost coincide in Figure 5.10a.

One can check that equations (5.233) are fulfilled for the considered data (5.237). Therefore, the bounds have the form (5.234)–(5.236). Substituting the data (5.237) into (5.231) we obtain the matrix (5.231). It is convenient to use the matrix (5.231) multiplied by f

$$f\boldsymbol{\psi}(\mathbf{D}^{\rm h}) = \begin{pmatrix} -0.36d_{11}^{\rm h} - 0.061d_{12}^{\rm h} + 1.06 & -0.36d_{12}^{\rm h} - 0.061d_{22}^{\rm h} - 0.061 \\ -0.061d_{11}^{\rm h} - 0.67d_{12}^{\rm h} + 0.18 & -0.061d_{12}^{\rm h} - 0.67d_{22}^{\rm h} - 0.67 \end{pmatrix}, \tag{5.240}$$

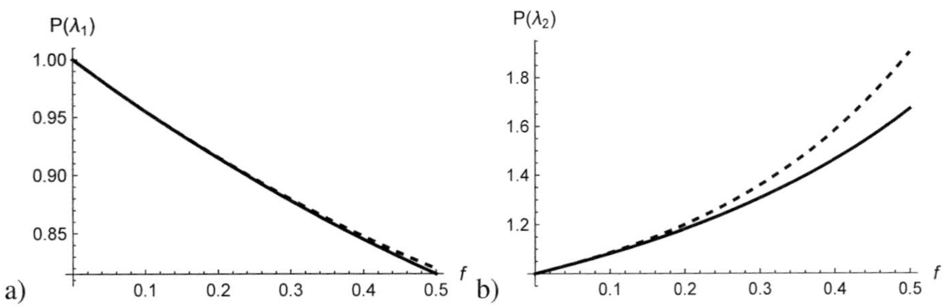

Figure 5.10 The dependence of the lower bounds (5.230) for $p(\varrho_j)$ (solid) and upper bounds (dashed) on concentration f: a) $P^L(\varrho_1')$ and $P^U(\varrho_1')$, b) $P^L(\varrho_2')$ and $p^U(\varrho_2')$.

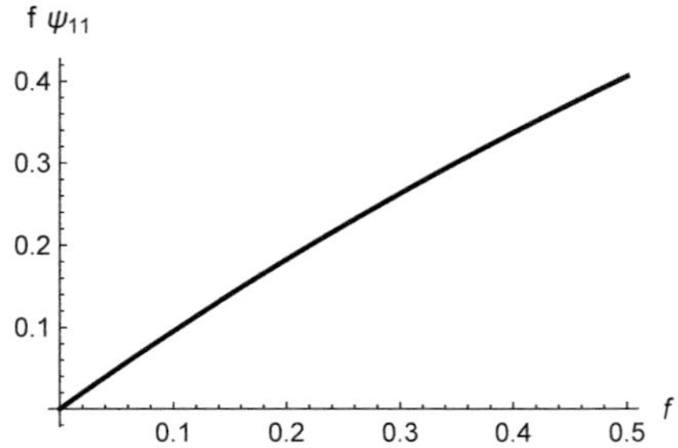

Figure 5.11 The coinciding bounds for the $(1,1)$-component of (5.240).

since it consists of a linear combination of the effective constants d^{h}_{11}, d^{h}_{12} and d^{h}_{22}. Substituting the data into the bounds (5.234)–(5.236) and using (5.240), we arrive at the numerical bounds. The lower and upper bounds for the $(1,1)$-component of (5.240) numerically coincide, see Figure 5.11. Hence, we take the following approximate equality instead of bounds

$$-0.36d^{\mathrm{h}}_{11} - 0.061d^{\mathrm{h}}_{12} + 1.06 = f\frac{1-2.7537f+2.5727f^2-0.707f^3-0.07527f^4-0.0554f^5-0.007575f^6}{1-2.2764f+1.2414f^2+0.5486f^3-0.3952f^4-0a.1466f^5-0.01136f^6}. \tag{5.241}$$

The bounds for the $(1,2)$- and $(2,2)$-components of (5.240) are displayed in Figure 5.12. It can be written numerically by the inequalities

$$b^L_{12}(f) \leq -0.36d^{\mathrm{h}}_{12} - 0.061d^{\mathrm{h}}_{22} - 0.061 \leq b^U_{12}(f), \tag{5.242}$$

where

$$b^L_{12}(f) = f^2\frac{4.38+0.91f+2.61f^2-8.48f^3-1.05f^4-0.33f^5-0.024f^6}{101.6-12.2f+1.9f^2-225.3f^3+21.8f^4+82.7f^5+19.4f^6+f^7}, \tag{5.243}$$

$$b^U_{12}(f) = f^2\frac{-3.79+6.58f-4.51f^2+1.16f^3+0.85f^4+0.12f^5}{-88+200.3f-109.3f^2-48.3f^3+34.8f^4+12.9f^5+f^6}, \tag{5.244}$$

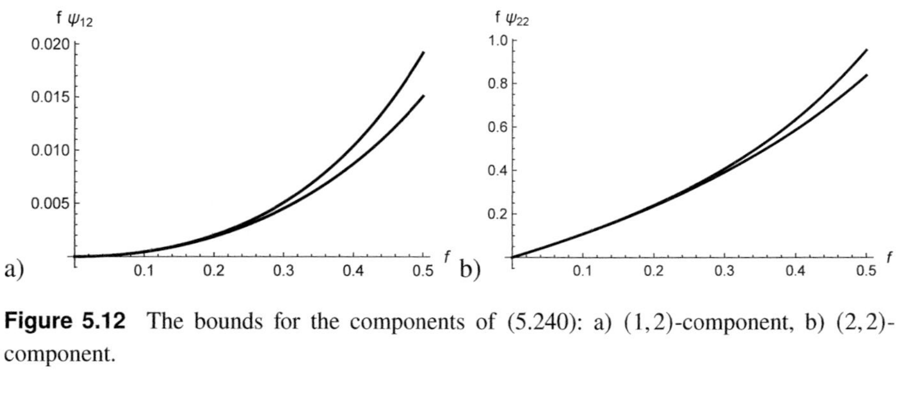

Figure 5.12 The bounds for the components of (5.240): a) $(1,2)$-component, b) $(2,2)$-component.

and

$$b_{22}^L(f) \le -0.061d_{12}^{\mathrm{h}} - 0.67d_{22}^{\mathrm{h}} - 0.67 \le b_{22}^U(f), \tag{5.245}$$

where

$$b_{22}^L(f) = f\frac{1+0.63f+0.58f^2-1.63f^3-1.05f^4-0.15f^5-0.00034f^6}{1-0.12f+0.019f^2-2.22f^3+0.21f^4+0.81f^5+0.19f^6}, \tag{5.246}$$

$$b_{22}^U(f) = f\frac{1-1.52f+0.436f^2+0.758f^3-0.451f^4-0.332f^5-0.0454f^6}{1-2.276f+1.241f^2+0.549f^3-0.395f^4-0.147f^5-0.0113f^6}. \tag{5.247}$$

Therefore, the estimated effective constants d_{12}^{h} and d_{22}^{h} satisfy the system of linear inequalities (5.242), (5.245). These inequalities determine a polygon for any fixed $0 < f < f_c = 0.505482$. In general, such a polygon can degenerate into a line or point.

Consider the case $f = 0.5$. The bounds in the inequalities (5.242) and (5.245) become

$$\begin{aligned} d_{12} &= -0.22 - 0.17d_{22}, \quad d_{12} = -0.23 - 0.17d_{22}, \\ d_{12} &= -25.17 - 11.17d_{22}, \quad d_{12} = -27 - 11.17d_{22}. \end{aligned} \tag{5.248}$$

They determine a thin parallelogram displayed in Figure 5.13. The effective constants d_{12}^{h} and d_{22}^{h} lie in this parallelogram. It follows from (5.241) that d_{11}^{h} is expressed through d_{12}^{h} by equation

$$d_{11}^{\mathrm{h}} = 1.818 - 0.169d_{12}^{\mathrm{h}}. \tag{5.249}$$

Ultimately, the effective constants d_{12}^{h} and d_{22}^{h} lie in the parallelogram displayed in Figure 5.13 and the constant d_{11}^{h} has the form (5.249).

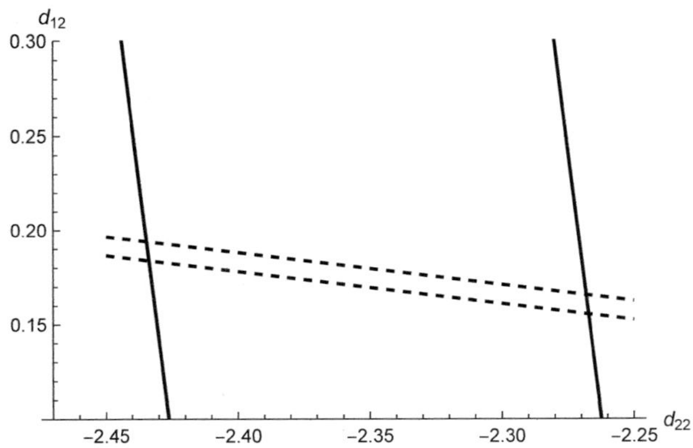

Figure 5.13 The first pare of equations (5.248) determines the solid lines; the second pare the dashed lines. The constants d_{12}^{h} and d_{22}^{h} lie in the parallelogram; its sides are defined by equations (5.248).

5.14 CONCLUSION AND DISCUSSION

The present chapter is devoted to the extension of the previous results to piezoelectric anti-plane problems by use of the vector-matrix $\mathbb{R}$–linear problem in a class of doubly periodic functions. We follow Filshtinsky's extension of Lekhnitskii's formalism to 2D MEE problems for composites. The difference between Filshtinsky's and Lekhnitskii's approach is clearly demonstrated in the example of a dielectric problem in Section 5.4. Using Filshtinsky's complex potentials, we reduce a boundary value problem for a hole and for an inclusion embedded in the MEE host material to the vector-matrix Riemann-Hilbert and to the $\mathbb{R}$-linear problems with a shift, respectively. A problem for a circular hole in the MEE bulk material is solved in closed form by the method of conformal mapping and Sochocki's formulas [79].

The general homogenization approach is applied to MEE composites. The anti-plane coupled problem for the skew-symmetric piezoelectric media is studied. The considered problem can be reduced to uncoupled permittivity (conductivity) equations by the matrix decomposition developed in [175,211] for two-phase composites. As a result, we derive exact and analytical approximate formulas for the effective piezoelectric tensor. It is demonstrated that any formula for the scalar permittivity of a 2D composite can be transformed and applied to a piezoelectric transversely isotropic fibrous composite.

The obtained decomposition is useful for numerical applications, too, since it is sufficient to solve uncoupled plane elastic problems and two uncoupled anti-plane problems instead of the vector-matrix problem expensive in computations.

The vector–matrix extension (5.147) of the Clausius–Mossotti approximation is obtained (cf [8] and [115]).

Sections 5.12 and 5.13 concern higher-order contrast bounds for piezoelectric fibrous composites. The analytical contrast bounds for dispersed composites can be divided into the lower-order bounds related to the 1- and 2- point correlation functions, e.g., Wiener [233], and Hashin-Shtrikman [103] bounds, and the higher-order contrast bounds based on the structural sums [67,93]. The 3-point correlation functions were computed for dispersed composites in [155,226].

The lower-order bounds are universal for composites. One needs to define structural geometrical parameters directly related to the effective constants to study the variety of dispersed composites. In the present chapter, we use the structural sums (4.65) as the set of structural parameters. The effective properties tensor can be expanded into the power series in concentration f. Every coefficient $\mathbf{A}_k$ on f^{k+1} is a finite linear combination of the structural sums $\mathbf{A}_k = \sum_{j=1}^{J(k)} \alpha_{kj} e_{\mathbf{m}(j)}$, see few first coefficients (4.69). The coefficients α_{kj} of every linear combination depend on the physical constants of components and do not depend on the location of inclusions. The infinite set of structural sums completely determines the distribution of inclusions, hence, the macroscopic properties of composites. The lower-order bounds do not take into account interactions among inclusions, and the higher-order contrast bounds do. Therefore, higher-order bounds describe the structure of dispersed composites and depend on the geometrical distribution of inclusion. In some special cases, the inclusions do not interact; hence, their location does not impact the effective constants, e.g., Hashin-Shtrikman assemblages and neutral inclusions [53,103,155].

The higher-order n-correlation functions were introduced by Beran [30] and applied to the contrast bounds on the theoretical level. Theory of bounds based on the contrast expansions was developed in [32, 33, 147, 152–154]. The contrast expansions for regular structures were derived in [145, 148, 204] by the method of Rayleigh [206]. These results were used in [224] to construct higher-order bounds of regular structures. The present chapter summarizes the first report [170], where the computationally effective method of structural sums is applied to higher-order contrast bounds. The analysis involves the decomposition of piezoelectric bounds into conductivity bounds, simplifying the verification process to phenomena governed by Laplace's equation.

The structural sums were initially used to construct polynomial approximations for random composites in [37, 162]. Construction of asymptotically equivalent rational approximations was performed in [145,148,204]. It was noted that the transformation of polynomial approximations to Padé approximations gives correct results, especially for percolation regimes [14,15]. It is worth noting that structural sums and the bounds presented here are computed in a couple of minutes on a regular laptop, including symbolic computations.

At the end of this section, consider the effective permittivity of the square array of disks. Figure 5.14 demonstrates that the exact expression (4.60) (red line) is between the 8-order bounds (solid blue lines). At the same time, Bruggeman's approximation lies above the upper bound (dashed-dotted line), and the "exact" formula from [207] gives the curve lying below the lower bound.

We adopt the contrast parameter $\varrho = 0.8497$ in Figure 5.14 for our comparisons, which aligns with the results [207]. Some clarifications concerning two different exact formulas for the same structure regarding Figure 5.14 are necessary. Note, that $|\varrho| < 1$. A truncated exact series (4.60) is used to employ a polynomial approximation up to $O(f^{27})$. For higher $|\varrho|$ and as f approaches $\frac{\pi}{4}$, the formula (6.8.90) from [93]

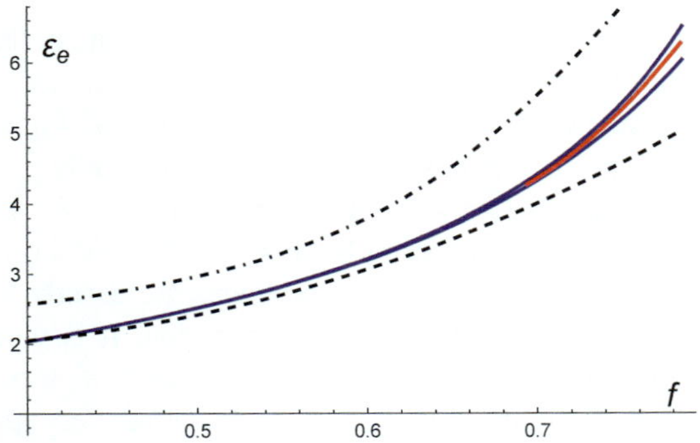

Figure 5.14 The effective permittivity of the square array. Data are for $\varrho = 0.8497$: the 8-order bounds (solid line), the exact formula from [93] (the red line between the bounds), the anti-plane shear modulus C_{44} from [207, Figure 3b] coinciding with the effective permittivity (dashed line below the bounds), Bruggeman's approximation (dashed-dotted line above the bounds).

should be used. Additionally, when $|\varrho| = 1$, the system passes into the percolation regime, and the formulas (6.6.61) and (6.8.90) from [93] similar to (5.221) become preferable.

As noted in [17], an approximate formula for the effective properties is sometimes called an "exact/closed form." The approach presented in a series of papers beginning from [207] involves expanding the unknown complex potentials for regular composites into series in elliptic functions. This method was initially proposed by V.Ya. Natanzon [194] in 1935 for 2D elasticity and by L.A. Filshtinsky [73] in 1964 for 2D conductivity (permittivity). It was rediscovered many times by various authors, see the historical review in [172]. The Natanzon-Filshtinsky method reduces the problem to an infinite system of linear algebraic equations equivalent to Rayleigh's system [206] for conductivity. It is worth noting that the Rayleigh and Natanzon-Filshtinsky methods differ but lead to the same system. The functional equations discussed in Section 4.1 can be considered as a continuous form of the infinite system by Rayleigh and Natanzon-Filshtinsky since the discrete form of the functional equation (for $N = 1$) after expansion into series leads to the same system. The infinite system was truncated to obtain numerical results in [73, 206], more precisely, a polynomial approximating the effective constant.

V.Ya. Natanzon justified the truncation method for the hexagonal array and didn't give numerical examples. L.A. Filshtinsky continued to study the infinite systems for various statements and developed a numerical truncation method for infinite systems. One can find a numerical example with 8 coefficients in the concentration expansion for the square array in the book [97, page 61].

The Natanzon-Filshtinsky method was applied in [207], where the obtained numerical results were called "exact." Comparison with the exact formula (4.60) first derived in [160, 165][2] shows that perhaps a polynomial of degree 6 was used [207] for the effective constants. A result from [207] is represented by the dashed line lying under the lower bound for $f > 0.4$ in Figure 5.14. Such a polynomial approximation was used in [100] for regular piezoelectric fibrous composites.

It is not surprising that polynomial approximations do not satisfy the higher-order bounds and even Hashin-Shtrikman bounds for high-contrast parameters since polynomials do not capture the percolation effect. Rayleigh [206], McPhedran [144, 146] intuitively used rational approximations of the derived analytically polynomials. After the derivation of bounds in terms of Padé approximants [147, 152, 154], their significance became clear. The exact formula (4.60) is obtained in the form of the power series in the concentration f, whose coefficients are exactly written. Calculating the infinite series (4.60), we have to use truncation. It is much easier than solving an infinite system, but the operation of truncation takes place. A truncation approximation yields the race to infinity [210] when the parameters approach critical values. It is worth noting that the application of Padé and more general quasi-rational approximations yields more accurate formulas in percolation regimes [91, 92]. Some accurate asymptotic expansions can be found in [67, 93].

---

[2]Unfortunately, we must divide formulas into exact and "exact" following [17].

Part IV

Image Analysis and Machine Learning

6 Digital Image Processing and Basic Granulometry

This chapter is split into two logical parts. The introduction to selected aspects of DIP (Digital Image Processing) will be presented in the first one. In the remaining part, we will employ this knowledge to construct image processing pipelines for specific images. We will illustrate all concepts using Python programming language. Therefore, we assume that the Reader has some experience with programming using this language. As a short introduction, we can recommend an appendix of our book [176]. You can also study free Scientific Python Lectures [4]. Complete examples of the code from this chapter can be downloaded from [1]. You can run it and experiment. We will use the OpenCV (Open Computer Vision) library for image processing. We will learn the specific algorithms in the library that are useful for image processing of tissues and selecting parts suspected to be cancer cells. Moreover, the same methods can be used in materials engineering for analyzing defects and identifying different phases in materials, and in biology, where one has to count bacteria cells in a microscope image. We focus mainly on cells in medical or biological applications or grains/granules in materials engineering. Therefore, this part will also be for nonspecialists who want to learn *granulometry* - a specific topic of DIP that focuses on characterizing granules in the image, e.g., counting them, defining their shape, color, classifying them, etc.

6.1 DIGITAL IMAGE PROCESSING

We begin with introducing fundamental concepts of DIP that will be used later in practical applications. We want to be manageable due to a specific aim in our head, and since DIP is a vast discipline, there are already excellent resources that treat this topic in great generality. You can consult [95] for very precise mathematical exposition or more concise yet still extremely precise [117].

We will illustrate concepts constructing Python code with the help of OpenCV library [2], a robust and well-optimized set of image processing tools for real-world applications. The library is successively amended by new algorithms and methods and is a de facto standard in Image Processing. The library is written in C language due to optimization and possible applications in embedded systems. However, a version for Python language allows us to use the benefits of high-level language if we do not need real-time responses. There are many excellent and practical books that introduce to the library, e.g., [108, 230], and reference books, e.g., [114].

DOI: 10.1201/9781032646633-6

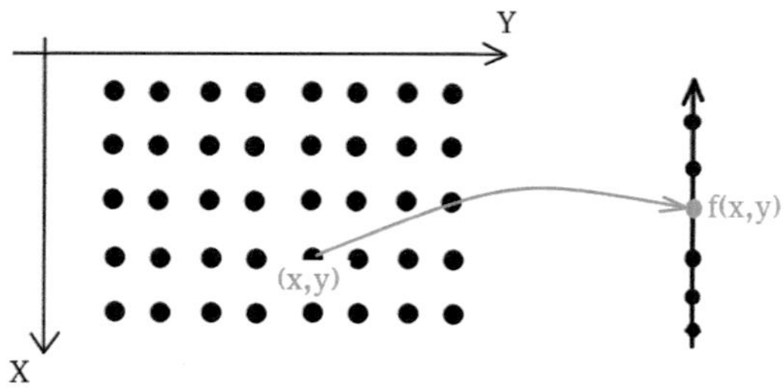

Figure 6.1 Schematic illustration of digital image. The $0X$ axis points down, and the $0Y$ axis points left. This is a typical axes orientation convention used in OpenCV.

In the following subsections, you will be introduced to all necessary topics from DIP to gain working knowledge for processing issues images, materials images, and biological microscopic images.

6.1.1 INTRODUCTION TO DIP

The main subject of Digital Image Processing is the *image*. We can define it as a as a mapping f from a connected and 'regular' subset U of $\mathbb{R}^2$ to an $\mathbb{R}^n$, $n > 0$, i.e., $f : U \longrightarrow \mathbb{R}^n$. Typically, U is a rectangle, and $n = 1$ if the image is grey, and $n = 3$ if the image is color. The imprecise word 'regular' is related to standard shapes such as square, rectangle, or ball - the shapes used for storing the image; however, we can imagine that the image is stored on the surface of some irregular set. The subset U can be treated as a 'photographic plate' or detector registering the image.

In practice, the domain U is not a continuum. It has a discrete structure of a finite rectangular lattice. This is related to the finite size of detectors that can register a single piece of image called *pixel* (picture element). This situation occurs in modern cameras, where the image is recorded in CCD (charge-coupled devices) - a fine array of capacitors that register photon hits. The *digital image* is a map f from a finite subset of $\mathbb{N}^2$ to the specific finite subset $\mathbb{N}^n$, see Figure 6.1.

The components of values $u = f(x, y)$ are restricted to the finite number of values in the quantization process. They are typically encoded in a specific number of bits. For example, for $n = 1$ (grayscale image), the 8-bit image means that u is a scalar integer value from the range from 0 to $2^8 - 1 = 255$. The pixel is the tuple $(x, y, f(x, y))$. Typically, 0 is displayed as a black pixel and 255 as a white pixel. We can also look at a digital image as an array $[f(x, y)]_{x,y}$, which opens effective image manipulation using matrix manipulations.

The color of a pixel can be encoded in various formats constructed in agreement with the natural perception of colors by humans[1]. We mention two such encodings

[1]Color, as perceived by a human, does not have to be a physical color related to the specific wavelength. The space of colors recognized by the human eye has the shape of a distorted triangle (simplex), and each color can be composed of the linear combination of three *primary colors* (from the colors we can construct

used later:

- RGB - colors are mixtures of three basic colors: Red, Green, and Blue; In OpenCV the order of colors is reversed to BGR.
- HSV/HSI/HSL (Hue, Saturation, Value/Intensity/Lightness) - *hue* defines the dominant wavelength in the mixture, *saturation* is related to the purity of color, and *value/intensity/lightness* is related to brightness.

There are well-known formulas for changing color coordinates from RGB to HSV and vice versa, see [95].

For color images, instead of one array, we get three for each *color channel.*

We now illustrate these basic concepts with Python code using OpenCV library [2]. We start from importing libraries which we will use later.

```
import numpy as np
import pandas as pd
import matplotlib.pylab as plt
import cv2
```

The last import is for the OpenCV library.

At the beginning, we should check the details of the library we will work with. We can do it using the following code:

```
print("OpenCV version:", cv2.__version__)
build_info = cv2.getBuildInformation()
print("OpenCV build information:")
print(build_info)
```

We get the current version of the library and build information.

In the next part, we have to work with some example pictures. Due to the suggestions from publishers [191], we will use standard public domain Mandrill image 6.2. OpenCV can load images in various formats into a numerical array as in the following code

```
image = cv2.imread("mandrill.png")
print("Shape = ", image.shape)
```

which produces

```
Shape =  (512, 512, 3)
```

We can see the shape $512x512$ and three color channels. Since it is an array, we can examine it

so-called *chromatic indices* that are barycentric coordinates in the simplex with respect to the edges). These primary colors are Red, Green, and Blue. The construction originates in three eye photoreceptors with sensitivity curves centered on primary colors. However, the light wave of a specific frequency representing a specific color is not equivalent to three waves of different frequencies. Moreover, the perception of 'color' changes with the intensity and the dominant color-to-background ratio. Therefore, in considering 'colors' as humans perceive, one has to distinguish psychological effects from physical ones. For details see [95, 117].

Figure 6.2 The mandrill image used for image processing.

```
 print("Image as an array: ")
print(image[0:5,0:5])
```

```
Image as an array:
[[[ 71 150 164]
  [ 31  57  63]
  [ 10  43  75]
  [ 46  94  95]
  [ 73 140 157]]

 [[ 62 125 120]
  [ 33  97 135]
  [ 23  35  55]
  [ 24  72  75]
  [ 46  86 116]]...
```

We can split the channels

```
blue, green, red = cv2.split(image)
print("Red channel: \n", red)
```

```
Red channel:
[[164  63  75 ... 117 141 179]
[120 135  55 ... 122 136 120]
[ 99 132  60 ... 118  87  96]
...
```

and draw plots

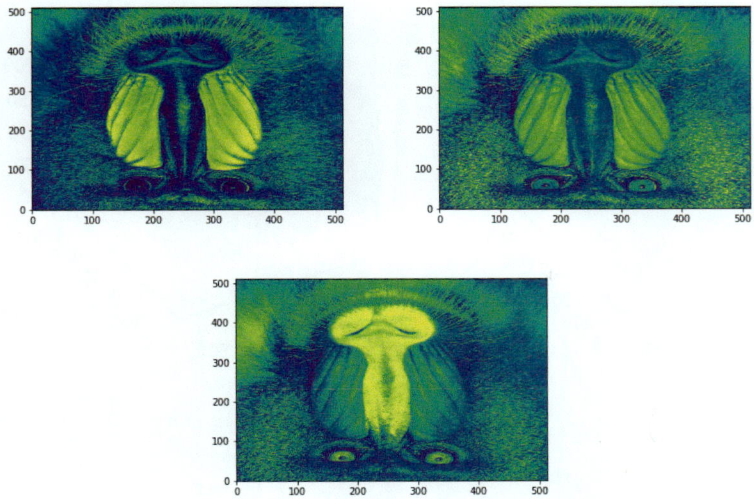

Figure 6.3 Blue, green, and red components.

```
x = np.linspace(0, image.shape[0], image.shape[0])
y = np.linspace(0, image.shape[1], image.shape[1])
X, Y = np.meshgrid(x, y)
plt.pcolor(X, Y, red)
plt.savefig("output/Red.png")
plt.show()

plt.pcolor(X, Y, green)
plt.savefig("output/Green.png")
plt.show()

plt.pcolor(X, Y, blue)
plt.savefig("output/Blue.png")
plt.show()
```

which produces Figs. 6.3.

We can plot them in 3D plots showing the image as a mapping f

```
for name, color in zip(("red", "green", "blue"), (red, green, blue)):
    ax = plt.figure().add_subplot(projection='3d')

    # Plot the surface.
    surf = ax.plot_surface(X, Y, color, linewidth=0, antialiased=
                                                     False)
    ax.set_xlabel("X")
    ax.set_ylabel("Y")
    ax.set_zlabel(name)
    plt.savefig("output/3D{}.png".format(name))
    plt.show()
```

that gives plots in Figure 6.4.

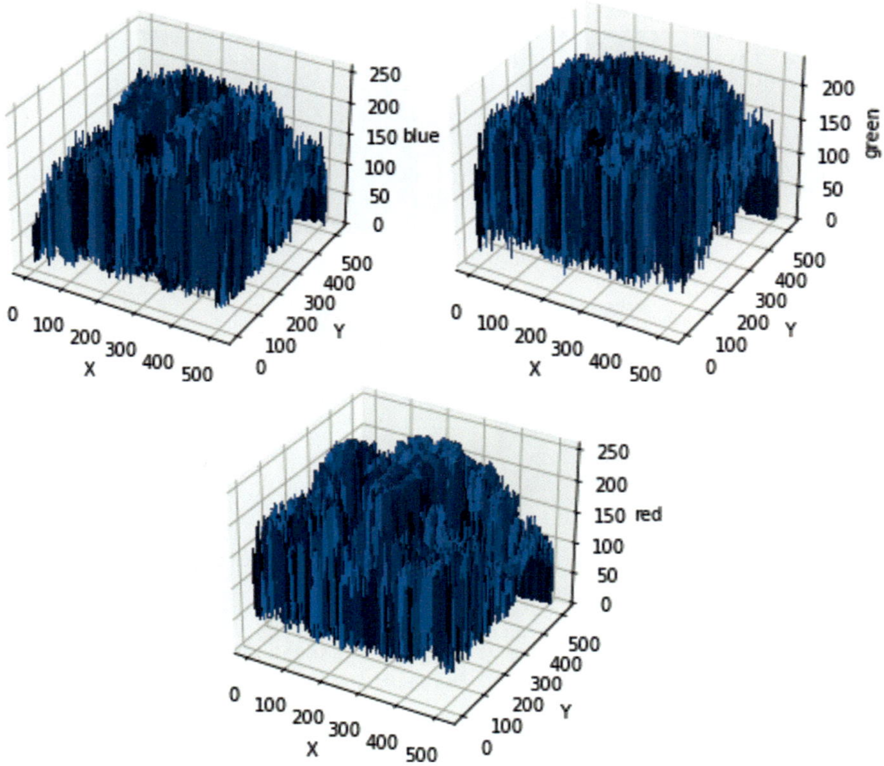

Figure 6.4 Blue, green, and red components in 3D plot.

Splitting into channels is useful in making transformations on them independently. In the end, we want to merge them into a single color image

```
img = cv2.merge([blue, green, red])
print("same :", (img == image).all())
```

which gives true value.

We will also need to convert images between different color spaces. For example, we can convert the image from BGR to HSV or a greyscale image

```
#convert to HSV
imageHSV = cv2.cvtColor(image, cv2.COLOR_BGR2HSV)
#convert to gray
imageGrey =  cv2.cvtColor(image, cv2.COLOR_BGR2GRAY)
```

We can save an image into file by

```
cv2.imwrite('output/MandrillBW.png', imageGrey)
```

OpenCV allows us to preview the image in a separate window

```
#window preview
cv2.imshow('Image', image)
```

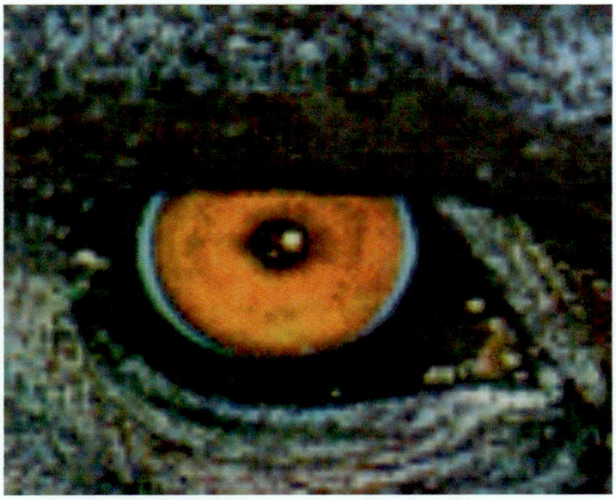

Figure 6.5 Cropped image.

```
##wait for user to press any key
cv2.waitKey(0)

## destroy all created windows
cv2.destroyAllWindows()
```

Since the image is an array, we can crop the image

```
cropped = image[20:100, 130:230, :]
cv2.imwrite("output/cropped.png", cropped)
```

which gives the image in Figure 6.5.

6.1.2 EQUALIZING HISTOGRAM OF IMAGE

The process of taking an image is imperfect and prone to register lighter and darker parts, not always as in reality. Therefore, after reading an image, it is good practice to adjust the contrast in the image by transforming the cumulative distribution for the histogram of intensities to the desired shape, usually the linear one [95]. Mathematically, the idea is the same as transforming probability distribution to another one using the cumulative distribution function.

In this part we discuss the practical way of equalizing histograms for gray and color images.

The intensity histogram will play the core issue in this and later parts. It can be constructed for each channel, representing the number of pixels with a specific intensity/value. Below, we will show you how to create it automatically using OpenCV.

We start practice from reading an image and making a gray copy for further processing

```
import numpy as np
import pandas as pd
```

Figure 6.6 Gray mandrill image converted from the color one.

```
import matplotlib.pylab as plt
import cv2

#read file
image = cv2.imread("mandrill.png")

# convert image to gray
imageGray = cv2.cvtColor(image, cv2.COLOR_BGR2GRAY)
cv2.imwrite("GrayMandrill.png", imageGray)
```

The gray mandrill image is presented in Figure 6.6. Next, we will equalize the grayscale image first since it has only one channel to deal with. First, let us construct the intensity histogram and cumulative sum distribution for it

```
#make histogram of gray image
histGray= cv2.calcHist([imageGray],[0],None,[256],[0,256])
plt.plot(histGray/max(histGray))
plt.xlabel("bin")
plt.ylabel("normalized cout")
plt.savefig("GrayHistogram.png")
plt.show()


#cumulative distribution
cumulativeSumGray = np.cumsum(histGray)
plt.plot(cumulativeSumGray/sum(histGray))
plt.xlabel("bin")
plt.ylabel("normalized sum")
plt.savefig("GrayHistogramCumulative.png")
plt.show()
```

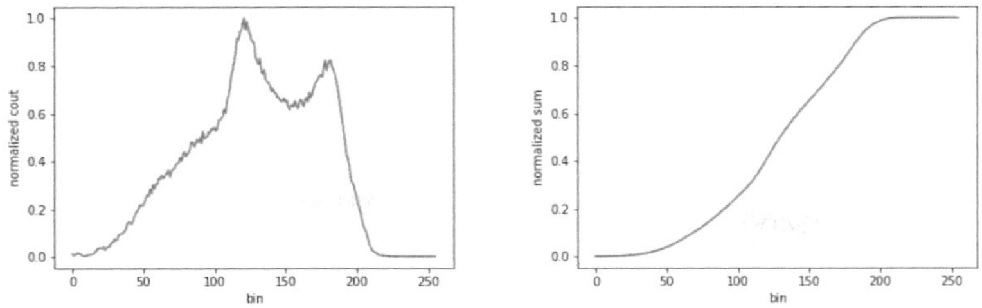

Figure 6.7 Intensity histogram and cumulative sum for equalized gray mandrill picture.

Figure 6.8 Equalized gray mandrill image. Compare this with Figure 6.6.

which produces plots from Figure 6.7. Note that we normalize the histograms by the maximal value, and cumulative sum by the number of pixels (sum of bins in the histogram).

Now we equalize the histogram for the gray image

```
imageGrayEqualized = cv2.equalizeHist(imageGray)
```

that produces Figure 6.8.

We can now repeat histogramming to see what changes were made. The following code

```
#make historgram of gray image
histGrayEqualized = cv2.calcHist([imageGrayEqualized],[0],None,[256],
                                  [0,256])
plt.plot(histGrayEqualized/max(histGrayEqualized))
```

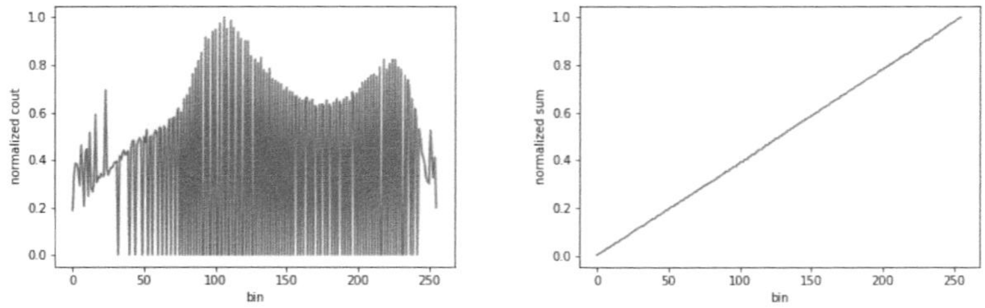

Figure 6.9 Intensity histogram and cumulative sum for histogram equalized gray mandrill picture. Compare with Figure 6.7.

```
plt.xlabel("bin")
plt.ylabel("normalized cout")
plt.savefig("GrayHistogramEqualized.png")
plt.show()

#cumulative distribution
cumulativeSumGrayEqualized = np.cumsum(histGrayEqualized)
plt.plot(cumulativeSumGrayEqualized/sum(histGrayEqualized))
plt.xlabel("bin")
plt.ylabel("normalized sum")
plt.savefig("GrayHistogramCumulativeEqualized.png")
plt.show()
```

that produces plots from Figure 6.9. We can see that the cumulative sum is a straight line, which precisely summarizes the idea of equalization.

We now focus on color image histogram equalization. We could equalize each of the BGR channels. However, we want to equalize intensity, so it is more natural to convert the image to HSV color space, equalize the value channel, and return to BGR color space. This is realized by the following code

```
src = cv2.cvtColor(image, cv2.COLOR_BGR2HSV)
h,s,v = cv2.split(src)
v = cv2.equalizeHist(v)
src = cv2.merge([h,s,v])
imageEqualized = cv2.cvtColor(src, cv2.COLOR_HSV2BGR)
cv2.imwrite("MandarillEqualized.png", imageEqualized)
```

which produces Figure 6.10.

We are now ready to manipulate images and extract information from them.

6.1.3 DRAWING AND WRITING ON IMAGE

Sometimes, we must draw some geometric shapes or write text on the image during processing. In this section, we show how to realize this using OpenCV.

We can draw simple geometric shapes providing typical geometric parameters:

- `cv2.line(image,(0,0),(200,200),(255,0,255),2)` - draw line from the point (0,0), to (200,200) using color (255,0,255) and line thickness 2 on *image*;

Figure 6.10 Mandrill image after histogram equalization. Compare this with Figure 6.2.

- `cv2.rectangle(image,(300,300),(400,400),(0,0,0),-1)` - draw a rectangle with opposite (top left and bottom right) corners located at (300,300), (400,400), with black color, and filled (by convention, the line thickens −1 means that the shape is filled);
- `cv2.circle(image,(400,100), 50, (0,0,255), 2)` - draw circle with center located at (400,100), radius 50, color red, and line thickens 2;
- `cv2.ellipse(image,(100,400),(100,50),0,0,270,(255,0,255), -1)` - draw 'partial' ellipse, with center (100,400), major and minor axes (100,50) with orientation angle 0, between angles from 0 to 270, color (255,0,255) and filled −1;
- `cv2.putText(image,'Example Text', (100,100), cv2.FONT_HERSHEY_SIMPLEX, 1, (0,0,0), 2, cv2.LINE_AA)` - write text 'Example Text', around the point (100,100), using font cv2.FONT _HERSHEY_ SIMPLEX, with scale 1, color black, thickness 2, and the line type cv2.LINE_AA;

We can also plot general polygons, however, basic shapes are sufficient for our purposes. Later, we will also need to detect contours and draw them.

The following code

```
import numpy as np
import pandas as pd
import matplotlib.pylab as plt
```

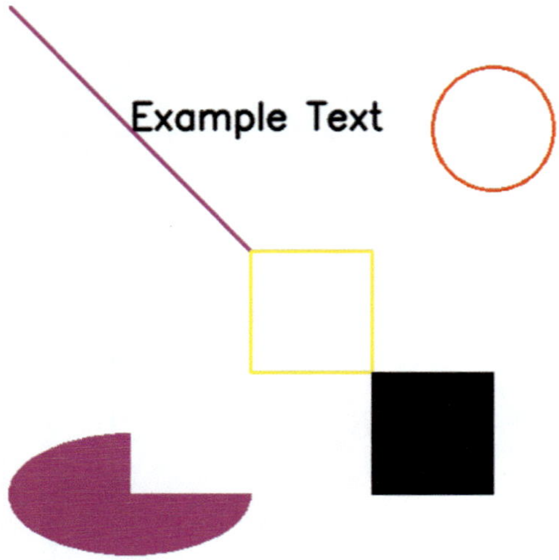

Figure 6.11 Example of geometric shapes and text.

```
import cv2

#create gray image
image = np.zeros((512,512,3), np.uint8)
#reverse colors
image = image + 255
#line
cv2.line(image,(0,0),(200,200),(255,0,255),2)
#rectangle with line width 2
cv2.rectangle(image,(200,200),(300,300),(0,255,255),2)
#filled rectangle (linewidth = -1)
cv2.rectangle(image,(300,300),(400,400),(0,0,0),-1)
#circle
cv2.circle(image,(400,100), 50, (0,0,255), 2)
#ellipse
cv2.ellipse(image,(100,400),(100,50),0,0,270,(255,0,255),-1)
#text
cv2.putText(image,'Example Text', (100,100), cv2.FONT_HERSHEY_SIMPLEX
                                , 1, (0,0,0), 2, cv2.LINE_AA)
cv2.imwrite("output/drawing.png", image)
```

produces Figure 6.11.

6.1.4 MORPHOLOGICAL OPERATIONS

Morphological operations are typically applied to grayscale images and are designed to extract specific structures from the image. They are defined by *structuring element/kernel*, which is an array of typical odd by odd size. Their definition in full generality can be very abstract mathematical concepts [95], although they represent intuitive concepts.

Our description follows [95]. We distinguish only basic operations for gray images represented as a function f. Consider the flat kernel b, which is a matrix of odd dimensions filled with ones. In order to compute specific morphological operation for (x,y) point, we center b at specific element (anchor) (x,y). Then b is treated as a mask that distinguishes points of the image that are taken into account.

We define *erosion* $f \ominus b$ by

$$(f \ominus b)(x,y) = min_{(s,t)\in b}(f(x+s,y+t)). \tag{6.1}$$

Intuitively, erosion expands the border pixels of the structure since they have a smaller value than the background white pixels. It can be helpful to expand small noise pints smaller to the size of the kernel.

Opposite operation to erosion is *dilation* defined by

$$(f \oplus b)(x,y) = max_{(s,t)\in b}(f(x-s,y-t)). \tag{6.2}$$

This operation removes border pixels that are darker than the background. Dilation is also helpful in eliminating noise that is represented by shapes smaller than the kernel size. Applying successively, we can remove small details and border points of the shape.

Since erosion removes small objects, we note that this operation cannot have the inverse operation, and therefore, dilation is definitely not the inverse. However, they 'go in opposite directions'.

When we have noisy images, we can remove noise using dilation. However, we also remove the border points of the interesting shapes. Therefore, to restore them, we apply erosion in the second step. We define *opening* operation as the successive application of dilation and then erosion. Dually, *closing* is the application of erosion and then dilation.

We note that opening and closing operators are idempotent, i.e., only one application affects the image. Two applications give the identity operation.

There are many more morphological operations; for details, see [95].

We now present examples in OpenCV. First, we construct a representative image using the following code

```
image = np.zeros((512,512,3), np.uint8)
image = image + 255
cv2.line(image,(0,0),(200,200),(0,0,0),2)
#filled rectangle
cv2.rectangle(image,(200,200),(400,400),(0,255,0),-1)
cv2.ellipse(image,(100,400),(100,50),0,0,270,(255,0,255),-1)
#generate random noise points
x = 512* np.random.rand(1000)
y = 512* np.random.rand(1000)
for i,j in zip(x,y):
    cv2.circle(image,(int(i),int(j)), 1, (0,0,0), -1)
image = cv2.cvtColor(image, cv2.COLOR_BGR2GRAY)
cv2.imwrite("output/initialImageMotphology.png", image)
```

that produces Figure 6.12.

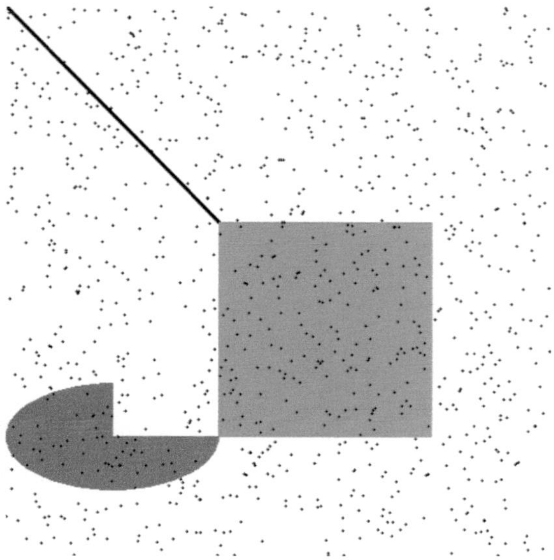

Figure 6.12 Initial shape for morphological operations.

Let us apply erosion

```
imageErode = cv2.erode(image, np.ones((3, 3), np.uint8))
cv2.imwrite("output/Erode.png", imageErode)
```

that gives Figure 6.13. As expected, the noise points and borders are enlarged.

Then we apply dilation on the initial image 6.12

```
imageDilate = cv2.dilate(image, np.ones((3, 3), np.uint8))
cv2.imwrite("output/Dilate.png", imageDilate)
```

that produces Figure 6.14. Since the kernel ($3x3$ matrix of ones) is greater than the noise points, they are removed. If the noise points are larger, we can increase the size of the kernel or apply dilation multiple times. One can also observe that the border pixels are removed, and the join of the ellipse and rectangle are deleted. Therefore, dilation can also be used to separate connected/overlapping shapes.

Next, we apply the opening to Figure 6.12

```
imageOpening = cv2.morphologyEx(image, cv2.MORPH_OPEN, np.ones((3, 3)
                                , np.uint8))
cv2.imwrite("output/Opening.png", imageOpening)
```

that produces Figure 6.15.

In order to remove noise only, we apply closing operation to Figure 6.12

```
imageClosing = cv2.morphologyEx(image, cv2.MORPH_CLOSE, np.ones((3, 3
                               ), np.uint8))
cv2.imwrite("output/Closing.png", imageClosing)
```

that gives Figure 6.16.

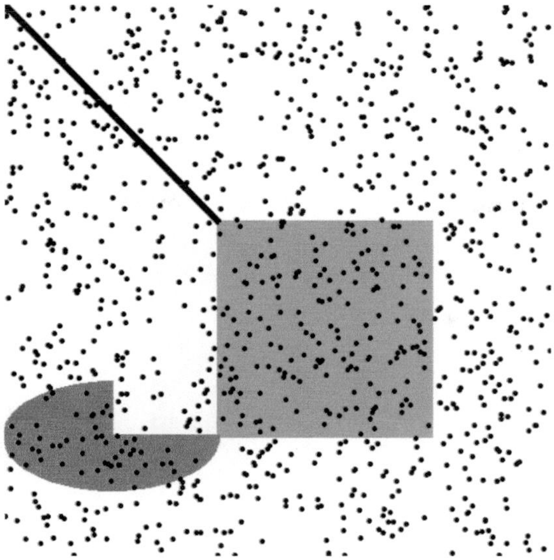

Figure 6.13 Erosion of image from Figure 6.12.

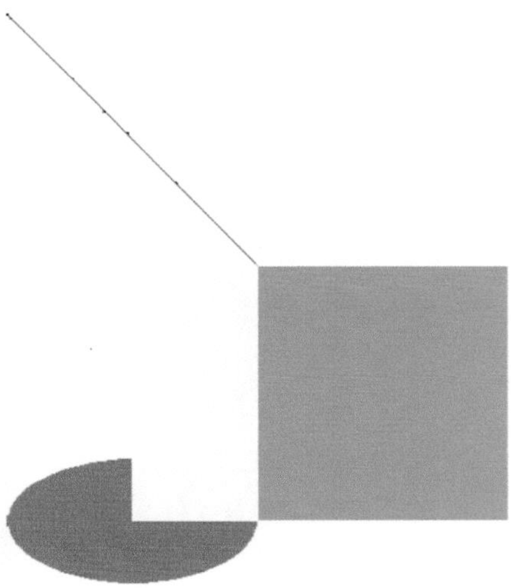

Figure 6.14 Dilation of the image from Figure 6.12.

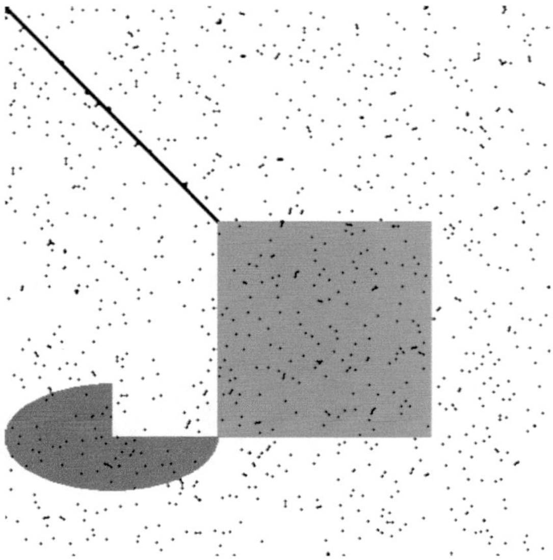

Figure 6.15 Opening of the image from Figure 6.12.

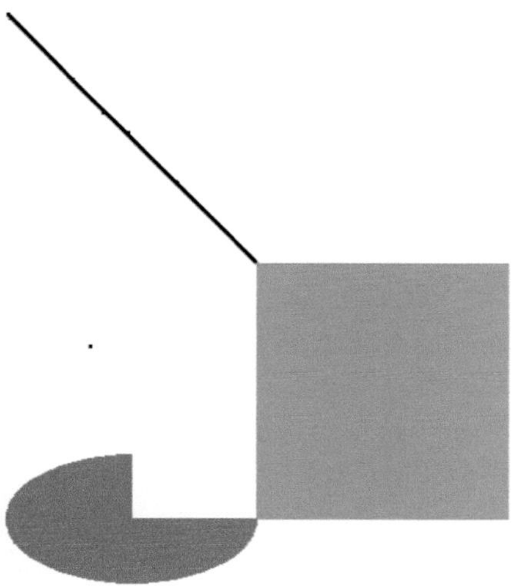

Figure 6.16 Closing of image from Figure 6.12.

Figure 6.17 Mask.

These operations are sufficient for separating shapes in an image and removing random noise points. However, as mentioned above, morphology is a large part of DIP and contains many operations that can be applied for advanced processing, see [95]. By using more advanced kernels, the possibilities are limitless.

6.1.5 MASKING

Selecting irregular parts of the image can be easily done using a mask. It is a grayscale image interpreted as 0 and 1 pixels (for positive values) that mark pixels of an image of the same image shape to be selected.

The mask can be constructed in various ways. The simplest case is to draw a grayscale image and use it. The other way, discussed later, is to extract from the original image mask and then apply it to the original image.

We first create a mask

```
import numpy as np
import cv2

#prepare mask
maskBGR = np.zeros((512,512,3), dtype='uint8')
cv2.circle(maskBGR,(200,100),100,(255,255,255),-1)
mask = cv2.cvtColor(maskBGR, cv2.COLOR_BGR2GRAY)
cv2.imwrite("output/Mask.png", mask)
```

which produces Figure 6.17. Then, we apply it to the mandrill image using the bitwise AND operator. We also reverse order for pixels that represent the background. We have

```
#read image
image = cv2.imread("mandrill.png")
#bitwise AND
```

Figure 6.18 Masked image (left), and with reversed background (right).

```
maskedImageAND = cv2.bitwise_and(image, image, mask=mask)
cv2.imwrite("output/imageMaskedAnd.png", maskedImageAND)
##reverse backround
maskedImageAND[mask == 0] = 255
cv2.imwrite("output/imageMaskedAndReversedBackground.png",
                                    maskedImageAND)
```

which produces Figure 6.18.

We can also add two images, where one is treated as foreground and the second one as a background

```
dst = cv2.add(maskBGR,image)
cv2.imwrite("output/imageAdd.png", dst)
```

that produces Figure 6.19.

6.1.6 THRESHOLDING

Having a grayscale image, we want to extract some shapes of a given intensity level. This can be done by selecting pixels above the fixed (intensity) threshold. We can use histograms to specify the threshold or use automatic methods.

The general idea of thresholding θ_t is to modify pixel-wise the value depending on the value of threshold t and maximal value V (typically the maximal intensity $V = 255$). In a simple binary thresholding we have

$$\theta_t^V(f)(x,y) = \begin{cases} 0 & \text{if} \quad f(x,y) < t \\ V & \text{if} \quad f(x,y) \geq t \end{cases} \tag{6.3}$$

We can also invert the values. This is the most useful in creating masks.

First, we create an image that contains two types of granules/circles of different sizes. We, moreover, add a disturbance to the background. We then convert the image to the grayscale. In real life, the complexity is larger

Figure 6.19 Addition of two images.

```
import numpy as np
import matplotlib.pylab as plt
import cv2

#prepare figure
image = np.zeros((512,512,3), np.uint8)
image = image + 255

#perturbation
cv2.rectangle(image,(100,0),(500,400),(240,240,240),-1)
cv2.rectangle(image,(0,100),(400,500),(0,240,240),-1)

#generate class 1
x = 512* np.random.rand(200)
y = 512* np.random.rand(200)
for i,j in zip(x,y):
    cv2.circle(image,(int(i),int(j)), 5, (100,100,100), -1)

#generate class 2
x = 512* np.random.rand(50)
y = 512* np.random.rand(50)
for i,j in zip(x,y):
    cv2.circle(image,(int(i),int(j)), 20, (10,10,10), -1)

imageGray = cv2.cvtColor(image, cv2.COLOR_BGR2GRAY)
cv2.imwrite("output/initialImageThresholding.png", imageGray)
```

which produces Figure 6.20. The next is to construct histogram of intensity distribution

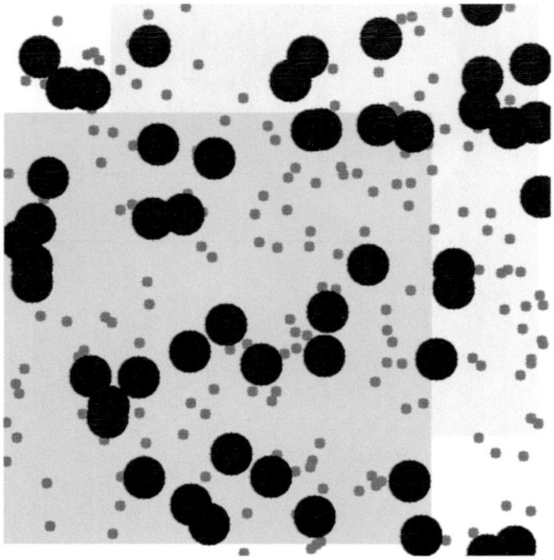

Figure 6.20 Initial image converted to grayscale.

```
histGray= cv2.calcHist([imageGray],[0],None,[256],[0,256])
plt.plot(histGray/max(histGray))
plt.xlabel("bin")
plt.ylabel("normalized cout")
plt.savefig("output/GrayHistogramThresholding.png")
plt.show()
```

that gives Figure 6.21. We can note in Figure 6.21 the bin close to zero representing bigger darker balls in Figure 6.20, and the second bin closer to 100 that represents smaller, lighter balls. These two kinds of balls are granules we want to extract. Moreover, we have three bigger peeks above 200: two light rectangles of big area and the last one at 255 represents the background white color.

The first kind of granules can be extracted easily by setting the threshold value to, e.g., 50. The pixels equal to or above this value will be set to white. We can do it using the following code

```
_ , mask1 = cv2.threshold(imageGray, 50, 255, cv2.THRESH_BINARY)
cv2.imwrite("output/MaskThresholding1.png", mask1)
```

which produces Figure 6.22, which can be used as a mask for selecting the first kind of balls.

For selecting the second kind of balls we have to subtract from Figure 6.20 the balls from image 6.22. That can be done by negating the mask 6.22 and using exclusive OR (XOR) binary operation with the original image 6.20. Then we can do thresholding below, e.g., 150

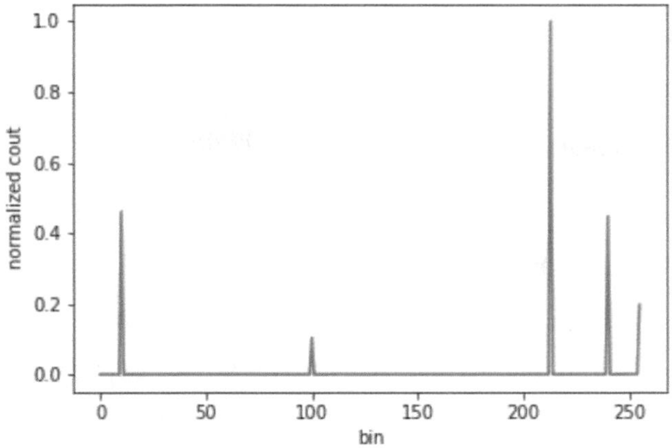

Figure 6.21 Intensity histogram for image 6.20.

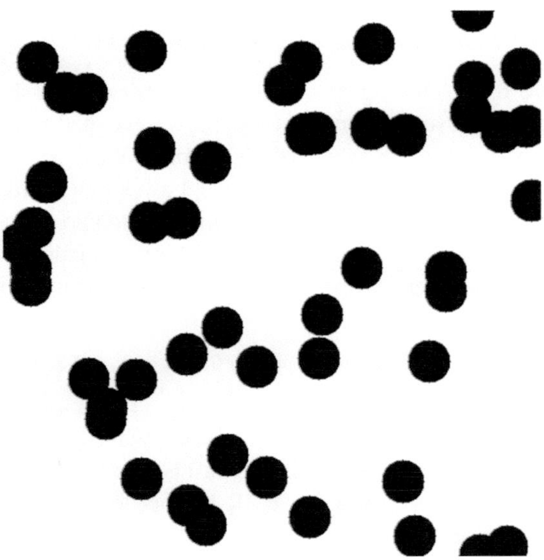

Figure 6.22 Extraction of larger granules from Figure 6.20.

Figure 6.23 Extraction of smaller granules from Figure 6.20.

```
_ , mask2 = cv2.threshold(cv2.bitwise_xor(imageGray, cv2.bitwise_not(
                                        mask1)) , 150, 255, cv2.
                                        THRESH_BINARY)
cv2.imwrite("output/MaskThresholding2.png", mask2)
```

that produces Figure 6.23. This figure can also be used as a mask for extracting smaller balls. One can observe that some of these balls were shadowed by the darker, bigger ones, and therefore, they are only partially restored. We can remove these parts using morphological operations as described above, however, we will not do it here. At this point we have desired *segmentation* of the image into two classes of granules.

The same extraction can be done using `cv2.inRange` function that is supplied with the boundaries of the intensity interval from which we extract pixels as in the following code

```
mask2 = cv2.bitwise_not(cv2.inRange(imageGray, 50, 150))
cv2.imwrite("output/MaskThresholding2.png", mask2)
```

We also inverted colors to have granules black and color white (opposite to the colors used in a mask). The function `cv2.inRange` can also be used to threshold color images, and then the boundary points are BGR vectors for minimal and maximal values of each color channel.

We can now focus on examining automatic thresholding. We first examine Otsu's method [201], which tries to split pixels into different classes and tries to maximize in-between-class variance, see, e.g., [95]. It is realized, but the following code

```
_, imageOtsu = cv2.threshold(imageGray, 0, 255, cv2.THRESH_OTSU)
cv2.imwrite("output/MaskThresholdingOtsu.png", imageOtsu)
```

that produces Figure 6.24 that removes background.

We can also use adaptive thresholding using the mean local value of pixels minus a constant. We have to provide the size of the local neighborhood. This is realized by

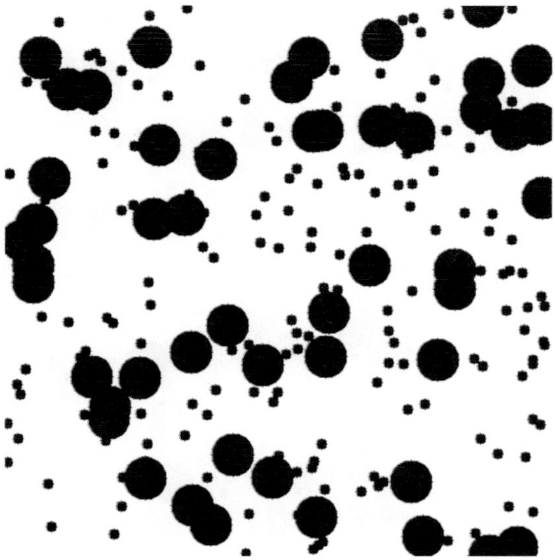

Figure 6.24 Otsu's method applied to Figure 6.20.

```
imageAdaptiveMean = cv2.adaptiveThreshold(imageGray,255,cv2.
                                    ADAPTIVE_THRESH_MEAN_C, cv2.
                                    THRESH_BINARY,51,1)
cv2.imwrite("output/MaskThresholdingAdaptiveMean.png",
                                    imageAdaptiveMean)
```

that produces Figure 6.25. The last two parameters in `cv2.adaptiveThreshold` are the size of the local neighborhood and the constant C that is subtracted from the local mean, giving the local threshold.

One can note that some artifacts resulting from removing background rectangles are present.

We can also apply Gaussian adaptive thresholding, which determines the threshold as Gaussian-weighted local neighborhood values minus a constant. This algorithm is applied in the following code

```
imageAdaptiveGaussian = cv2.adaptiveThreshold(imageGray,255,cv2.
                                    ADAPTIVE_THRESH_GAUSSIAN_C, cv2.
                                    THRESH_BINARY,51,1)
cv2.imwrite("output/MaskThresholdingAdaptiveGaussian.png",
                                    imageAdaptiveGaussian)
```

that produces Figure 6.26. In this case, also, some artifacts are visible.

Summing up, the basic automatic methods are less helpful in extracting granules in such a simple example.

6.1.7 CONTOURS

Having extracted granules, we want to characterize them. In this section we will learn how to extract contours, fit geometric shapes to them, and compute moments of contours.

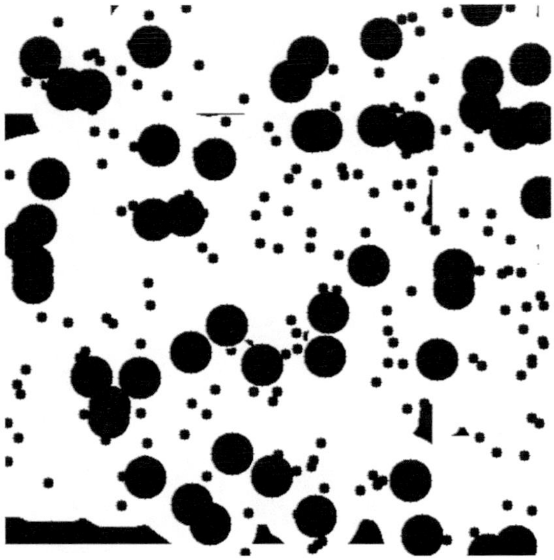

Figure 6.25 Adaptive mean thresholding applied to Figure 6.20.

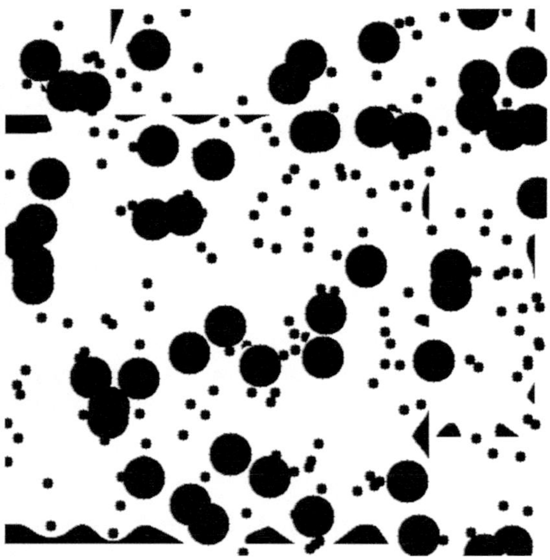

Figure 6.26 Adaptive gaussian thresholding applied to Figure 6.20.

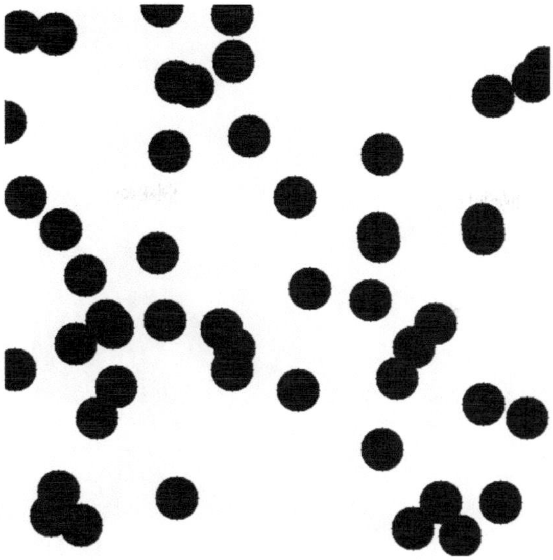

Figure 6.27 Initial image for detecting contours.

First, we must prepare the image. We use an academic example of random balls. The difficulty will be with overlapping balls. The following code

```
import numpy as np
import pandas as pd
import matplotlib.pylab as plt
import cv2

#prepare figure
image = np.zeros((512,512,3), np.uint8)
image = image + 255

#generate granules
np.random.seed(1)
x = 512* np.random.rand(50)
y = 512* np.random.rand(50)
for i,j in zip(x,y):
    cv2.circle(image,(int(i),int(j)), 20, (255,0,0), -1)

cv2.imwrite("output/initialImageContours.png", image)
```

creates Figure 6.27.

First, we create a grayscale mask

```
imageGray = cv2.cvtColor(image, cv2.COLOR_BGR2GRAY)
imageGray = cv2.bitwise_not(imageGray)
cv2.imwrite("output/maskImageContours.png", imageGray)
```

that gives Figure 6.28. Since some granules are merged we can try to disconnect them using erosion multiple times. However, overlapping is too big in some cases

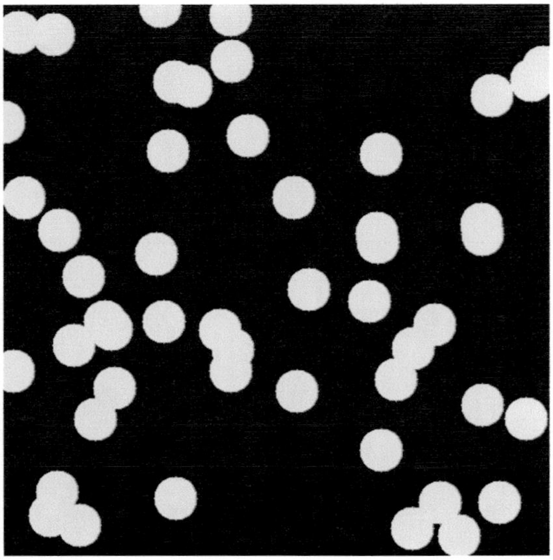

Figure 6.28 Grayscale image of Figure 6.27.

to separate them by erosion without significantly removing white pixels from well-separated images. Moreover, in real applications, granules have irregular shapes and are connected to such an extent that they are impossible to distinguish even by the human eye.

OpenCV has many specialized methods that can detect contours. Contours are lists of points that sit at the edge of shapes. In the simplest case, the algorithm tries to find the contours(borders) of connected components. We use a general one for detecting contours on grayscale image

```
contours, _ = cv2.findContours(imageGray, cv2.RETR_LIST, cv2.
                                        CHAIN_APPROX_TC89_L1 )
#draw contours
imageGrayContours =  image.copy()
#imageGrayContours =  np.zeros(image.shape)+255
cv2.drawContours(imageGrayContours, contours, -1, (0, 0, 255), 3)
cv2.imwrite('output/contoursImageContours.png', imageGrayContours)
```

which gives Figure 6.29. One can observe that some blobs are treated as a single connected component.

The function `cv2.findContours` has various options and types of returned data. For detecting granules we need only a flat structure of the list (option `cv2.RETR_LIST`). In general, it can return a hierarchical structure of contours using `cv2.RETR_TREE` instead. We can also use various approximation methods that optimize the number of points. Since we need many points in order to match ellipses to the contours, we used option `cv2.CHAIN_APPROX_TC89_L1` which uses the Teh-Chin chain approximation algorithm [222]. We can use other Teh-Chin algorithm by option

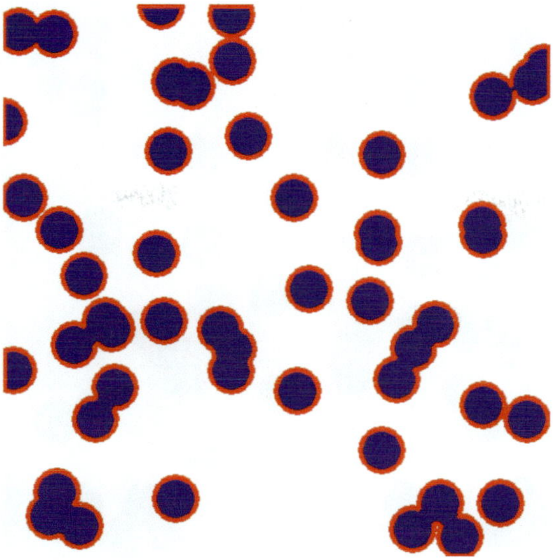

Figure 6.29 Detected contours (red) from Figure 6.27.

cv2.CHAIN_APPROX_TC89_KCOS, or simple approximation with a small number of points cv2.CHAIN_APPROX_SIMPLE, or no approximation cv2.CHAIN_APPROX_NONE - stores all points of the contours.

The contour is drawn on the image imageGrayContours all (−1 contours of all indices) detected contours using a red line of thickness 3.

Since we know that the contours are balls, we can apply more specialized algorithms to detect circles that are available in OpenCV. We use the method based on Hough transform [68]

```
img = imageGray.copy()
colorimg = image.copy()
circles = cv2.HoughCircles(img,cv2.HOUGH_GRADIENT, 1, 20, param1=20,
                                    param2=10, minRadius=2, maxRadius=
                                    50)
circles = np.uint16(np.around(circles))
for circle in circles[0,:]:
    # draw the outer circle
    cv2.circle(colorimg,(circle[0],circle[1]),circle[2],(0,0,255),2)
    # draw the center of the circle
    cv2.circle(colorimg,(circle[0],circle[1]),3,(0,0,0),3)
cv2.imwrite('output/HoughCirclesDetectorImageContours.png', colorimg)
```

that produces Figure 6.30. One can note that more balls are detected, however, the algorithm does not spot some overlapping granules.

The list of circles returned by the cv2.HoughCircles method represented by x and y coordinates (indices 0 and 1) of the center and the radius (index 2).

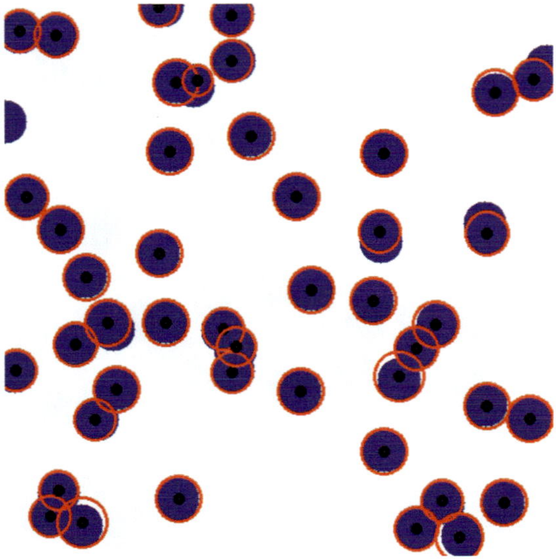

Figure 6.30 Hough algorithm for detecting circles applied to Figure 6.27.

Let us return to contours, as they are more general types of closed curves used for bonding regions in the images. We want to draw around the contour bounding rectangles. Moreover, we cut the rectangular image enclosing the contours and save them in images, moreover, we enumerate detected contours. This is realized by the following code

```
img = image.copy()
for idx, contour in enumerate(contours):

    #draw minimum area rectangle
    minRect = cv2.minAreaRect(contour)
    box = cv2.boxPoints(minRect)
    box = np.int0(box)
    cv2.drawContours(img, [box], -1, (0,0,255), 1)
    #enumarate contours
    cv2.putText(img, '{}'.format(idx), (box[0][0], box[0][1]), cv2.
                                        FONT_HERSHEY_SIMPLEX, 0.5, (0,
                                        0,0), 1, cv2.LINE_AA)

    #extract image from bounding rectangle and write them to files
    x, y, w, h = cv2.boundingRect(contour)
    cropped = image[y:y + h, x:x + w, :].copy()
    feature = cv2.resize(cropped, (20, 20))
    cv2.imwrite("./output/figs/{}.png".format(idx), feature)
cv2.imwrite('output/MinAreaRectangleImageContours.png', img)
```

which produces Figure 6.31. Extracted images can be used for (unsupervised) training Machine Learning algorithms to discriminate their types if there are more types of them. We will return to Machine Learning later in this chapter.

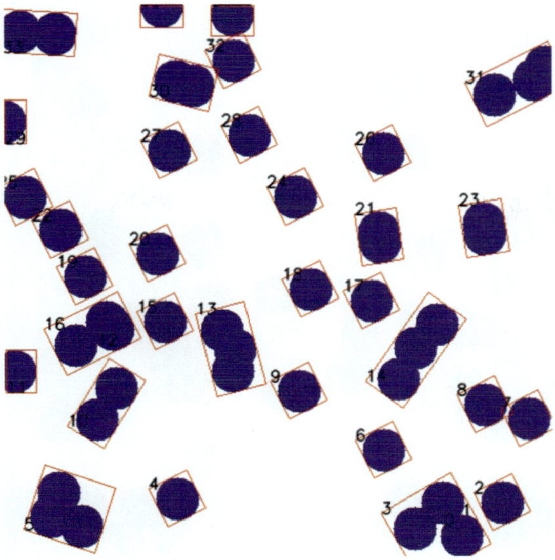

Figure 6.31 Minimal area rectangle for detected contours from Figure 6.29.

When the granules have no circular symmetry, we sometimes want to know their orientation and non-symmetry. The easiest way is to fit ellipses and extract their parameters. OpenCV has a function that can fit an ellipse using at least five points of contours using the least square method. We do it in the following code

```
img = image.copy()
ellipses = [] #list for storing ellipsis parameters
for idx, contour in enumerate(contours):
    if contours[idx].size < 10 :
        continue
    ellipse = cv2.fitEllipse(contour) # ellypse is ((x,y), (majorAxis
                                        , minorAxis), angle)
    #remove bigger ellipses
    if max(ellipse[1]) < 200:
        ellipses.append({"id": idx, "x":ellipse[0][0], "y":ellipse[0]
                                             [1], "majorAxis": ellipse[
                                             1][0], "minorAxis":
                                             ellipse[1][1], "angle":
                                             ellipse[2]})
        cv2.ellipse(img, ellipse, (0,0,255), 3, cv2.LINE_AA)
        cv2.putText(img, '{}'.format(idx), (int(ellipse[0][0]), int(
                                             ellipse[0][1]) ), cv2.
                                             FONT_HERSHEY_SIMPLEX, 0.2,
                                              (0, 0, 0), 1, cv2.LINE_AA
                                             )
#save parameters of ellipses
df_ellipses = pd.DataFrame.from_records(ellipses)
df_ellipses.to_csv("output/ellipses.csv")
#write image with elipses
cv2.imwrite('output/EllipsesImageContours.png', img)
```

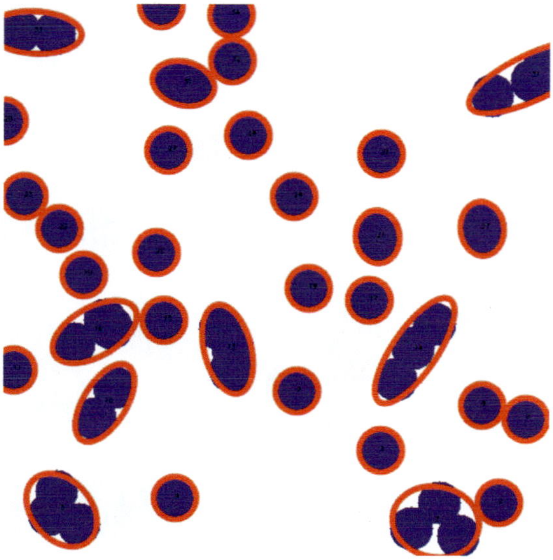

Figure 6.32 Fitted ellipses to contours from Figure 6.29.

that produces the image in Figure 6.32. The ellipse is characterized by the center (x, y), major axis, minor axis, and orientation given by the angle between the positive $0X$ axis and the major axis. In the code, these data are packed into a Python `DataFrame` structure and saved in `ellipses.csv` file[2]

Sometimes, we want to characterize contours more precisely. We can use moments of distribution. For a contour $c(x, y)$ we can get *static real moments* (sometimes called '*spatial or raw moments*')

$$m_{ij} = \sum_{x,y} c(x, y)x^i y^j, \tag{6.4}$$

central moments

$$\mu_{ij} = \sum_{x,y} c(x, y)(x - \bar{x})^i (y - \bar{y})^j, \tag{6.5}$$

where the 'mass center' is characterized by

$$\bar{x} = \frac{m_{10}}{m_{00}}, \quad \bar{y} = \frac{m_{01}}{m_{00}}, \tag{6.6}$$

and *normalized central moments* ($\nu ='' nu''$) defined as

$$\nu_{ij} = \frac{\mu_{ij}}{m_{00}^{(i+j)/2+1}}. \tag{6.7}$$

[2]CVS (comma-separated values) is a simple and portable file format for saving tabular data in a text file. Each row of the table corresponds to the row of the file, and columns are represented by data separated by a comma or other separator. These files can be easily imported into a spreadsheet program.

Since $\mu_{00} = m_{00}$, $\nu_{00} = 1$, $\nu_{10} = \mu_{10} = \nu_{01} = \mu_{01} = 0$, so these values are not returned.

The considered static real moments correspond to the static complex moments (2.95) introduced in Chapter 2. The complex Green's formulas (2.88) allow us to treat line integral over a closed contour as a surface area over the area enclosed by the contour. The same holds for the static real moments.

The contours with self-intersections can have zero m_{00} even if the area is not zero due to the implicit orientation of areas in Green's formula. Moreover, the value depends on rasterization, and therefore, the computed values can be treated only as approximations. Moreover, the more points of the contour yield a more accurate approximation.

The contours (viewed as continuous curves/before discretization) can be transformed by scaling, translations, and rotations (after undoing discretization on the raster lattice). We are interested in having the same moments for two of the same contours, even if one is a transformed version of the second one. Therefore, in analyses, we should use moments that are invariant under specific symmetry groups. A well-known fact is that the central moments are invariant with respect to translations. The moments ν_{ij} are invariant with respect to translations as defined by central moments, and also with respect to scaling. The natural next step is to define the invariant moments concerning scaling, rotations, and translations. These seven moments were defined by Hu [109], and then amended [77] and generalized for spatial objects [78]. *Hu moment invariants* are defined as

$$
\begin{aligned}
I_1 &= \nu_{20} + \nu_{02}, \\
I_2 &= (\nu_{20} - \nu_{02})^2 + 4\nu_{11}^2, \\
I_3 &= (\nu_{30} - 3\nu_{12})^2 + (3\nu_{21} - \nu_{03})^2, \\
I_4 &= (\nu_{30} + \nu_{12})^2 + (\nu_{21} - \nu_{03})^2, \\
I_5 &= (\nu_{30} - 3\nu_{12})(\nu_{30} + \nu_{12})[(\nu_{30} + \nu_{12})^2 - 3(\nu_{21} + \nu_{30})^2] + \\
&\quad +(3\nu_{12} - \nu_{03})(\nu_{21} + \nu_{03})[3(\nu_{30} + \nu_{12})^2 - (\nu_{21} + \nu_{03})^2], \\
I_6 &= (\nu_{20} - \nu_{02})[(\nu_{30} + \nu_{12})^2 - (\nu_{21} + \nu_{03})^2] + 4\nu_{11}(\nu_{30} + \nu_{12})(\nu_{21} + \nu_{03}), \\
I_7 &= (3\nu_{21} - \nu_{03})(\nu_{30} + \nu_{12})[(\nu_{30} + \nu_{12})^2 - 3(\nu_{12} + \nu_{03})^2] - \\
&\quad (\nu_{30} - 3\nu_{12})(\nu_{21} + \nu_{03})[3(\nu_{30} + \nu_{12})^2 - (\nu_{21} + \nu_{03})^2].
\end{aligned}
\tag{6.8}
$$

They are not all independent. Moreover, I_7 is reflection antisymmetric and the remaining ones reflection symmetric, they can be used to detect mirror symmetry of the contour. In addition, there are more invariants of fourth order [77].

The Hu moment invariants can be computed in OpenCV using `cv2.HuMoments` function using the result of `cv2.moments` that computes m_{ij}, μ_{ij} and ν_{ij} moments. The following code, being a continuation of the previous code, computes moments and Hu moment invariants and saves it to a CSV file

```
import math
huMomentsList = []
for idx, contour in enumerate(contours):
    #calculate moments
    moments = cv2.moments(contour)
    print("contour nr ", idx, ", moments:", moments)
    #calculate Hu moments
    huMoments = cv2.HuMoments(moments)
    #scaling Hu moments
    for i in range(0,7):
       if huMoments[i] == 0:
           huMoments[i] = 0.0
       else:
           huMoments[i] = -1* math.copysign(1.0, huMoments[i]) * math
                                            .log10(abs(huMoments[i]
                                            ))

    huMomentsList.append({'h0':huMoments[0], 'h1':huMoments[1],
                          'h2':huMoments[2], 'h3':huMoments[3],
                          'h4':huMoments[4], 'h5':huMoments[5],
                          'h6':huMoments[6]})

df_HuMoments = pd.DataFrame.from_records(huMomentsList)
df_HuMoments.to_csv("output/HuMoments.csv")
```

Since Hu moment invariants can have arbitrary values, we adjust their range using log_{10} transform and preserve their sign. It is standard operation, and it can be helpful when we use moments for Mchine Learning algorithms that require data to be properly normalized/standardized to have a similar range. This code generates the file presented in Figure 6.33.

In the present section, the discussed static moments are used for the image analysis. It is worth noting that the static moments affect the physical properties in combination with structural sums. This is explicitly shown by the formulas (2.112)–(2.114).

6.1.8 CLASSIFYING

Once we 'segmented' image into granules and identify potentially interesting ones, the next step is to identify them. The most reliable is the 'hand-made' classification by a skilled human expert. On the other hand, this is the slowest and least automatic way of classification. In some important cases where the error should be minimized, like cancer cell detection, the human expert should have the last word in such classification. However, automatic methods can help in such classification and speed up the process.

In this part, we make a short review of Machine Learning (ML) methods that can be applied to the classification of images. We focus mostly on the methods where no expert insight is available (unsupervised learning). We also explain how to use other types of Machine Learning where expert data is present for learning before classification (supervised learning). The basics of ML in the context of mathematical modeling can be learned in our book [176]. Moreover, a more formal yet practical introduction to Machine Learning are the books [86,205]. This part cannot be treated

	h0	h1	h2	h3	h4	h5	h6
0	[0.]	[0.]	[0.]	[0.]	[0.]	[0.]	[0.]
1	[0.]	[0.]	[0.]	[0.]	[0.]	[0.]	[0.]
2	[0.79814845]	[26.81188262]	[24.41845927]	[23.47695713]	[47.42470121]	[-36.88289946]	[49.31560298]
3	[0.70538293]	[2.69093786]	[2.58304728]	[4.38886088]	[-7.8800969]	[-5.73709618]	[-8.68443074]
4	[0.79814845]	[26.85279981]	[24.42416307]	[24.02403254]	[48.2512061]	[-37.47078364]	[-49.17406799]
5	[0.74091863]	[2.46926156]	[2.9935646]	[4.60553714]	[-8.84461097]	[-6.23294822]	[8.43585707]
6	[0.79814845]	[0.]	[25.18292176]	[24.10978444]	[-48.98377938]	[0.]	[48.8498556]
7	[0.79814845]	[26.81178664]	[23.9546497]	[23.7333268]	[-48.094335]	[-37.32975677]	[-47.59838247]
8	[0.79814845]	[27.41384663]	[24.16494398]	[23.87036614]	[-48.0824341]	[-37.63872767]	[-48.00203766]
9	[0.79814845]	[28.01652069]	[24.77083858]	[24.68023155]	[49.49318743]	[-38.7606507]	[49.64558229]
10	[0.67523629]	[1.79176439]	[25.59654951]	[25.54615833]	[51.11772537]	[26.92941533]	[52.6216787]
11	[0.76012002]	[2.48727811]	[3.35418212]	[4.80944317]	[-8.89125582]	[-6.05308223]	[0.]
12	[0.]	[0.]	[0.]	[0.]	[0.]	[0.]	[0.]
13	[0.70056534]	[1.91643844]	[3.56724453]	[4.68988083]	[-8.82139728]	[-5.65043331]	[9.75310793]
14	[0.58037321]	[1.39557297]	[3.782052]	[4.72392509]	[-9.11828063]	[-5.50417661]	[9.13697904]
15	[0.79814845]	[29.21941298]	[26.6618193]	[26.11570643]	[52.50449099]	[-40.7895659]	[54.5047043]
16	[0.6857275]	[1.89187407]	[3.40205217]	[4.3795409]	[8.2942191]	[5.35392537]	[8.76151372]
17	[0.79814845]	[28.01652069]	[24.51249657]	[24.61855795]	[49.28468392]	[-38.85433458]	[-49.39952709]
18	[0.79814845]	[0.]	[26.02654629]	[25.700419]	[-51.80239727]	[0.]	[-51.65197979]
19	[0.79814845]	[27.7213525]	[25.73798887]	[25.60301727]	[51.27462302]	[-39.46378692]	[-52.42122381]
20	[0.79814845]	[27.82192699]	[26.37755539]	[25.60589352]	[53.07603682]	[-39.58323299]	[-51.59785794]
21	[0.78997163]	[3.06604495]	[25.77686473]	[25.79253896]	[52.34820172]	[27.85054456]	[-51.58356701]
22	[0.79814845]	[29.02657497]	[26.31685638]	[26.26031951]	[52.54917525]	[-40.7864668]	[54.00352057]
23	[0.78905807]	[3.00941968]	[25.17852668]	[26.13276919]	[52.34125909]	[27.75796314]	[-51.80614589]
24	[0.79814845]	[0.]	[26.69939242]	[25.85411951]	[53.36018816]	[0.]	[-52.13163211]
25	[0.79804425]	[5.66204838]	[6.19735014]	[9.43787949]	[-17.25549431]	[-12.26890368]	[-25.32645678]
26	[0.79814845]	[27.82245498]	[24.73266168]	[24.70911716]	[49.430966]	[-38.63688209]	[-50.60786036]
27	[0.79814845]	[0.]	[29.45039529]	[27.1276657]	[56.57571274]	[0.]	[55.41774276]
28	[0.79814845]	[27.91895058]	[25.97686679]	[25.63567101]	[51.4647807]	[-39.61840774]	[51.94228181]
29	[0.69109948]	[1.88105976]	[2.92548302]	[4.13412645]	[-7.66393119]	[-5.07465633]	[-18.75625795]
30	[0.77060218]	[2.50929185]	[27.41107951]	[27.55199248]	[55.03535016]	[28.84551973]	[-56.07258]
31	[0.60488307]	[1.57827313]	[3.01942336]	[4.14726545]	[-7.81259446]	[-4.94539935]	[7.98182496]
32	[0.79814845]	[29.02657497]	[26.31685638]	[26.26031951]	[52.54917525]	[-40.7864668]	[-54.00352057]
33	[0.69716981]	[1.92795111]	[4.89335164]	[6.67083021]	[13.07703084]	[7.90232684]	[12.46554216]
34	[0.76012002]	[2.48727811]	[3.35418212]	[4.80944317]	[-8.89125582]	[-6.05308223]	[-22.07892886]
35	[0.69109948]	[1.88105976]	[2.92548302]	[4.13412645]	[-7.66393119]	[-5.07465633]	[18.75278852]

Figure 6.33 Hu moments for contours from Figure 6.29.

as an introduction to the vast area of ML. However, you will learn the idea and see an example that can be generalized to any other, more advanced ML method.

Since computers are good only at crunching numbers, all data, before processing, must be converted into numbers or vectors of numbers. The general idea of ML methods is to distinguish separate classes in the data and associate class labels to them. We distinguish two main cases:

- **supervised learning** - when class labels are present in the training data; Then ML algorithm learns on examples to associate labels to the new data on training. This type of algorithms focus on making precise borders between areas with points associated with different class labels (decision regions). In this class we have algorithms like SVM (Support Vector Machines) and various architectures of Artificial Neural Networks.
- **unsupervised learning** - when that data does not contain class labels, the ML algorithm must infer how many classes there are and label them. The classes are typically determined as clusters in the data space. All kinds of clustering algorithms like k-Means or Gaussian Mixture Model represent this class.

Supervised learning can be made with the data that are labeled. If no such data are available for our case, the way to detect differences in data is to use unsupervised learning. In the later part we will focus on unsupervised learning. Already trained supervised learning algorithms can be treated as a replacement in the code encountered in this part.

Many ML algorithms are implemented in the Scikit Learn library [5]. The work with classifiers is rather standard and consists of two steps:

1. `fit(...)` - learn classifier;
2. `predict(...)` - use a classifier to predict classes; usually used for new data not used in the learning process;

A prominent example of unsupervised ML algorithms is the k-means method. It tries associating data to k clusters, minimizing WSSE (clusters within the sum of square error). The input is the number of clusters, however, it can be determined. In order to understand the idea, we present the following toy example, where we create artificial data with some clusters and then apply the k-Means algorithm. More details can be found in our book [176].

```
import matplotlib.pylab as plt

#preparing data
from sklearn.datasets import make_blobs
X, y = make_blobs(n_samples=1000, centers=3, n_features=3,
                                  random_state=0)

# plot data
plt.scatter(X[:,0], X[:,1], c=y)
plt.xlabel("$x1$")
plt.ylabel("$x2$")
plt.savefig("output/data-kMeans.png")
```

which produces three clusters in Figure 6.34.

Three clusters are clearly visible, and we expect to distinguish them. The easiest way to determine the number of clusters is to plot WSSE vs the number of clusters

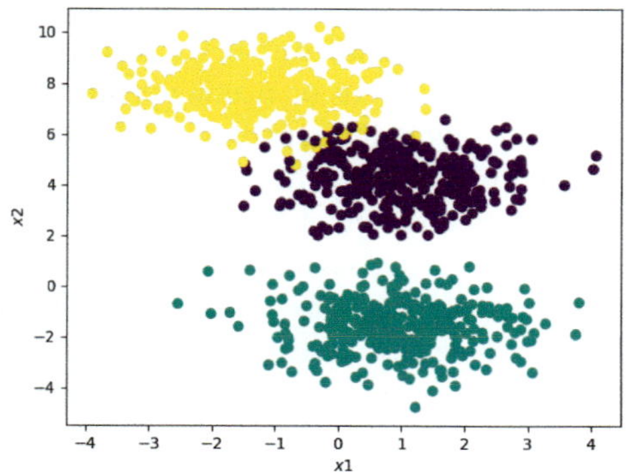

Figure 6.34 Initial data for k-Means algorithm.

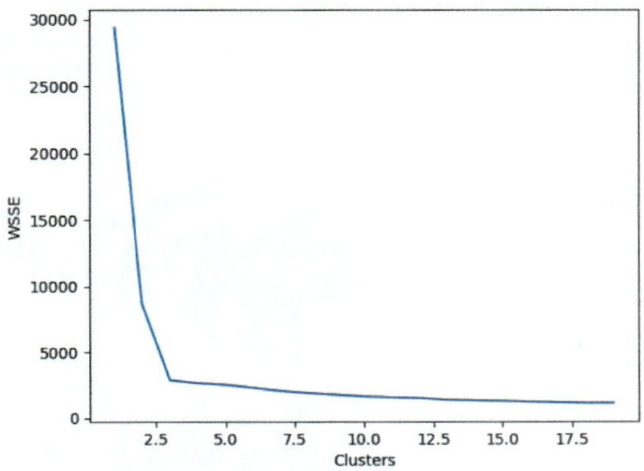

Figure 6.35 Scree plot.

(the scree plot) and identify characteristic elbow. The following code

```
from sklearn.cluster import KMeans
wsse = []
for k in range(1, 20):
    kmeans = KMeans(n_clusters=k, n_init=1)
    kmeans.fit(X)
    wsse.append(kmeans.inertia_)

#make plot
plt.plot(range(1, 20), wsse)
plt.xlabel("Clusters")
plt.ylabel("WSSE")
plt.savefig("output/kMenas_Scree.png")
```

gives Figure 6.35. One can see 'elbow' with 'joint' at $k = 3$ clusters. Above this number of clusters, increase does not improve WSSE much.

We can now make k-Means clustering using $k = 3$ clusters

```
kmeans = KMeans(n_clusters=3, n_init=1)
kmeans.fit(X)
##clusters labels
labels = kmeans.predict(X)
##locations of centroids
centroids  = kmeans.cluster_centers_

#plot
plt.scatter(X[:,0], X[:,1], c=labels)
plt.scatter(centroids[:,0],centroids[:,1], c='r', marker='v', label="
                                            centers")
plt.legend()
plt.xlabel("x1")
plt.ylabel("x2")
plt.savefig("output/kMeans-clustering.png")
plt.show()
```

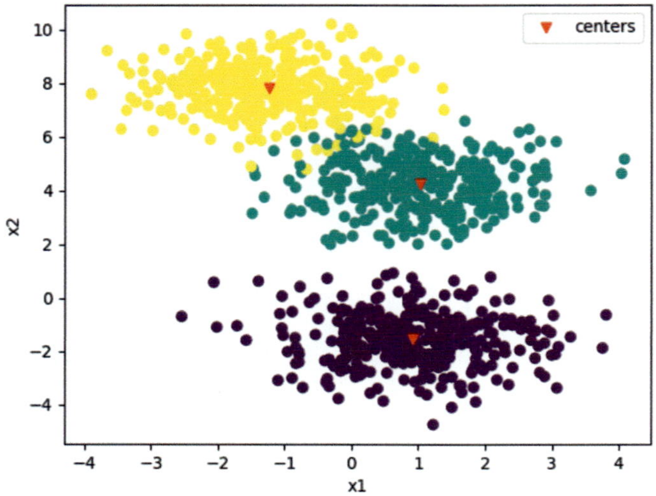

Figure 6.36 Clustering result.

that gives plot in Figure 6.36. The above example was an ideal one. Real-world data are typically not so well separable.

The next example will present the k-means algorithm for image classification, which will be useful in the next part of this chapter. We use handwritten digits [7] as input data. They are loaded by the following code

```
import numpy as np
import matplotlib.pylab as plt

from sklearn.datasets import load_digits

digits = load_digits()
X = digits.data
Y = digits.target

#reverse color
X = 255-X

#example digit
plt.imshow(X[100].reshape(8,8))
plt.savefig("output/digitSingleOne.png")
plt.show()
```

that produces Figure 6.37. The image is encoded as a vector of 64 entries and has to be converted to 8×8 grayscale array.

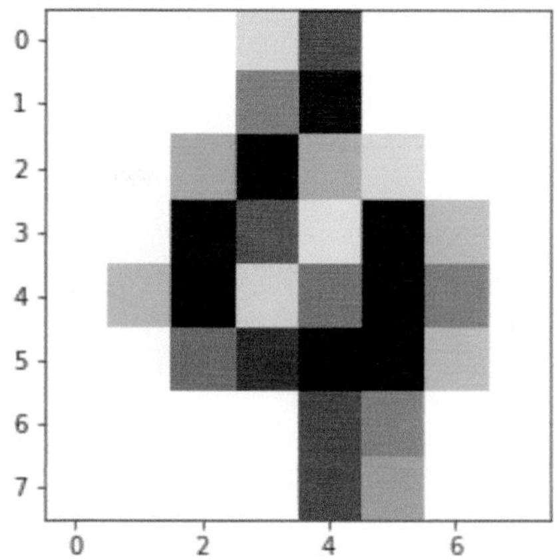

Figure 6.37 An example of handwritten digit.

The next step is to estimate the number of clusters

```
from sklearn.cluster import KMeans
k = range(1,20)
WSSE = []
for i in k:
    model = KMeans(n_clusters = i, n_init='auto')
    model.fit_predict(X)
    WSSE.append(model.inertia_)

plt.plot(k,WSSE)
plt.xlabel('K')
plt.ylabel('Sum of Squared Errors')
plt.savefig("output/digitsScree.png")
plt.show()
```

that gives Figure 6.38. One can note that the elbow joint is about 10, which is expected. Therefore in the next step we can use the k-Means classifier to cluster digits into similar sets

```
clusters = 10
kmeans = KMeans(n_clusters=clusters, n_init='auto')
kmeans.fit(X)
Z = kmeans.predict(X)

#print selected class
##get images of numbers from given class
clusterIndex = 2
#get images with a specific number of cluster
imageIndex = np.where(Z==clusterIndex)[0]
#plt.figure(figsize=(10,10))
```

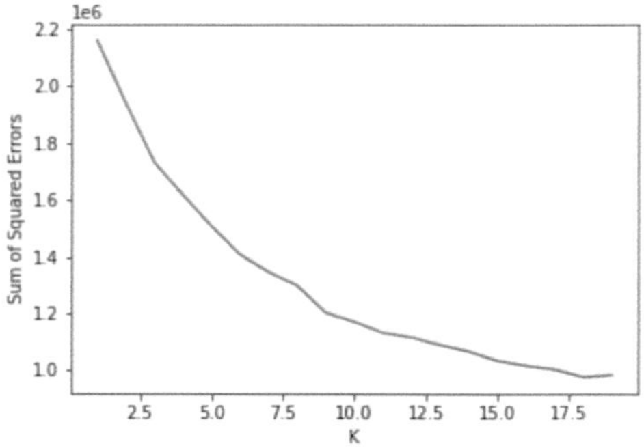

Figure 6.38 Scree plot for handwritten digits data.

```
for k in range(0, imageIndex.shape[0]):
    image = X[imageIndex[k],:]
    image = image.reshape(8, 8)
    plt.imshow(image, cmap='gray')
    plt.savefig("output/digits/{}_{}.png".format(clusterIndex, k))
    plt.show()
```

The results assigned to the class 2 can be inspected visually. It is not perfect, however, many images represents the same digit.

Instead of the k-means algorithm, one can use any other unsupervised ML algorithm or trained, supervised ML algorithm.

6.1.9 SEGMENTATION USING K-MEANS ALGORITHM

One of the standard techniques of segmentation of an image is related to similar colors of objects we want to extract. If this is the case, we can do clusterization in color space, and then extract objects belonging to the same cluster (similar color). Clusterization can be performed using an arbitrary unsupervised ML algorithm. We present an example that uses the k-means algorithm.

We start by preparing an image

```
import numpy as np
import pandas as pd
import matplotlib.pylab as plt
import cv2

#prepare figure
image = np.zeros((512,512,3), np.uint8)
image = image + 255

#perturbation
cv2.rectangle(image,(100,0),(500,400),(240,240,240),-1)
cv2.rectangle(image,(0,100),(400,500),(0,240,240),-1)

#set reproducibility of initial image
```

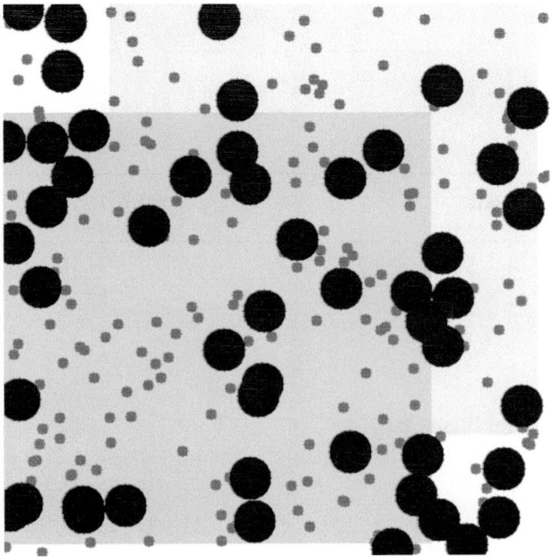

Figure 6.39 Initial image.

```
np.random.seed(1)

#generate class 1
x = 512* np.random.rand(200)
y = 512* np.random.rand(200)
for i,j in zip(x,y):
    cv2.circle(image,(int(i),int(j)), 5, (100,100,100), -1)

#generate class 2
x = 512* np.random.rand(50)
y = 512* np.random.rand(50)
for i,j in zip(x,y):
    cv2.circle(image,(int(i),int(j)), 20, (10,10,10), -1)
```

that produces an image from Figure 6.39.

Next, we localize the optimal number of classes

```
from sklearn.cluster import KMeans

k = range(1,5)
WSSE = []

for i in k:
    model = KMeans(n_clusters = i, n_init='auto')
    model.fit_predict(df)
    WSSE.append(model.inertia_)

plt.plot(k,WSSE)
plt.xlabel('K')
plt.ylabel('WSSE')
```

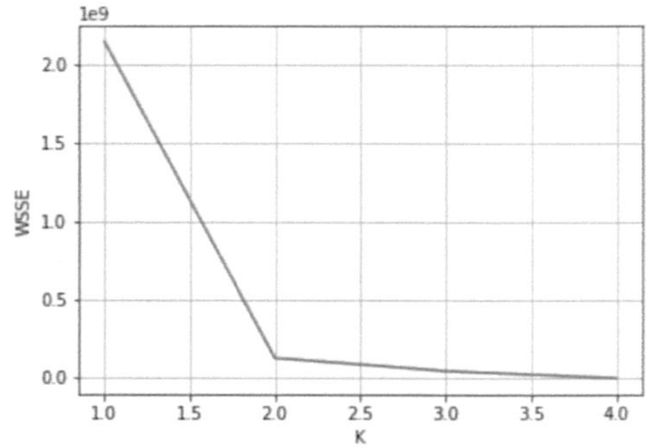

Figure 6.40 Scree plot for Figure 6.39.

```
plt.grid(True)
plt.savefig("output/kmeansSegmentationColor.png")
plt.show()
```

that produces Figure 6.40.

One can observe that 2 clusters are enough. The following code clusters data with respect to 2 clusters

```
NClusters = 2

km = KMeans( n_clusters= NClusters, n_init='auto')
labels = km.fit_predict(df)
#create mask from class labels
mask = labels.reshape(xShape, yShape, 1)

imgLabelled = []
masks = []
for label in range(NClusters):
    maskClass = np.where(mask == label, 255, 0)
    maskClass = maskClass.astype(np.uint8)
    masks.append(maskClass)
    cv2.imwrite("output/SegmentationColorMask{}.png".format(label),
                                             maskClass)
    imgLabel = cv2.bitwise_and(imageGray, imageGray, mask = maskClass
                                             )
    imgLabelled.append(imgLabel)
    cv2.imwrite("output/SegmentationColorLabel{}.png".format(label),
                                             imgLabel)
```

that produces segments in Figure 6.41, which were made using masks Figure 6.42 applied to Figure 6.39. We can see that the masks are inverse to each other.

It is instructive to consider a higher number of clusters. Figure 6.43 represents masks for `NClusters = 4` clusters in the k-Means algorithm. We observe that the k-Means algorithm separates small and big granules, as well as elements in the shaded region and its complement.

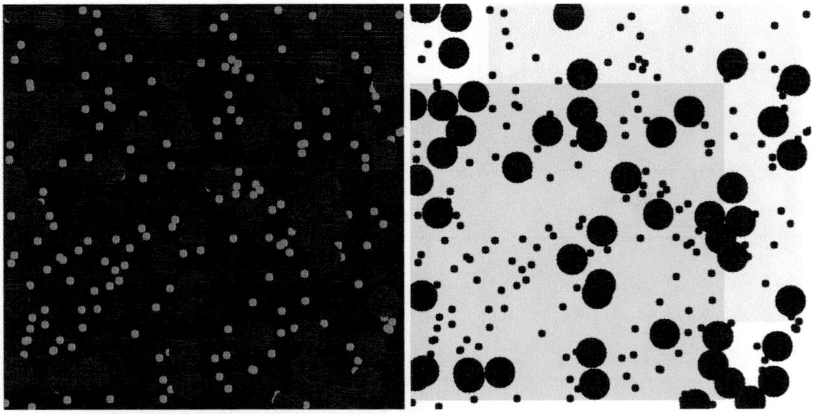

Figure 6.41 Segmentated image for Figure 6.39.

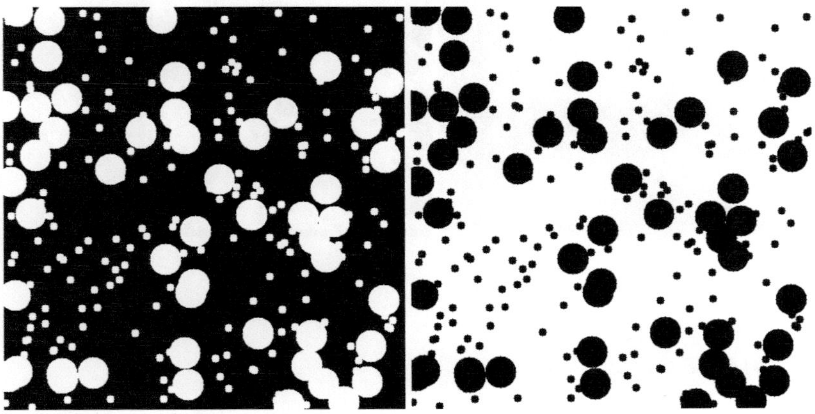

Figure 6.42 Masks that applied to Figure 6.39 result in Figure 6.41.

6.1.10 SUMMARY

Until now, we have presented many essential methods of image manipulation with the OpenCV library that can be used for extracting and classifying granules in an image. They mustn't be used as a black box. They must be used in a specific order, and parameters must be adjusted to a particular situation/type of image. Some experience is needed, however, thanks to them you can automatize granulometry.

6.2 EXAMPLES

In this section, we will describe more advanced examples that use methods from the previous part. We present four examples.

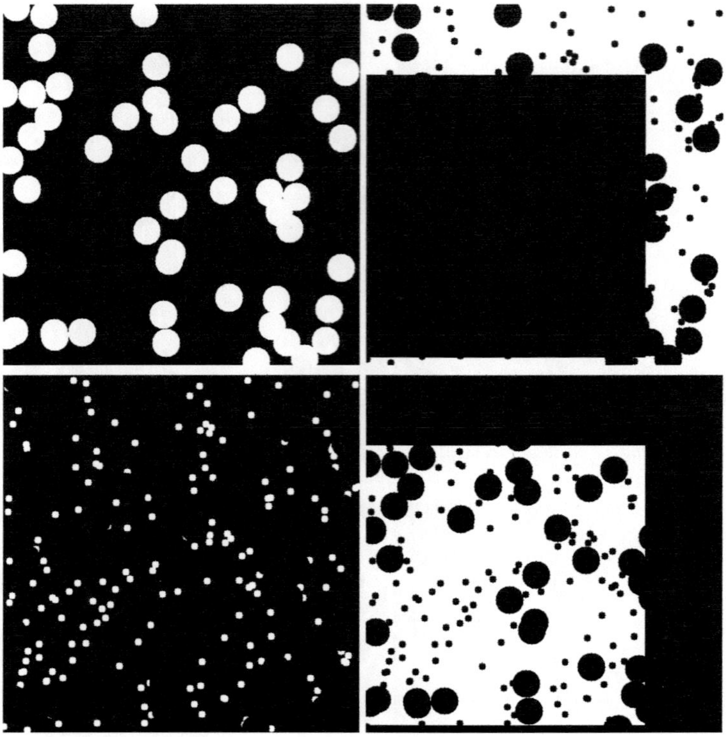

Figure 6.43 Masks for $k = 4$ clusters.

6.2.1 EXAMPLE 1

In the first example we want to select darker granules from Figure 6.44. This is a microscope image of a tissue.

In the first step we read the image and cut out the top part of the image

```
import numpy as np
import pandas as pd
import matplotlib.pylab as plt
import cv2

img1 = cv2.imread("Picture1.png")

#cut out printout
img1 = img1[50:-10,10:]
cv2.imwrite("output/Example1Cut.png", img1)
```

which gives Figure 6.45.

We now improve contrast by histogram equalization

```
#histograms of colors
hist= cv2.calcHist([img1],[0],None,[256],[0,256])
```

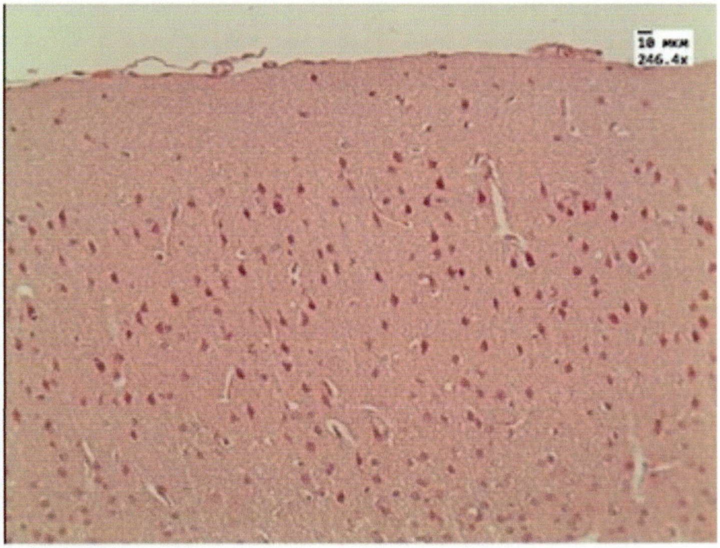

Figure 6.44 Initial image. Microscope image of cancerous mouse brain biopsy.

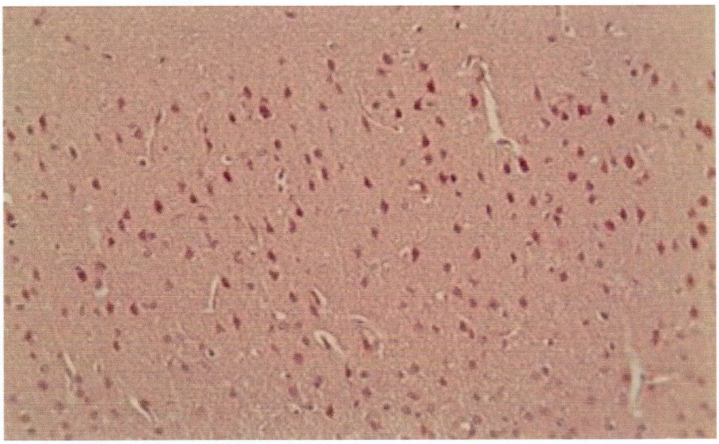

Figure 6.45 Cropped image that will be further processed.

```
hist2=cv2.calcHist([img1],[1],None,[256],[0,256])
hist3=cv2.calcHist([img1],[2],None,[256],[0,256])
##plot histograms
plt.subplot(1, 3, 1)
plt.plot(hist/max(hist))
plt.title('Hist Blue : ')
plt.subplot(1, 3, 2)
plt.plot(hist2/max(hist2),color="green")
plt.title('Hist Green : ')
plt.subplot(1, 3, 3)
plt.plot(hist3/max(hist3),color="red")
plt.title('Hist Red : ')
plt.savefig("output/Example1ColorHistograms")
plt.show()

#histogram equalization
src = cv2.cvtColor(img1, cv2.COLOR_BGR2HSV)
h,s,v = cv2.split(src)
v = cv2.equalizeHist(v)
src = cv2.merge([h,s,v])
img1 = cv2.cvtColor(src, cv2.COLOR_HSV2BGR)
cv2.imwrite("output/Example1Equalization.png", img1)

#histograms of colors after normalization
hist= cv2.calcHist([img1],[0],None,[256],[0,256])
hist2=cv2.calcHist([img1],[1],None,[256],[0,256])
hist3=cv2.calcHist([img1],[2],None,[256],[0,256])
plt.subplot(1, 3, 1)
plt.plot(hist/max(hist))
plt.title('Hist Blue : ')
plt.subplot(1, 3, 2)
plt.plot(hist2/max(hist2),color="green")
plt.title('Hist Green : ')
plt.subplot(1, 3, 3)
plt.plot(hist3/max(hist3),color="red")
plt.title('Hist Red : ')
plt.savefig("output/Example1ColorHistogramsEqualization.png")
plt.show()
```

that apart of color histograms, produces histogram-equalized Figure 6.46.

In order to enhance contours, we apply bilateral filtering, convert to grayscale, and compute histogram in order to determine the threshold. The following code

```
#scale figure
img1 = cv2.resize(img1,None,fx=4, fy=4, interpolation = cv2.
                                            INTER_CUBIC)
# convert to BW
img1_BW = cv2.cvtColor(img1, cv2.COLOR_BGR2GRAY)
cv2.imwrite("output/Example1BW.png", img1_BW)

#bilateral filtering
img1_BW = cv2.bilateralFilter(img1_BW,9,75,75)
cv2.imwrite("output/Example1BWFiltering.png", img1_BW)

#compute gray histogram
hist1 = cv2.calcHist([img1_BW], [0], None, [256], [0,256])
```

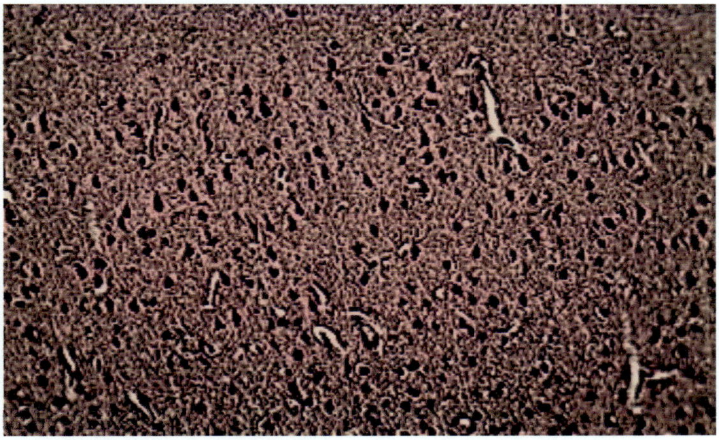

Figure 6.46 Result of histogram equalization applied to Figure 6.45.

```
plt.plot(hist1.ravel()/hist1.max())
plt.savefig("output/Example1HistogramBW.png")
plt.show()
```

which produces Figs. 6.47, 6.48. One can note that the black dots are represented by the pixels with intensity below 40. We use this information in thresholding and below. We also do morphological operations to delete small granules in the following code

```
ret1m, thresh1 = cv2.threshold(img1_BW, 40, 256, cv2.THRESH_BINARY)

#remove small dots
thresh1 = cv2.dilate(thresh1, np.ones((5, 5), np.uint8))
thresh1 = cv2.erode(thresh1, np.ones((5, 5), np.uint8))

cv2.imwrite("output/Example1BWThresholding.png", thresh1)
```

that produce Figure 6.49.

We are ready to detect contours and draw minimum area rectangle

```
#reverse colors in mask
thresh2 = abs(thresh1 - 255)

#find contours
contours, _ = cv2.findContours(thresh2, cv2.RETR_LIST, cv2.
                                   CHAIN_APPROX_TC89_L1 )
blank = np.ones(thresh2.shape)*255
cv2.drawContours(blank, contours, -1, (0, 0, 0), 3)
print("# of contours found: ", len(contours))
cv2.imwrite("output/Example1BWCountour.png", blank)

#minmimal area rectangle
img1_tmp1 = img1.copy()
img1_tmp2 = img1.copy()
```

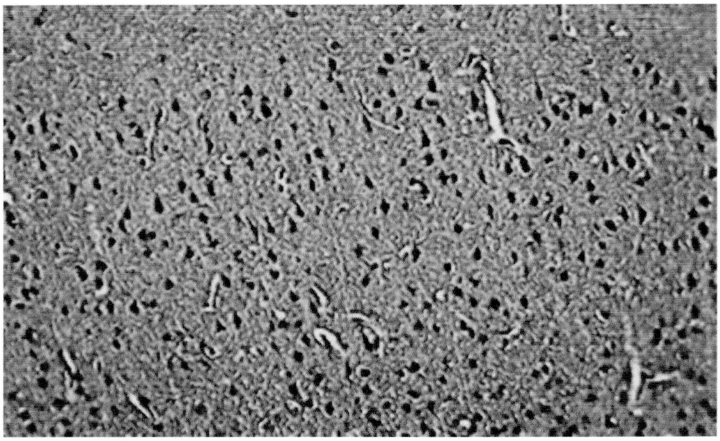

Figure 6.47 Grayscale image after bilateral filtering to enhance gradients.

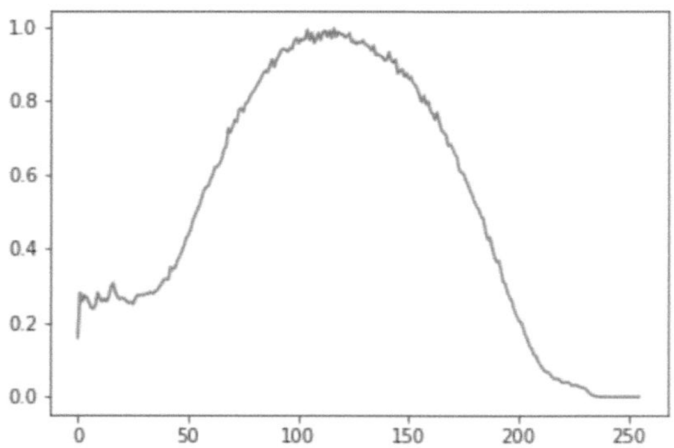

Figure 6.48 Histogram for Figure 6.47.

Figure 6.49 The result of thresholding in Figure 6.47.

```
for idx, cnt in enumerate(contours):
    minRect = cv2.minAreaRect(cnt)
    box = cv2.boxPoints(minRect)
    box = np.int0(box)
    cv2.drawContours(img1_tmp1, [box], 0, (0,0,255), 2)
    cv2.drawContours(img1_tmp2, [box], 0, (0,0,255), 2)
    cv2.putText(img1_tmp2, '{}'.format(idx), (int(box[0][0]), int(box
                                            [0][1]) ), cv2.
                                            FONT_HERSHEY_SIMPLEX, 0.5, (
                                            255, 255, 255), 1, cv2.LINE_AA
                                            )
cv2.imwrite("output/Example1BWMinAreaRectangle.png", img1_tmp1)
cv2.imwrite("output/Example1BWMinAreaRectangleWithIndices.png",
                                        img1_tmp2)
```

That produces Figure 6.50. We can see that most of the granules were detected by this (semi) automatic procedure.

The final step is to fit ellipses to the contours, and save their parameters in CSV file, that is realized by the following code

```
blank = np.ones(thresh2.shape)*255
ellipsesData = []
img_tmp = img1.copy()
for cnt in contours:
    #if less points preventing fitting ellipse
    if cnt.size < 10 :
        continue
    x,y,w,h = cv2.boundingRect(cnt)
    #remove small conoturs
    if w < 2 or h < 2:
        continue
    ellipse = cv2.fitEllipse(cnt)
    cv2.ellipse(blank, ellipse, (0,255, 255), 1, cv2.LINE_AA)
    cv2.ellipse(img_tmp, ellipse, (0,255, 255), 1, cv2.LINE_AA)
```

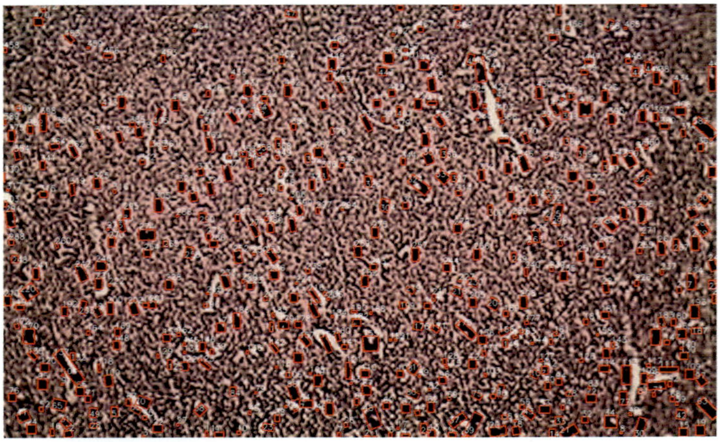

Figure 6.50 Minimal area rectangles enclosing detected contours with enumeration.

```
    ellipsesData.append({"id": idx, "x":ellipse[0][0], "y":ellipse[0]
                                            [1], "majorAxis": ellipse[1][0
                                            ], "minorAxis": ellipse[1][1],
                                             "angle": ellipse[2]})

cv2.imwrite("output/Example1BWEllipses.png", blank)
cv2.imwrite("output/Example1BWEllipsesColor.png", img_tmp)

df_ellipses = pd.DataFrame.from_records(ellipsesData)
df_ellipses.to_csv("output/ellipses.csv")

plt.hist(np.array(df_ellipses.angle)%90)
plt.xlabel("angle[deg]  mod 90")
plt.ylabel("No. of ellipses")
plt.savefig("output/Example1AngleDistibution.png")
plt.show()
```

This produces Figure 6.51. Moreover, when we are looking for the orientation of ellipses that can represent the orientation of non-spherical granules, we get from the above code Figure 6.52.

6.2.2 EXAMPLE 2 - CLASSIFICATION USING IMAGES

We now want to classify contours from the previous example using the k-means algorithm. As mentioned above, the classification results must be consulted with an expert. The first part of the code related to preparing images and creating masks is the same as before. We provide it for completeness

```
import numpy as np
import pandas as pd
import matplotlib.pylab as plt
import cv2
```

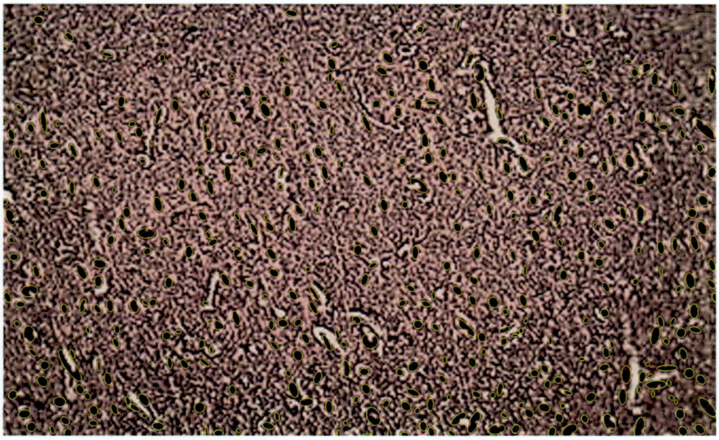

Figure 6.51 Ellipses fit to the contours.

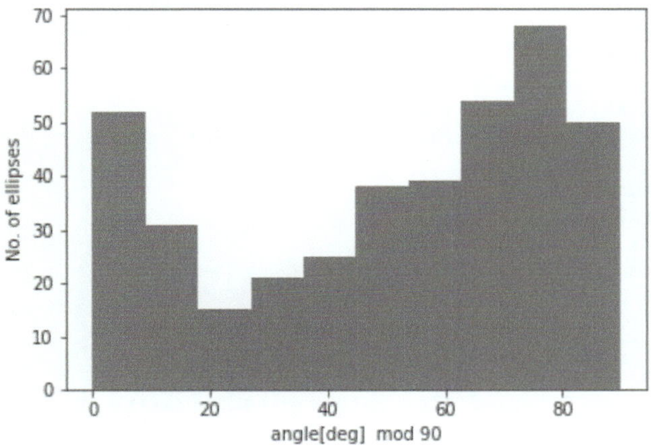

Figure 6.52 Angle distributions for ellipses.

```
img1 = cv2.imread("Picture1.png")

#cut out printout
img1 = img1[50:-10,10:]
cv2.imwrite("output/Example2Cut.png", img1)

#histograms of colors
hist= cv2.calcHist([img1],[0],None,[256],[0,256])
hist2=cv2.calcHist([img1],[1],None,[256],[0,256])
hist3=cv2.calcHist([img1],[2],None,[256],[0,256])
##plot histograms
plt.subplot(1, 3, 1)
plt.plot(hist/max(hist))
plt.title('Hist Blue : ')
plt.subplot(1, 3, 2)
plt.plot(hist2/max(hist2),color="green")
plt.title('Hist Green : ')
plt.subplot(1, 3, 3)
plt.plot(hist3/max(hist3),color="red")
plt.title('Hist Red : ')
plt.savefig("output/Example2ColorHistograms")
plt.show()

#histogram equalization
src = cv2.cvtColor(img1, cv2.COLOR_BGR2HSV)
h,s,v = cv2.split(src)
v = cv2.equalizeHist(v)
src = cv2.merge([h,s,v])
img1 = cv2.cvtColor(src, cv2.COLOR_HSV2BGR)
cv2.imwrite("output/Example2Equalization.png", img1)

#histograms of colors after normalization
hist= cv2.calcHist([img1],[0],None,[256],[0,256])
hist2=cv2.calcHist([img1],[1],None,[256],[0,256])
hist3=cv2.calcHist([img1],[2],None,[256],[0,256])
plt.subplot(1, 3, 1)
plt.plot(hist/max(hist))
plt.title('Hist Blue : ')
plt.subplot(1, 3, 2)
plt.plot(hist2/max(hist2),color="green")
plt.title('Hist Green : ')
plt.subplot(1, 3, 3)
plt.plot(hist3/max(hist3),color="red")
plt.title('Hist Red : ')
plt.savefig("output/Example2ColorHistogramsEqualization.png")
plt.show()

#scale figure
img1 = cv2.resize(img1,None,fx=4, fy=4, interpolation = cv2.
                                                INTER_CUBIC)
# convert to BW
img1_BW = cv2.cvtColor(img1, cv2.COLOR_BGR2GRAY)
cv2.imwrite("output/Example2BW.png", img1_BW)

#bilateral filtering
img1_BW = cv2.bilateralFilter(img1_BW,9,75,75)
```

```
cv2.imwrite("output/Example2BWFiltering.png", img1_BW)

#compute gray histogram
hist1 = cv2.calcHist([img1_BW], [0], None, [256], [0,256])
plt.plot(hist1.ravel()/hist1.max())
plt.savefig("output/Example2HistogramBW.png")
plt.show()

#tresholding - we select 40 for selecting black granules
ret1m, thresh1 = cv2.threshold(img1_BW, 40, 256, cv2.THRESH_BINARY)

#remove small dots
thresh1 = cv2.dilate(thresh1, np.ones((5, 5), np.uint8))
thresh1 = cv2.erode(thresh1, np.ones((5, 5), np.uint8))

cv2.imwrite("output/Example2BWThresholding.png", thresh1)

#reverse colors in mask
thresh2 = abs(thresh1 - 255)

#find contours
contours, _ = cv2.findContours(thresh2, cv2.RETR_LIST, cv2.
                                    CHAIN_APPROX_TC89_L1 )
blank = np.ones(thresh2.shape)*255
cv2.drawContours(blank, contours, -1, (0, 0, 0), 3)
print("# of contours found: ", len(contours))
cv2.imwrite("output/Example2BWCountour.png", blank)

#minmimal area rectangle
img1_tmp1 = img1.copy()
img1_tmp2 = img1.copy()
for idx, cnt in enumerate(contours):
    minRect = cv2.minAreaRect(cnt)
    box = cv2.boxPoints(minRect)
    box = np.int0(box)
    cv2.drawContours(img1_tmp1, [box], 0, (0,0,255), 2)
    cv2.drawContours(img1_tmp2, [box], 0, (0,0,255), 2)
    cv2.putText(img1_tmp2, '{}'.format(idx), (int(box[0][0]), int(box
                                        [0][1]) ), cv2.
                                        FONT_HERSHEY_SIMPLEX, 0.5, (
                                        255, 255, 255), 1, cv2.LINE_AA
                                        )
cv2.imwrite("output/Example2BWMinAreaRectangle.png", img1_tmp1)
cv2.imwrite("output/Example2BWMinAreaRectangleWithIndices.png",
                                    img1_tmp2)
```

We now crop from the initial image as bounding rectangles surrounding contours. These images will be reshaped into vectors to be an input for the ML algorithm. The following code prepares the data

```
image = img1.copy()
img_tmp = img1.copy()
filtered_contours = []
selected_img = []
for idx, contour in enumerate(contours):
    #draw selected contour
```

```
    minRect = cv2.minAreaRect(contour)
    box = cv2.boxPoints(minRect)
    box = np.int0(box)
    cv2.drawContours(img_tmp, [box], -1, (0,0,255), 1)

    #extract image from bounding rectangle
    x, y, w, h = cv2.boundingRect(contour)
    cropped = image[y:y + h, x:x + w, :].copy()
    feature = cv2.resize(cropped, (20, 20))
    cv2.imwrite("output/figs/{}.png".format(idx), feature)

    #scale cropped image to 1D array
    feature = feature.reshape(-1)
    selected_img.append(feature)
cv2.imwrite("output/Exmaple2SelectedContours.png", img_tmp)
df = pd.DataFrame(selected_img)
```

We now use these data to determine optimal number of clusters (similar granules)

```
from sklearn.cluster import KMeans
k = range(1,20)
WSSE = []
for i in k:
    model = KMeans(n_clusters = i, n_init='auto')
    model.fit_predict(df)
    WSSE.append(model.inertia_)
plt.plot(k,WSSE)
plt.xlabel('K')
plt.ylabel('WSSE')
plt.grid(True)
plt.savefig("output/Example2kmeans.png")
plt.show()
```

that produces Figure 6.53. From the 'elbow curve', we can observe that $k = 4$ clusters are sufficient. We therefore do classification using this value. Moreover, we fit ellipses to the contours and assign them to the class labels. The following lengthy code does the mentioned operations

```
km = KMeans( n_clusters=4, n_init='auto')
df['label'] = km.fit_predict(df)

df.to_csv("output/LabelledContours.csv")

img_tmp = image.copy()
img_ellipse = image.copy()
img_ellipse_closed = image.copy()
ellipsesData = []
for label, df_grouped in df.groupby('label'):
    Ngranules = len(df_grouped)
    print(Ngranules, " in label ", label)
    mask = np.zeros_like(image[:, :, 0])
    cv2.drawContours(mask, [contours[i] for i in df_grouped.index], -
                                            1, (255), -1)
    masked_image = cv2.bitwise_and(image, image, mask=mask)
    masked_image[mask==0] = (255,255,255)
```

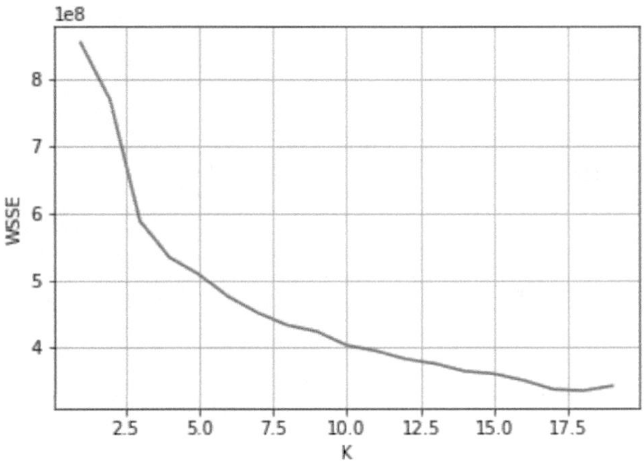

Figure 6.53 Angle distributions for ellipses.

```
cv2.imwrite("output/Example2label{}.png".format(label),
                                        masked_image)

print("fitting for label = ", label)
for  idx in df_grouped.index:
    minRect = cv2.minAreaRect(contours[idx])
    box = cv2.boxPoints(minRect)
    box = np.int0(box)
    cv2.drawContours(img_tmp, contours[idx], -1, (0,0,label*50),
                                             1)

    #fit ellypses
    if contours[idx].size < 10 :
        continue
    if w < 2 or h < 2:
        continue
    ellipse = cv2.fitEllipse(contours[idx])
    ellipsesData.append({"id": idx, "x":ellipse[0][0], "y":
                                             ellipse[0][1], "majorAxis"
                                             : ellipse[1][0], "
                                             minorAxis": ellipse[1][1],
                                              "angle": ellipse[2], "
                                             class": label})
    if max(ellipse[1]) < 50:
        cv2.ellipse(img_ellipse, ellipse, (0,0,label*50), 1, cv2.
                                                 LINE_AA)
        cv2.putText(img_ellipse, '{}_{}'.format(idx, label), (int
                                                 (box[0][0]), int(box[0
                                                 ][1]) ), cv2.
                                                 FONT_HERSHEY_SIMPLEX,
                                                 0.5, (255, 255, 255),
                                                 1, cv2.LINE_AA)
```

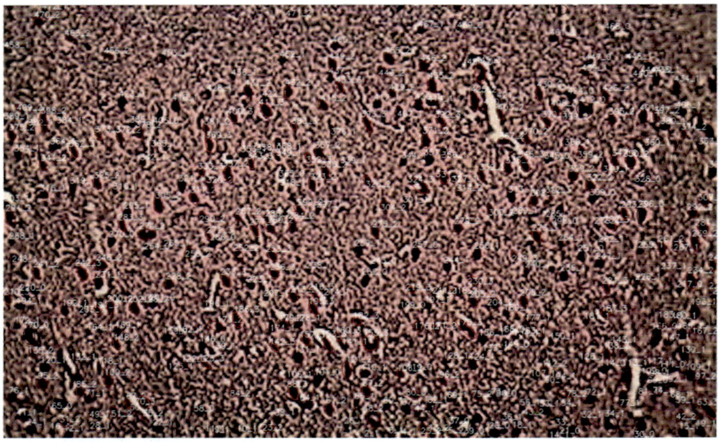

Figure 6.54 Fitted ellipses with numeration and class in the format: number_label.

```
            cv2.ellipse(img_ellipse_closed, ellipse, (0,0,label*50),
                                               -1, cv2.LINE_AA) #open
                                                ellipses
            cv2.putText(img_ellipse_closed, '{}_{}'.format(idx, label
                                               ), (int(box[0][0]),
                                               int(box[0][1]) ), cv2.
                                               FONT_HERSHEY_SIMPLEX,
                                               0.5, (255, 255, 255),
                                               1, cv2.LINE_AA)

df_ellipses = pd.DataFrame.from_records(ellipsesData)
df_ellipses.to_csv("output/ellipses.csv")

cv2.imwrite('output/Example2_color_segmentation.png', img_tmp)
cv2.imwrite('output/Example2_color_ellypses.png', img_ellipse)
cv2.imwrite('output/Example2_color_ellypses_closed.png',
                                   img_ellipse_closed)
```

which produces figures marking classified ellipses, as Figure 6.54.

6.2.3 EXAMPLE 3

In this example, we do a similar analysis to the previous one using Figure 6.55. After reading and cropping the image by the following code snippet

```
img1 = img1[50:-80,10:]
```

that result in Figure 6.56. The histogram of intensity for a grayscale image, Figure 6.57 suggests the threshold at about 50. The identified contours are presented in Figure 6.58.

Finally, on classifying and fitting ellipse we get Figure 6.59.

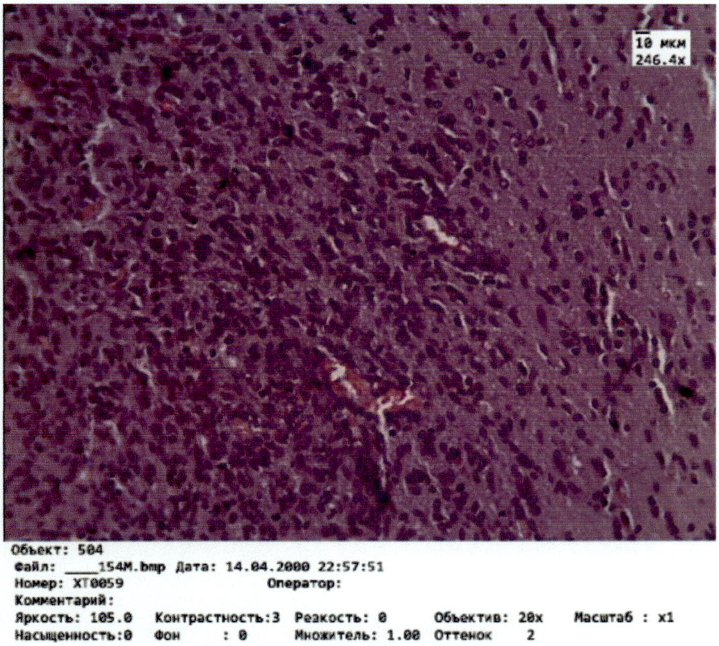

Figure 6.55 Initial image. Microscope image of cancerous mouse brain biopsy.

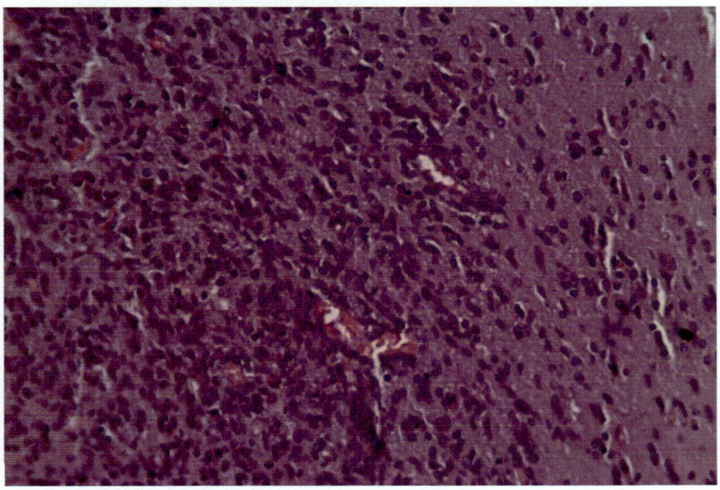

Figure 6.56 Cropped image.

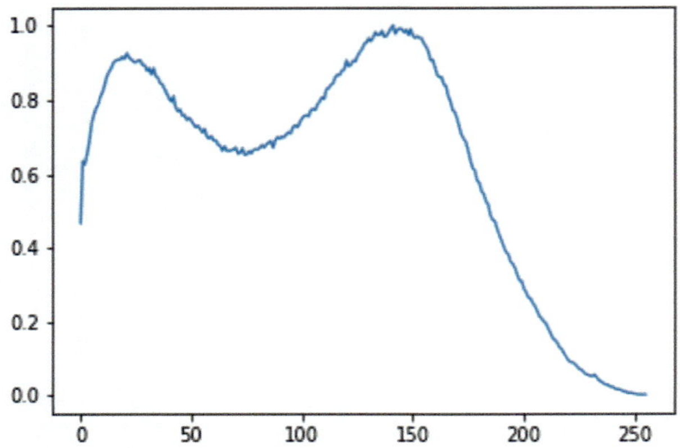

Figure 6.57 Intensity histogram.

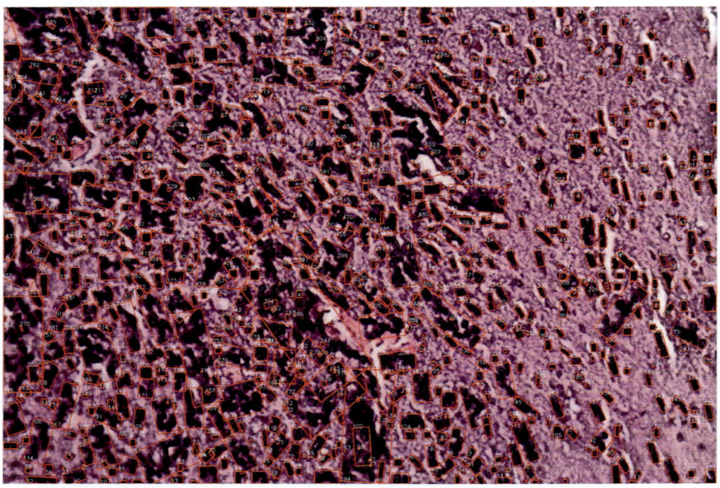

Figure 6.58 Contours marked by minimal area rectangles and enumerated.

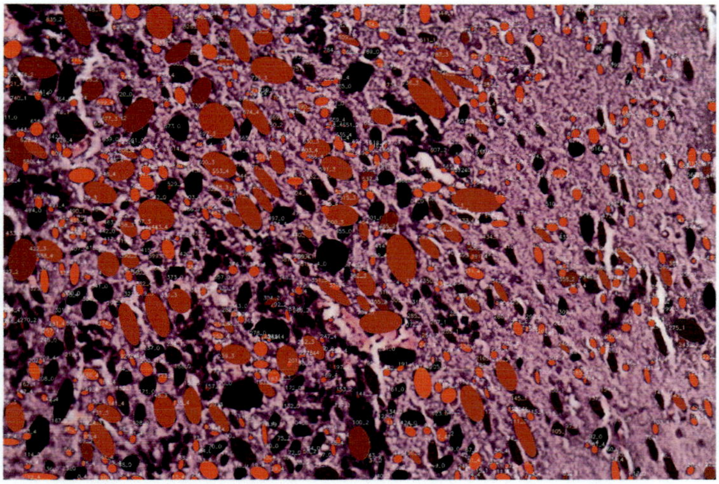

Figure 6.59 Ellipses fitted to contours with classes marked by different colors and enumeration in the format: number_label.

6.2.4 EXAMPLE 4 - CLASSIFICATION USING MEAN COLOR AND MOMENTS OF CONTOURS

We use again the image from Figure 6.55 and the code from the previous example. However, instead of using a full image representing a granule cropped from the original image as data for the classifier, we use mean color (each color channel) and Hu moment invariants. These parameters can be important if we expect that the geometry of granules, as well as their (average) color, are parameters that are important to discriminating different classes.

The change in the previous code that prepares the data is as follows

```
import math
image = img1.copy()
img_tmp = img1.copy()
filtered_contours = []
selected_data = []
for idx, contour in enumerate(contours):
    # get mean color of contour:
    masked = np.zeros_like(image[:, :, 0])
    cv2.drawContours(masked, [contour], 0, 255, -1)

    B_mean, G_mean, R_mean, _ = cv2.mean(image, mask=masked)

    #calculate moments
    moments = cv2.moments(contour)
    # Calculate Hu Moments
    huMoments = cv2.HuMoments(moments)

    #scaling Hu moments
    for i in range(0,7):
        #print("hu moment ", i, " ", huMoments[i])
```

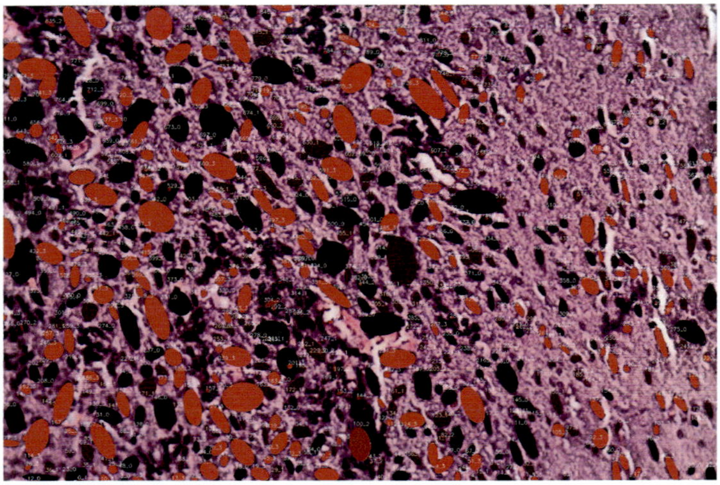

Figure 6.60 Ellipses fitted to contours with classes marked by different colors and enumeration in the format: number_label.

```
        if huMoments[i] == 0:
            huMoments[i] = 0.0
        else:
            huMoments[i] = -1* math.copysign(1.0, huMoments[i]) *
                                                   math.log10(abs(
                                                   huMoments[i]))

    selected_data.append({'B_mean': B_mean, 'G_mean': G_mean, 'R_mean
                                          ': R_mean, 'h0':huMoments[0],
                                          'h1':huMoments[1], 'h2':
                                          huMoments[2], 'h3':huMoments[3
                                          ], 'h4':huMoments[4], 'h5':
                                          huMoments[5], 'h6':huMoments[6
                                          ]})
cv2.imwrite("output/Exmaple2SelectedContours.png", img_tmp)
df = pd.DataFrame(selected_data)
```

For comparison, fitted ellipses are presented in Figure 6.60.

7 Cancer cells detection using Neural Networks

In this chapter, we construct a simple artificial neural network (NN) to classify cancer in histopathologic scans. It is more technical and demanding since we expect the reader to be familiar with the Digital Image Processing methods described in the previous chapter and have at least basic knowledge of the construction of Artificial Neural Networks described, e.g., in [86, 205]. In the previous chapter, we used unsupervised learning that allows an algorithm for clustering similar pixels. We obtained some groups of cells, which experts should examine since the algorithm k-Means, which was used, has no internal knowledge of detecting cancer cells. To make the classification more automatic, we must invest additional nowledge in preparing data and then use supervised learning algorithms to learn these examples. The NN approach is one of the most universal approaches to supervised learning that can detect complex relations in the data. We will not focus on the broad theory and practice of specific NN architectures since many resources cover these topics in detail, e.g., [86, 205]. In the chapter, we show an example of how to construct and train a model and then use it for classification. This chapter contains two sections. In the first one, we describe the preparation of the classifier; in the second section, we use it to make the classification. Complete examples of the code from this chapter can be downloaded from [1]. You can run it and experiment.

7.1 CLASSIFIER PREPARATION

The essential step in classifier preparation is collecting suitable data representing the problem under consideration. We will consider cancer cells in a tissue. The expert learns to spot cancer cells using theoretical aspects and experience in the analysis of many examples. One of the typical shapes of a cancer cell is the so-called 'signet ring cell' (due to its appearance), which is a cell with a large vacuole. However, it is not the only marker. Also, the surrounding tissue can have an effect. This leads to the great complexity of features that suggest that a given cell is a cancer cell. Therefore, we should use a large set of example images to train neural networks.

We choose the publicly available PatchCamelyon (PCam) dataset [3, 26, 228]. 'It consists of 327.680 color images (96 x 96px) extracted from histopathologic scans of lymph node sections. Each image is annotated with a binary label indicating the presence of metastatic tissue. PCam provides a new benchmark for machine learning models: bigger than CIFAR10, smaller than imagenet, trainable on a single GPU.' [3]. It is also important to note that: 'A positive label indicates that the center 32x32px

DOI: 10.1201/9781032646633-7

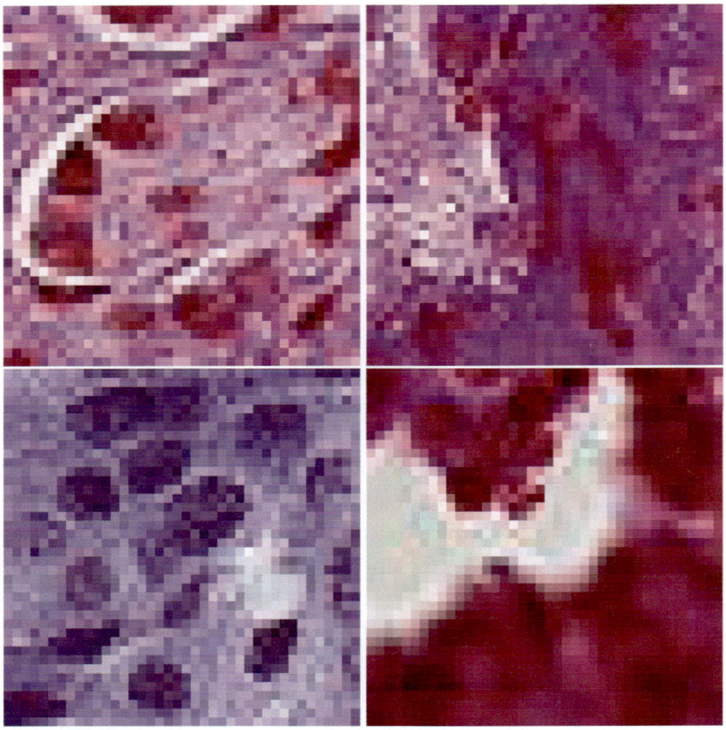

Figure 7.1 Example images containing cancer cells from PCam dataset [3]. 40×40 pixels cropped from the center of the images labeled as containing cancer.

region of a patch contains at least one pixel of tumor tissue.' [3], and therefore, we must extract at least such central area of the image for classification.

The dataset is split into three parts:

- training set - 262144 (2^{18}) examples;
- validation set - 32768 (2^{15}) examples;
- test set - - 32768 (2^{15}) examples;

These sets are balanced because 50% of images are labeled as containing cancer cells. We present a few examples of cancer-labelled images in Figure 7.1.

To speed up the training process, we use only combined test and validation data from [3], namely, the files:

- `camelyonpatch_level_2_split_valid_x.h5.gz` - zipped HDF5 file containing images for validation;
- `camelyonpatch_level_2_split_valid_y.h5.gz` - zipped HDF5 file containing labels fo validation images (0- no cancer; 1 - cancer);
- `camelyonpatch_level_2_split_test_x.h5.gz` - zipped HDF5 file containing images for testing;

- `camelyonpatch_level_2_split_valid_y.h5.gz` - zipped HDF5 file containing labels fo testing images (0- no cancer; 1 - cancer);

For preparation, we crop the 40×40 pixels image around the center to ensure that in the case of cancer cells, the subimage contains some pixels belonging to the cancer cells. We operate on unzipped files. The following code is self-explanatory

```
import numpy as np
import h5py
import pylab as plt
import cv2

#read X data - images
print("reading X data")

x = []
for fileName in ['./camelyonpatch_level_2_split_valid_x.h5', './
                                    camelyonpatch_level_2_split_test_x
                                    .h5']:
    print("... reading file: ", fileName)
    fx = h5py.File(fileName, 'r')

    for idx,image in enumerate(fx['x']):
        #32 pixels around center containes at least one pixel of
                                            cancer cell
        imageRescaled = image[int(image.shape[0]/2.-20):int(image.
                                            shape[0]/2.+20), int(image
                                            .shape[1]/2.-20) : int(
                                            image.shape[1]/2.+20) ]
        imageRescaled = cv2.resize(imageRescaled, (40,40))
        #save image for later processing
        x.append(imageRescaled)
        #save samples to files
        #cv2.imwrite("image/{}.png".format(idx), image)
x = np.array(x)
print("DONE")


#preparing y data - cancer/no cancer
print("reading Y data")
y = []
for fileName in ['./camelyonpatch_level_2_split_valid_y.h5', './
                                    camelyonpatch_level_2_split_test_y
                                    .h5']:
    fy = h5py.File(fileName, 'r')
    for idx,datay in enumerate(fy['y']):
        y.append(datay.flatten())

y = np.array(y)
print("DONE")

print("dataset shape X: ", x.shape)
print("dataset shape Y: ", y.shape)
print("cancer cases: ", np.sum(y))
```

If needed, we can also extract the cancer-labelled images only

```
positive = x[y.reshape(-1)==1]
for idx,image in enumerate(positive):
   cv2.imwrite("positive/{}.png".format(idx), image)
```

The next step is to shuffle data to make their appearance more random than in the file. Moreover, we split the whole dataset into three parts:

- **train set** - used for training; we use 75% of dataset;
- **validation set** - used for validation during training; we use 15% of dataset;
- **test set** - used for testing after training process; we use remaining 10% of the dataset;

The above percentage is rather arbitrary and standard. It is reflected in the following code

```
#shufle data randomly
from sklearn.utils import shuffle
x, y = shuffle(x, y)

#split into train, test and validation data
from sklearn.model_selection import train_test_split

train_ratio = 0.75
validation_ratio = 0.15
test_ratio = 0.10

x_train, x_test, y_train, y_test = train_test_split(x, y, test_size=1
                                       - train_ratio)
x_val, x_test, y_val, y_test = train_test_split(x_test, y_test,
                                   test_size=test_ratio/(test_ratio +
                                    validation_ratio))
```

In the next step, we must prepare our neural network, which will be trained and validated. We will use feed-forward architecture as the most popular and most straightforward to describe. No general mathematical framework describes the work of large neural networks. The general theorem is related to the Cybenko's paper [61], saying that the simple NN works as universal approximator of functions, and its generalizations. Moreover, there is no strict knowledge of what types of architectures are suitable for specific data. The experts make educated guesses guided by experiments and experience.

Our dataset contains images, and therefore, the first hint is that the 2-dimensional convolutional neural networks are the most suitable choice for the first layers. However, the layer works as an aggregator for neighborhood pixels, and the distance is determined by the shape of the convolutional kernel. This kernel is typically small, like 3×3 pixels. In our case, the cancer cell, if present and in the ideal case, will occupy almost all central parts of the picture and, therefore, the large size of the kernel or a few successive convolutional layers. Consequently, the complexity grows.

The above description suggests that the better approach is to use a top layer to analyze the 'whole picture at once' to extract the features of possible cancer cells.

In other words, the potential cancer cell occupies almost the whole image, so the suggested approach is the global view of the entire image by the first layer. The natural choice is a dense layer aggregating information in pixels, not necessarily neighbor ones, so it should have fewer neurons than the pixels in the image. The image is flattened to the vector, and then the dense layer is fed. The next layer is again dense, analyzing features extracted from the top dense layer. The number of units is selected by the trial-and-error approach. Finally, the output is a single unit with the sigmoid activation function. After quantization, this unit gives an output, which provides information on whether there is a cancer cell (1) in the image or not (0). Before feeding the top dense layer, we applied random reflection and rotation of images since our model should classify cancer cells in every orientation, and we do not have much control over the orientation as we operate on prepared pictures by technicians.

We use TensorFlow library [6] to construct the neural network

```
from tensorflow.keras.models import Sequential
from tensorflow.keras.layers import Dense, Flatten
from tensorflow.keras.layers import Dropout, BatchNormalization,
                                   Input
from tensorflow.keras.layers import RandomFlip, RandomRotation

import tensorflow as tf

model = Sequential([Input(shape=x.shape[1:]),
                    RandomFlip("horizontal_and_vertical"),
                    RandomRotation(0.5),
                    Flatten(),
                    Dense(40, activation='relu'),
                    BatchNormalization(),
                    Dense(1000, activation='relu'),
                    Dropout(0.2),
                    BatchNormalization(),
                    Dense(1, activation='sigmoid')
                    ])
```

We can also optimize the architecture, which is a more advanced topic [86, 205], and we do not follow it here.

Then, we can compile the model and train it. In the following code, we are interested in accuracy, and we use 30 epochs for training

```
#compile model
model.compile(
    loss=tf.keras.losses.BinaryCrossentropy(),
    optimizer='adam',
    metrics=['accuracy']
)

#train model
history = model.fit(x_train, y_train, epochs=30, validation_data=(
                                        x_val, y_val))
```

We can plot train history

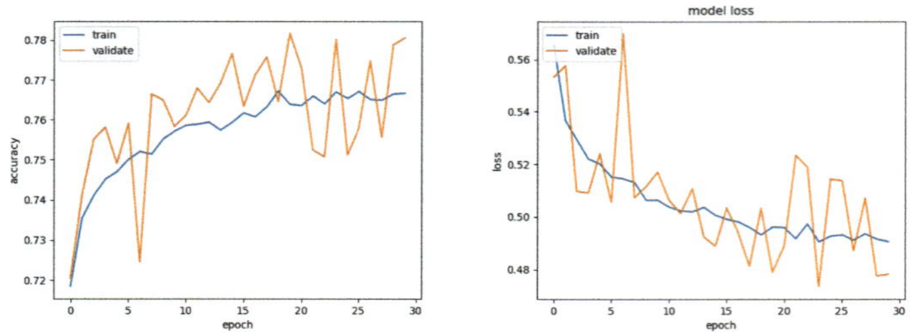

Figure 7.2 Accuracy and loss plots for training and validation dataset. The accuracy saturates at about 77%.

```
#accuracy plot
plt.plot(history.history['accuracy'])
plt.plot(history.history['val_accuracy'])
plt.ylabel('accuracy')
plt.xlabel('epoch')
plt.legend(['train', 'validate'], loc='upper left')
plt.savefig("output/accuracy.png")
plt.show()
# loss plot
plt.plot(history.history['loss'])
plt.plot(history.history['val_loss'])
plt.title('model loss')
plt.ylabel('loss')
plt.xlabel('epoch')
plt.legend(['train', 'validate'], loc='upper left')
plt.savefig("output/loss.png")
plt.show()
```

presented in Figure 7.2.

The following code realizes the validation on test data

```
results = model.evaluate(x_test, y_test, batch_size=128)
print("test loss, test acc:", results)
```

that produces

```
test loss, test acc: [0.46777990460395813,
                      0.7879157662391663]
```

This confirms that the model is about 77% accurate. The more detailed test involves cross-validation [86, 205], however, we will no do this here since the basic measures we obtained are consistent enough to estimate accuracy.

Finally, we save the model for later use in image classification in the next section, using

```
model.save('model/NNModel.h5')
```

7.2 IMAGE ANALYSIS

In this section, we use the NN model trained in the previous section to classify granules selected by thresholding, as in the previous chapter.

We will start from the initial image for analysis Figure 6.55. We crop it and histogram-equalize as in the following code

```
import numpy as np
import pandas as pd
import matplotlib.pylab as plt
import cv2

img = cv2.imread("Picture2.png")

#cut out printout
img = img[50:-80,10:]
cv2.imwrite("output/Picture2Cut.png", img)

#histograms of colors
hist= cv2.calcHist([img],[0],None,[256],[0,256])
hist2=cv2.calcHist([img],[1],None,[256],[0,256])
hist3=cv2.calcHist([img],[2],None,[256],[0,256])
##plot histograms
plt.subplot(1, 3, 1)
plt.plot(hist/max(hist))
plt.title('Hist Blue : ')
plt.subplot(1, 3, 2)
plt.plot(hist2/max(hist2),color="green")
plt.title('Hist Green : ')
plt.subplot(1, 3, 3)
plt.plot(hist3/max(hist3),color="red")
plt.title('Hist Red : ')
plt.savefig("output/Picture2ColorHistograms")
plt.show()

#histogram equalization
src = cv2.cvtColor(img, cv2.COLOR_BGR2HSV)
h,s,v = cv2.split(src)
v = cv2.equalizeHist(v)
src = cv2.merge([h,s,v])
img1 = cv2.cvtColor(src, cv2.COLOR_HSV2BGR)
cv2.imwrite("output/Picture2Equalization.png", img1)

#histograms of colors after normalization
hist= cv2.calcHist([img1],[0],None,[256],[0,256])
hist2=cv2.calcHist([img1],[1],None,[256],[0,256])
hist3=cv2.calcHist([img1],[2],None,[256],[0,256])
plt.subplot(1, 3, 1)
plt.plot(hist/max(hist))
plt.title('Hist Blue : ')
plt.subplot(1, 3, 2)
plt.plot(hist2/max(hist2),color="green")
plt.title('Hist Green : ')
plt.subplot(1, 3, 3)
plt.plot(hist3/max(hist3),color="red")
plt.title('Hist Red : ')
```

```
plt.savefig("output/Picture2ColorHistogramsEqualization.png")
plt.show()

#scale figure
img1 = cv2.resize(img1,None,fx=4, fy=4, interpolation = cv2.
                                        INTER_CUBIC)
img = cv2.resize(img,None,fx=4, fy=4, interpolation = cv2.INTER_CUBIC
                                        )
# convert to BW
img1_BW = cv2.cvtColor(img1, cv2.COLOR_BGR2GRAY)
cv2.imwrite("output/Picture2BW.png", img1_BW)

#bilateral filtering
img1_BW = cv2.bilateralFilter(img1_BW,9,75,75)
cv2.imwrite("output/Picture2BWFiltering.png", img1_BW)

#compute gray histogram
hist1 = cv2.calcHist([img1_BW], [0], None, [256], [0,256])
plt.plot(hist1.ravel()/hist1.max())
plt.savefig("output/Picture2HistogramBW.png")
plt.show()

#tresholding - we select 40 for selecting black granules
ret1m, thresh1 = cv2.threshold(img1_BW, 50, 256, cv2.THRESH_BINARY)

#remove small dots
thresh1 = cv2.dilate(thresh1, np.ones((5, 5), np.uint8))
thresh1 = cv2.erode(thresh1, np.ones((5, 5), np.uint8))

cv2.imwrite("output/Picture2BWThresholding.png", thresh1)

#reverse colors in mask
thresh2 = abs(thresh1 - 255)

#find contours
contours, _ = cv2.findContours(thresh2, cv2.RETR_LIST, cv2.
                                        CHAIN_APPROX_TC89_L1 )
blank = np.ones(thresh2.shape)*255
cv2.drawContours(blank, contours, -1, (0, 0, 0), 3)
print("# of contours found: ", len(contours))
cv2.imwrite("output/Picture2BWCountour.png", blank)

#minmimal area rectangle
img1_tmp1 = img.copy()
img1_tmp2 = img.copy()
for idx, cnt in enumerate(contours):
    minRect = cv2.minAreaRect(cnt)
    box = cv2.boxPoints(minRect)
    box = np.int0(box)
    cv2.drawContours(img1_tmp1, [box], 0, (0,0,255), 2)
    cv2.drawContours(img1_tmp2, [box], 0, (0,0,255), 2)
    cv2.putText(img1_tmp2, '{}'.format(idx), (int(box[0][0]), int(box
                                        [0][1]) ), cv2.
                                        FONT_HERSHEY_SIMPLEX, 0.5, (
                                        255, 255, 255), 1, cv2.LINE_AA
                                        )
```

```
cv2.imwrite("output/Picture2BWMinAreaRectangle.png", img1_tmp1)
cv2.imwrite("output/Picture2BWMinAreaRectangleWithIndices.png",
                                        img1_tmp2)
```

The results of this code are the same as in the previous chapter when we analyzed `Picture2.png`.

Having found contours that contain possibly suspected structures that can be cancer cells, we now cut them out from the whole image, adding 20 pixels from all four sides since the contours have different spans. We then rescale all cropped parts to 40×40 pixels in accordance with the NN model input. We use non-histogram-equalized images since we use the non-histogram-equalized images for training. We collect subimages in the following code

```
 image = img.copy()
img_tmp = img1.copy()
filtered_contours = []
selected_img = []
for idx, contour in enumerate(contours):
    #draw selected contur
    minRect = cv2.minAreaRect(contour)
    box = cv2.boxPoints(minRect)
    box = np.int0(box)
    cv2.drawContours(img_tmp, [box], -1, (0,0,255), 1)

    #extract image from bouding rectangle
    x, y, w, h = cv2.boundingRect(contour)
    #cropped = image[y:y + h, x:x + w, :].copy()
    a = 20
    cropped = image[y-a if y-a >=0 else 0 :y + h+a, x-a if x-a >= 0
                                  else 0:x + w+a, :].copy()
    feature = cv2.resize(cropped, (40, 40))
    cv2.imwrite("output/figs/{}.png".format(idx), feature)

    #scale cropped image to 1D array
    #feature = feature.reshape(-1)
    selected_img.append(feature)

cv2.imwrite("output/Picture2SelectedContours.png", img_tmp)
selected_img = np.array(selected_img)
print("selected_img = ", selected_img.shape)
```

where selected contours are marked by red rectangles in Figure 7.3.

As a final part, we read the NN model trained in the previous section and applied it to selected sub-images containing granules. We focus on granules with contours large enough to fit ellipses (number of contour points no less than 10 points) and not too big (major axis of ellipse less than 150 pixels). We use the following code

```
#load NN model
import tensorflow as tf
model = tf.keras.models.load_model('model/NNModel.h5')

#classify sub-images
y = model.predict(selected_img)
```

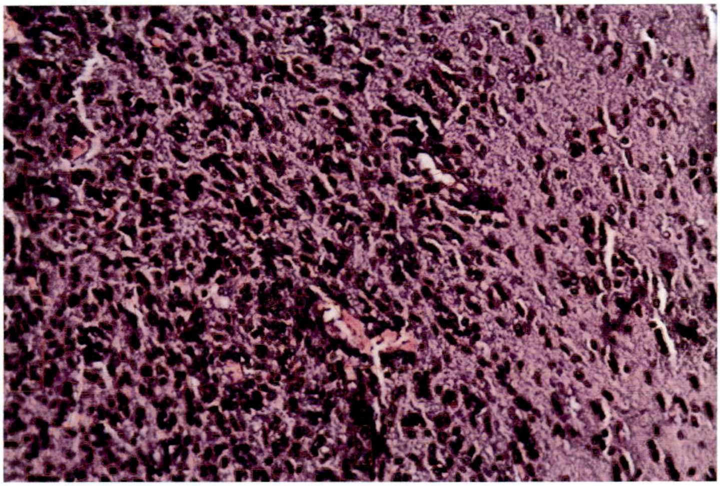

Figure 7.3 Selected granules for classification. Minimal area rectangle surrounding contours were used.

```
#quantize output to 0 or 1
y = np.where(y>0.5, 1, 0)

print("number of cancer cells: ", np.sum(y))

img_tmp = img1.copy()
img_ellipse = img1.copy()
img_ellipse_closed = img1.copy()

ellipsesData = []
for idx, contour in enumerate(contours):

    minRect = cv2.minAreaRect(contour)
    box = cv2.boxPoints(minRect)
    box = np.int0(box)
    cv2.drawContours(img_tmp, contour, -1, (50,50,50), 1)

    #fit ellipses
    if contour.size < 10 :
        continue
    ellipse = cv2.fitEllipse(contour)
    if max(ellipse[1]) < 150:
        label = y[idx][0]
        ellipsesData.append({"id": idx, "x":ellipse[0][0], "y":
                                           ellipse[0][1], "majorAxis"
                                           : ellipse[1][0], "
                                           minorAxis": ellipse[1][1],
                                            "angle": ellipse[2], "
                                           class": label})

        cv2.putText(img_ellipse, '{}'.format(label), (int(box[0][0]),
                                            int(box[0][1]) ), cv2.
```

```
                                    FONT_HERSHEY_SIMPLEX, 0.5,
                                     (255, 255, 255), 1, cv2.
                                    LINE_AA)
        cv2.putText(img_ellipse_closed, '{}'.format(label), (int(box[
                                    0][0]), int(box[0][1]) ),
                                    cv2.FONT_HERSHEY_SIMPLEX,
                                    0.5, (255, 255, 255), 1,
                                    cv2.LINE_AA)
        cv2.putText(img_tmp, '{}'.format(label), (int(box[0][0]), int
                                    (box[0][1]) ), cv2.
                                    FONT_HERSHEY_SIMPLEX, 0.5,
                                     (255, 255, 255), 1, cv2.
                                    LINE_AA)
        if label == 1:
            cv2.drawContours(img_tmp, [box], -1, (0,0,250), 1)
            cv2.ellipse(img_ellipse, ellipse, (0,0,255), 1, cv2.
                                        LINE_AA)
            cv2.ellipse(img_ellipse_closed, ellipse, (0,0,250), -1,
                                        cv2.LINE_AA)
        else:
            cv2.drawContours(img_tmp, [box], -1, (0,250,0), 1)
            cv2.ellipse(img_ellipse, ellipse, (0,255,0), 1, cv2.
                                        LINE_AA)
            cv2.ellipse(img_ellipse_closed, ellipse, (0,250,0), -1,
                                        cv2.LINE_AA)

df_ellipses = pd.DataFrame.from_records(ellipsesData)
df_ellipses.to_csv("output/ellipses.csv")

cv2.imwrite('output/Picture2_color_segmentation.png', img_tmp)
cv2.imwrite('output/Picture2_color_ellypses.png', img_ellipse)
cv2.imwrite('output/Picture2_color_ellypses_closed.png',
                                 img_ellipse_closed)
```

The parameters of ellipses and their labels from the NN model are saved in CSV file `ellipses.csv`.

We obtain ellipses presented in Figure 7.4. For better visibility, we also plot the filled ellipses in Figure 7.5.

A note of warning is in order now. The classification results are only a hint since the algorithm is only 77% accurate, and some specific cases may not be reflected in training data or be too complicated to learn in such a simple architecture. Therefore, these results must be cross-checked by an expert. Nevertheless, these results can simplify and speed up the diagnosis process.

The approach presented here can be improved in various directions, however, the general idea of the approach is very general.

After reading this and previous chapters you should be well-prepared to construct image data analysis and classification Python script independently.

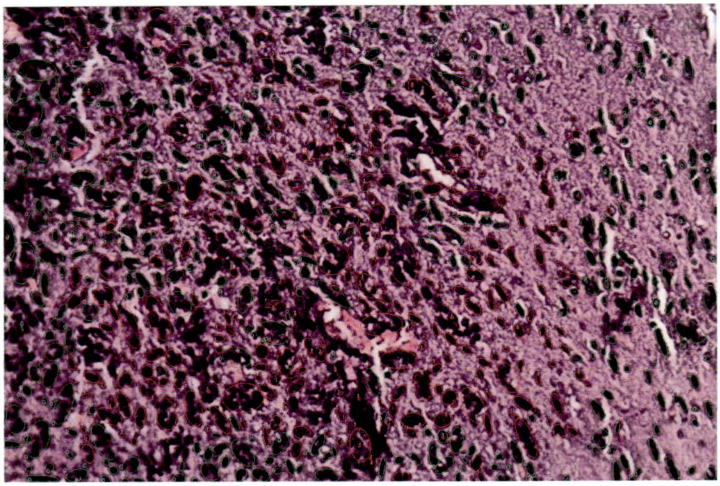

Figure 7.4 Fitted ellipses. Color corresponds to classification: 1/red - cancer cell(s); 0/green - no-cancer cell(s);

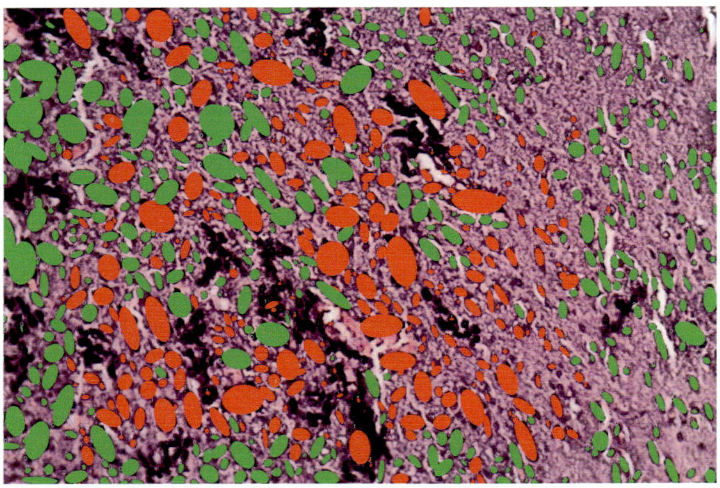

Figure 7.5 Fitted ellipses with filling for better visualization. Color corresponds to classification: 1/red - cancer cell(s); 0/green - no-cancer cell(s);

8 Applications

First, image analysis of one cancerous mouse brain biopsy microscope image is applied to detect the inclusions (in the terminology of composite media) or cancerous and noncancerous cells (in the medicine terminology). Next, the derived analytical formulas for three-phase composites are applied to determine the macroscopic permittivity for various frequencies. Different variants of shapes are considered, in particular, elliptic for glioma and circular for neurons. The contributions of the Maxwell-Garnett term, the **J**-shape term, and the interaction term $\boldsymbol{\wp}$ into the macroscopic permittivity tensor are analyzed. The results for the Maxwell-Garnett and Bruggemen approximations, as well as derived higher-order approximations, coincide only for very low frequencies. Low frequency corresponds to small absolute values of contrast parameters. The aRVE theory is applied to study the distribution of all the inclusions (cells) in the following way. A small sliding box $B(\mathbf{x})$ centered at $\mathbf{x}$ with a side less than 10 times the side of the large square Q is considered. The structural sums e_2, e_{22} and e_{44} are calculated over $B(\mathbf{x})$ as functions of $\mathbf{x}$. According to the decomposition theorem, these structural sums characterize the local geometrical distribution of inclusions in Q, which impacts the macroscopic permittivity tensor.

8.1 BIOLOGICAL RANDOM MEDIUM IMPACTED BY THE MORPHOLOGY OF THE TUMOR CELLS

In the present section, cancerous (glioma) cells are modeled by elliptic inclusion, and noncancerous (neuron) cells by disks. The permittivity ε_1 of glioma and ε_2 of a neuron depends on the frequency ω [96]

$$\varepsilon_1 = -\frac{3.5\left(2.45678\times 10^{10}\mathrm{i}\omega+\omega^2+2.52773\times 10^{19}\right)}{\omega(3\omega-10^{11}\mathrm{i})}, \tag{8.1}$$

$$\varepsilon_2 = \frac{540\left(-1.5428\times 10^{11}\mathrm{i}\omega+\omega^2-2.69803\times 10^{20}\right)}{\omega(9\omega-10^{12}\mathrm{i})}. \tag{8.2}$$

We now apply the formulas derived in Section 4.2 to the structure in Figure 8.1. This figure is similar to Figure 6.55, but represents another image of a cancerous mouse.

The centers of inclusions, the permittivities (8.1)–(8.2) are fixed, and the total concentration and shapes change. Consider the unit square periodicity cell Q with N inclusions G_k as shown in Figure 8.2.

The general formula (4.99) is used for the computations. It can be written in the form

$$\boldsymbol{\varepsilon}_{\perp} = \frac{1+\varrho f}{1-\varrho f}\mathbf{I}+\mathbf{J}+\boldsymbol{\wp}+O(f^3\varrho^3), \tag{8.3}$$

DOI: 10.1201/9781032646633-8

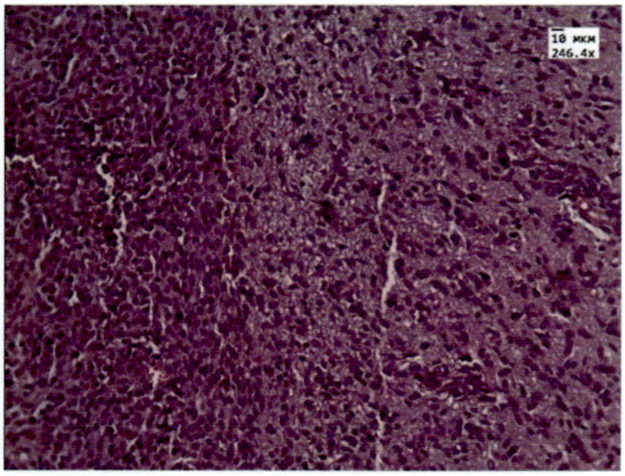

Figure 8.1 Microscope image of cancerous mouse brain biopsy.

where the first term is the Maxwell-Garnett approximation of order $O(|\varrho|f)$, $\mathbf{J}$ the shape term of order $O(\varrho^2 f)$

$$\mathbf{J} = 2\varrho^2 f_1 \alpha \frac{1}{N_1} \sum_{k=1}^{N_1} \begin{pmatrix} \cos 2\varphi_k & -\sin 2\varphi_k \\ -\sin 2\varphi_k & -\cos 2\varphi_k \end{pmatrix}, \tag{8.4}$$

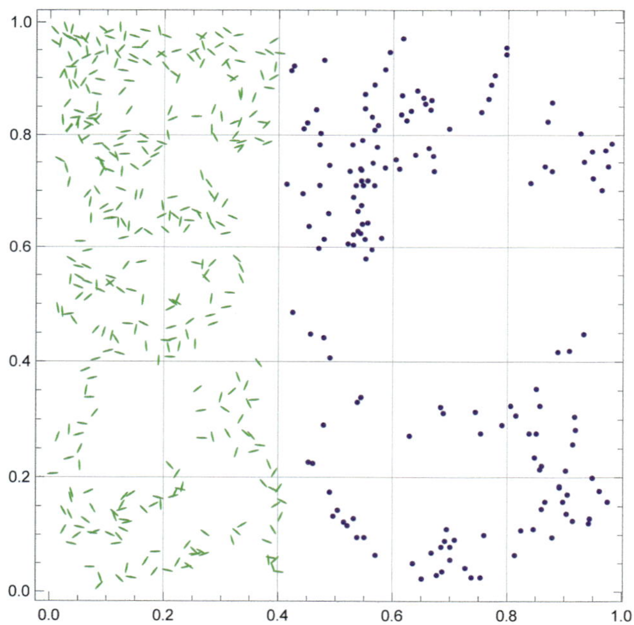

Figure 8.2 The digitized image corresponding to Figure 8.1, where the square part from the left side of the rectangular is taken. The side of the square is normalized to unity. $N_1 = 356$ uniformly oriented elliptic inclusions and $N_2 = 159$ disks in the square periodicity cell.

and the $\wp$ term of order $O(\varrho^2 f^2)$ is given by

$$\boldsymbol{\wp} = \frac{2f^2\varrho^2}{\pi N^2}\sum_{k,m=1}^{N}\begin{pmatrix} \mathrm{Re}\,\wp(a_m-a_k) & -\mathrm{Im}\,\wp(a_m-a_k) \\ -\mathrm{Im}\,\wp(a_m-a_k) & -\mathrm{Re}\,\wp(a_m-a_k)\end{pmatrix}. \tag{8.5}$$

We recall that N is the total number of inclusions per unit square Q and N_1 the number of ellipses. The radius of $(N-N_1)$ circular inclusions is r, the semi-axes of ellipses are $r_1(1\pm\alpha)$, and the angle φ_k is the inclination of the kth ellipses to the x_1-axis. It is convenient to use the concentration of ellipses $f_1 = \pi r_1^2(1-\alpha^2)$ and the total concentration $f = f_1 + (N-N_1)\pi r^2$ in the final formulas instead of r and r_1.

If ellipses are reduced to circles, then $f_1 = 0 \Leftrightarrow N_1 = 0$ and the shape term $\mathbf{J}$ vanishes. The interaction $\boldsymbol{\wp}$ term (8.5) accumulates the first-order interactions between all the inclusions. Here, the interactions are defined by Schwarz's method discussed in Chapter 2 and illustrated in Figure 2.2.

The contribution of the terms to the effective permittivity is shown in Figures 8.9 and 8.10.

It can be observed from Figures 8.3-8.8, that the effective permittivity components reach the highest values in the elliptical inclusions. It should be mentioned that the permittivity peaks correspond to the presence of the cancerous zones in the sample shown in Figure 8.1. Doing so, the highest permittivity values in the case of the elliptical inclusions stand as the perfect evidence that cancerous cells possess an ellipsoidal shape rather than a circular one.

Figures 8.11 and 8.12 complement the discussion of Chapter 3 on the Maxwell-Garnett and Bruggeman formulas for three-phase composites, giving a good

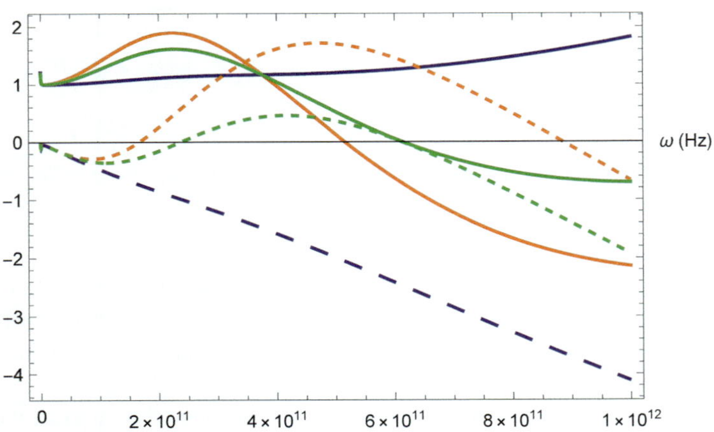

Figure 8.3 The dependence of the component ε_{11} on frequency ω for the concentration $f = 0.1$. Data for circular inclusions ($\alpha = 0$) are shown by blue lines Re ε_{11} (blue solid), Im ε_{11} (blue dashed). Data for unidirectional elliptical inclusions parallel to the x_1-axis ($\alpha = 0.8$, $\phi_k = 0$) and circular inclusions are shown by orange lines Re ε_{11} (orange solid), Im ε_{11} (orange dashed). Data for random inclinations of elliptical inclusions ($\alpha = 0.8$, $\phi_k = \frac{\pi}{4}$) and circular inclusions are shown by green lines Re ε_{11} (green solid), Im ε_{11} (green dashed).

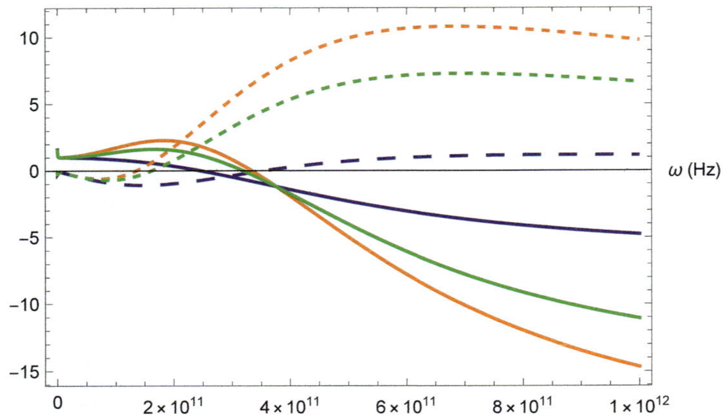

Figure 8.4 The dependence of the component ε_{11} on frequency ω for $f = 0.25$. The designations are the same as in Figure 8.3.

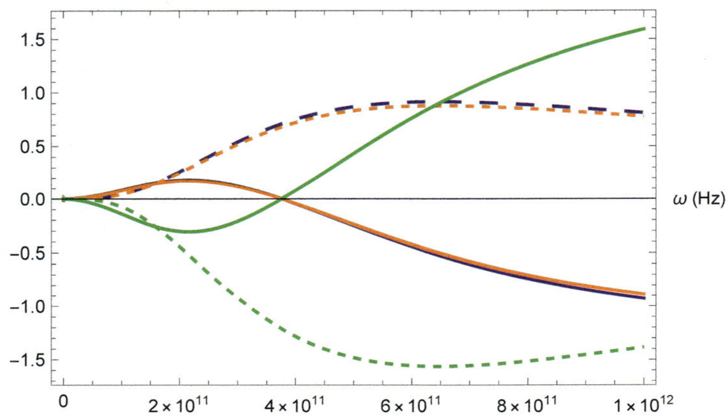

Figure 8.5 The dependence of the component ε_{12} on frequency ω for $f = 0.1$. The designations are the same as in Figure 8.3.

approximation for small frequencies and divergent for high frequencies. It is worth noting that the Maxwell-Garnett and Bruggeman formulas do not change with the change in the location of inclusions, while the tensor $\boldsymbol{\varepsilon}_{\perp}$ calculated by the higher-order formula (8.3) does.

The convergence estimation of the graphs for the considered numerical data $|\varrho|$ can be checked by Figure 4.4. The bound $f|\varrho(\omega)| < 1$ holds for $\omega < 10^{12}$ if $f \leq 0.77$; for $\omega < 5 \cdot 10^{11}$ if $f \leq 0.1$ and for $\omega < 1.5 \cdot 10^{11}$ if $f \leq 0.25$. Figure 8.13 shows the dependence of ε_{11} on ω for various f. One can see that the graphs for $f|\varrho(\omega)| > 1$ in the considered interval $0 < \omega < 10^{12}$ do not drastically change their behavior. This means that their slight continuation to $0 < \omega < 10^{12}$ holds. Figures 8.13-8.10 show the dependence of the real and imaginary parts of the components ε_{ij} on the frequency ω.

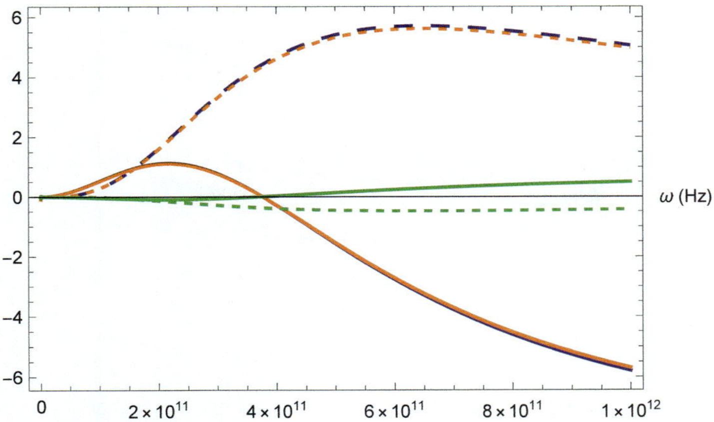

Figure 8.6 The dependence of the component ε_{12} on frequency ω for $f = 0.25$. The designations are the same as in Figure 8.3.

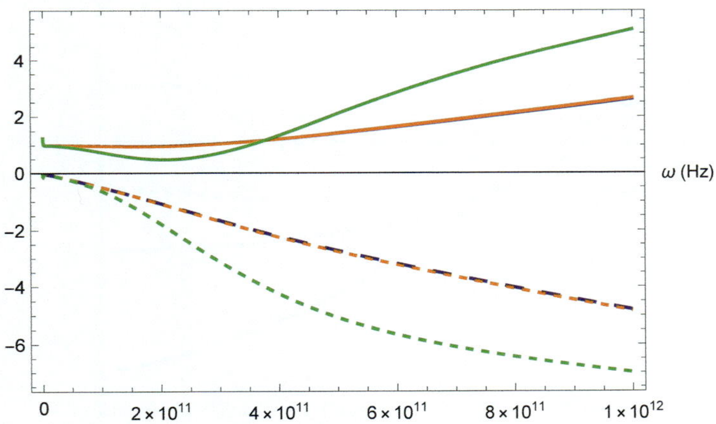

Figure 8.7 The dependence of the component ε_{22} on frequency ω for $f = 0.1$. The designations are the same as in Figure 8.3.

8.2 APPLICATION OF ANALYTICAL RVE

The theory of aRVE and its constructive application to random composites are outlined in Section 4.1.7. A dispersed random medium is characterized by a set of its structural sums.

Consider Figures 8.1 and 8.2 with $N_1 = 356$ glioma cells (green) and $N_2 = 159$ neuron cells (blue). Below, the union of these sets $\mathcal{U}$ is investigated by three local structural sums $e_2(\mathbf{x})$, $e_{22}(\mathbf{x})$ and $e_{44}(\mathbf{x})$ ($\mathbf{x} = (x_1, x_2)$) calculated as follows. First, it is convenient to change the scale for simulations and enlarge the unit cell Q in 10 times. Consider the sliding square box of size 1×1 in the enlarged square Q

$$B(\mathbf{x}) = \{\mathbf{y} = (y_1, y_2) \in Q : \ |y_j - x_j| < 1 \ (j = 1, 2)\}. \tag{8.6}$$

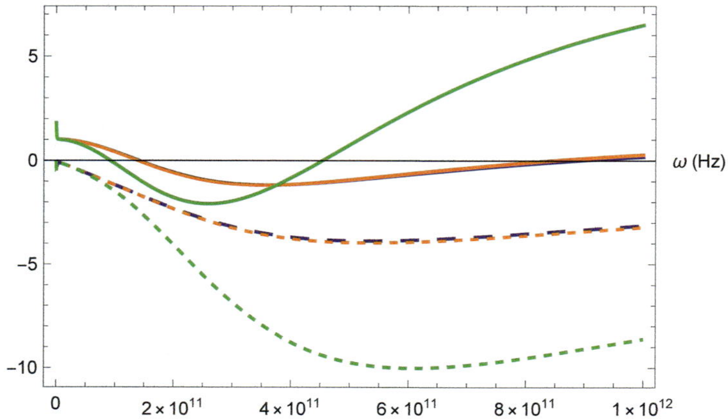

Figure 8.8 The dependence of the component ε_{12} on frequency ω for $f = 0.25$. The designations are the same as in Figure 8.3. The blue lines coincide with the orange lines.

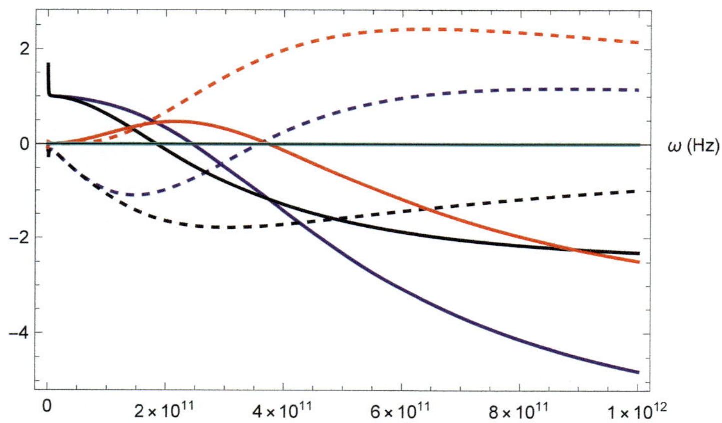

Figure 8.9 The dependence of ε_{11} on frequency ω for $f = 0.25$ in the case of circular inclusions. Re ε_{11} and Im ε_{11} are shown by blue lines, solid and dashed, respectively. The contributions into ε_{11} of the terms: the Maxwell-Garnett term (black), the J-shape term (cyan), and the $\wp$ term (red). In the case of circular inclusions, the J-shape term vanishes.

Calculate the structural sums over the unit square $B(\mathbf{x})$. Figures 8.14-8.15 represent the results excluding the domain with too high dominated values corresponding to the left-top small square in Figure 8.2.

The values $e_2(\mathbf{x})$ in Figure 8.14 characterize the anisotropic behavior of the local permittivity tensor. The permittivity tensor becomes isotropic for $e_2(\mathbf{x}) = \pi$. It is worth noting that the computation of $e_2(\mathbf{x})$ corresponds to the computation of 2-point correlation function. The values $e_{22}(\mathbf{x})$ and $e_{44}(\mathbf{x})$ in Figure 8.15 are always positive. They characterize the behavior of the considered structure, which can be described by 4- and 8-point correlation functions. It is difficult to say what exactly characterizes

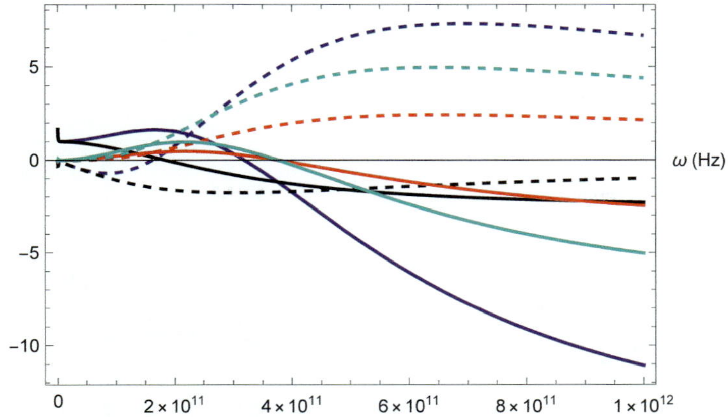

Figure 8.10 The dependence of ε_{11} on frequency ω for $f = 0.25$ for elliptical ($\alpha = 0.8$, $\phi_k = \frac{\pi}{4}$) and circular inclusions shown in Figure 8.2 by green and blue, respectively. The designations are the same as in Figure 8.9.

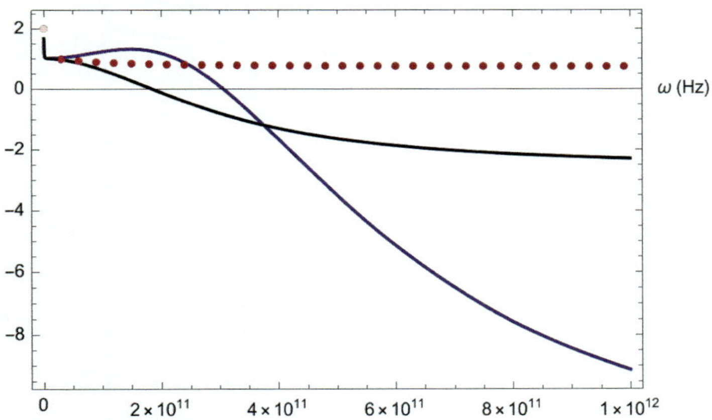

Figure 8.11 The comparison of Re ε_{11} for $f = 0.25$ calculated by the Maxwell-Garnett formula (black), Bruggeman formula (dotted) and (4.99) (blue).

the higher-order structural sums like skewness and kurtosis used in one-dimensional statistics. It can certainly be asserted that the infinite set of structural sums completely describes the location of inclusions that impact the effective constants of composites.

The considered example demonstrates applications of structural sums to characterize the structural properties of the set of points displayed in Figures 8.1 and 8.2. Similar investigations concerning the collective behavior of bacteria can be found in [131]. The papers [203, 221] concern the interactions of different sets of inclusions and their applications to the classification of clouds. As mentioned in the Preface, huge amounts of statistical data should be processed by the methods developed in the present book in the frame of the aRVE theory.

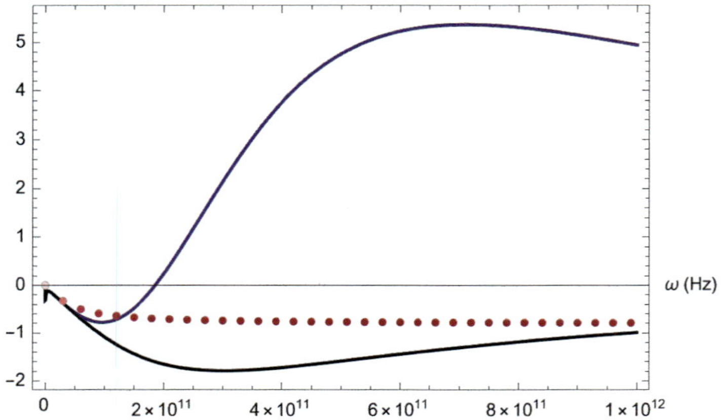

Figure 8.12 The comparison of Im ε_{11} for $f = 0.25$ calculated by the Maxwell-Garnett formula (black), Bruggeman formula (dotted) and (4.99) (blue).

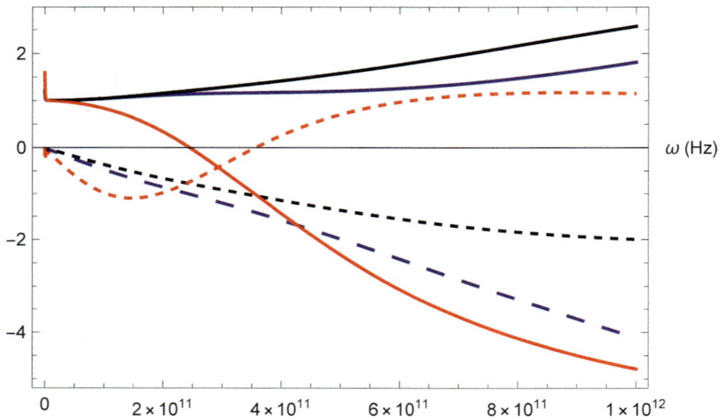

Figure 8.13 The dependence of Re ε_{11} (solid) and Im ε_{11} (dashed) on ω for circular inclusions. Data are for: $f = 0.1$ (blue); $f = 0.077$ (black); $f = 0.25$ (red).

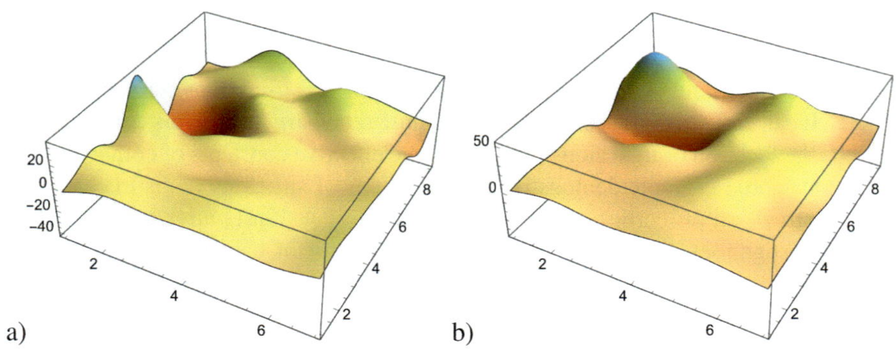

Figure 8.14 a) Re e_2 and b) Im e_2.

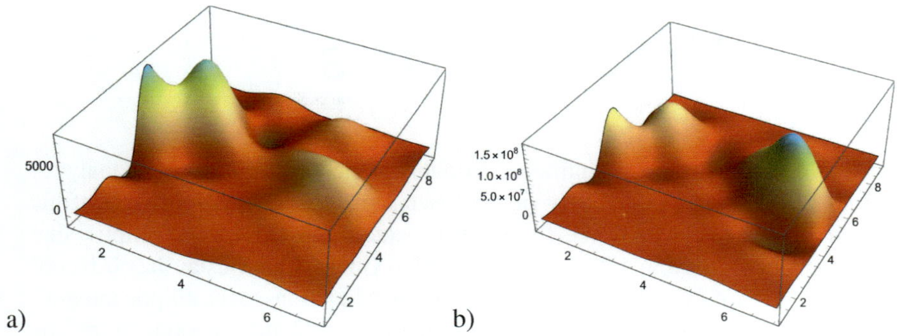

Figure 8.15 a) e_{22} and b) e_{44}.

A Elliptic Functions

In 1847, Gotthold Eisenstein introduced doubly periodic meromorphic (analytic) functions [70]. The modern theory of the Eisenstein functions and analysis of his approach can be found in [231]. It should be noted that the special elliptic sine function was introduced in 1799 by Carl Friedrich Gauss. Carl Gustav Jacob Jacobi introduced the elliptic functions in 1827 as the inverse functions to the elliptic integral. In 1863, Karl Weierstrass introduced doubly periodic analytic functions by means of absolutely convergent series. The absence of a uniform convergence argument in Eisenstein's study was the reason why Weierstrass and Rayleigh (1892) later ignored his work. Some historical notes can be found in [172, 231].

In the present Appendix, we select the relations for the Weierstrass and Eisenstein functions used in the theory of composites.

A.1 LATTICE SUMS

Consider the unit cell Q determined by two fundamental translation vectors expressed by two complex numbers ω_1 and ω_2. Without loss of generality we assume that $\omega_1 > 0$, $\text{Im}\ \omega_2 > 0$ and the area

$$|Q| = \omega_1 \text{Im}\ \omega_2 = 1. \tag{A.1}$$

Let m_1, m_2 run over the integer numbers $\mathbb{Z}$. The Eisenstein–Rayleigh lattice sums are defined as

$$S_m = \sum_{m_1,m_2}{}' (m_1\omega_1 + im_2\omega_2)^{-m}, \quad m = 2,3,\ldots, \tag{A.2}$$

The prime means that the term $m_1 = m_2 = 0$ is skipped in the summation. The double series (A.2) is conditionally convergent for $m = 1$ and $m = 2$, hence, its value depends on the order of summation. Eisenstein proposed the iterated summation method

$$\sum_{m_1,m_2}{}^{e} := \lim_{M_2\to+\infty} \lim_{M_1\to+\infty} \sum_{m_2=-M_2}^{M_2} \sum_{m_1=-M_1}^{M_1}, \tag{A.3}$$

where the symbol $\sum^e$ stands for the Eisenstein summation. The lattice sums $S_m = 0$ for odd m.

The following formula was proved in [158]

$$S_2 = \frac{2}{\omega_1}\zeta\left(\frac{\omega_1}{2}\right), \tag{A.4}$$

where $\zeta(z)$ denotes the Weierstrass ζ-function considered in the next section. Application of Akhiezer's formula for the ζ-function [10, p. 204] yields the computationally effective formula

$$S_2 = \left(\frac{\pi}{\omega_1}\right)^2 \left(\frac{1}{3} - 8\sum_{m=1}^{\infty} \frac{q^{2m}}{(1-q^{2m})^2}\right), \quad q = \exp\left(\pi i\, \frac{\omega_2}{\omega_1}\right). \tag{A.5}$$

DOI: 10.1201/9781032646633-A

Hurwitz's formula from [110, p. 210] yields

$$S_2 = \left(\frac{\pi}{\omega_1}\right)^2 \left(\frac{1}{3} - 8\sum_{m=1}^{\infty} \frac{mq^{2m}}{1-q^{2m}}\right). \tag{A.6}$$

The lattice sums S_4 and S_6 for a general lattice are found from the quickly convergent series

$$\begin{aligned} S_4 &= \left(\frac{\pi}{\omega_1}\right)^4 \left(\frac{1}{45} + \frac{16}{3}\sum_{m=1}^{\infty} \frac{m^3q^{2m}}{1-q^{2m}}\right), \\ S_6 &= \left(\frac{\pi}{\omega_1}\right)^6 \left(\frac{2}{945} - \frac{16}{15}\sum_{m=1}^{\infty} \frac{m^5q^{2m}}{1-q^{2m}}\right). \end{aligned} \tag{A.7}$$

The next lattice sums S_{2n} $(n \geq 4)$ can be calculated by the recurrence formula

$$S_{2n} = \frac{3}{(2n+1)(2n-1)(n-3)} \sum_{m=2}^{n-2} (2m-1)(2n-2m-1)\, S_{2m} S_{2(n-m)}. \tag{A.8}$$

The end of the present subsection concerns the square array used in the bulk of the book. In order to calculate S_2 for $\omega_1 = 1$ and $\omega_2 = \mathrm{i}$ Rayleigh [206] replaced the lattice sum S_2 with the double integral

$$S_2 = \int\int_{\mathbb{R}^2} \frac{dx_1\, dx_2}{(x_1 + \mathrm{i}x_2)^2} \tag{A.9}$$

and extended the Eisenstein method by performing integration

$$S_2 = 2\int_v^{\infty} \left[\int_{-v}^{v} \frac{dx_2}{(x_1 + \mathrm{i}x_2)^2}\right] dx_1 = \pi. \tag{A.10}$$

Conversion of the double integral (A.9) into the iterated integral (A.10) corresponds to the Eisenstein summation of the series.

The direct application of the Eisenstein summation (A.3) to the lattice sum S_2 from (A.2) for the square array requires the following transformations

$$\begin{aligned} &S_2 = \sum_{m_1 \neq 0} \frac{1}{m_1^2} + \sum\nolimits^{e}_{\substack{m_1, m_2 \\ m_2 \neq 0}} \frac{1}{(m_1 + \mathrm{i}m_2)^2} = \sum_{m_1 \neq 0} \frac{1}{m_1^2} - \sum_{m_2 \neq 0} \frac{1}{m_2^2} \\ &+ \lim_{N \to +\infty} \sum_{\substack{m_2 = -N \\ m_2 \neq 0}}^{N} \sum_{m_1 = 1}^{\infty} \left(\frac{1}{(m_1 + \mathrm{i}m_2)^2} - \frac{1}{(m_1 - \mathrm{i}m_2)^2}\right) = \\ &\lim_{N \to +\infty} \sum_{\substack{m_2 = -N \\ m_2 \neq 0}}^{N} 2\sum_{m_1 = 1}^{\infty} \frac{m_1^2 - m_2^2}{(m_1^2 + m_2^2)^2} = 2\sum_{m_2 = 1}^{\infty} \left(\frac{1}{m_2^2} - \frac{\pi^2}{\sinh^2 \pi m_2}\right) = \pi. \end{aligned}$$

The known formula $\sum_{m_2=1}^{\infty} \sinh^{-2} \pi m_2 = \frac{1}{6} - \frac{1}{2\pi}$ is used to find the final equality. Rayleigh did not explain the application of the Eisenstein summation. It was accomplished in [144, 146] and [157]. Further discussion and exact formulas for S_2 for various arrays can be found in [172, 236]. We have for the square array, $S_{4n+2} = 0$ $(n = 1, 2, \ldots)$. The values S_{4n} are calculated by (A.8) in terms of the Γ-function

$$S_4 = \frac{\Gamma^8(\frac{1}{4})}{960\pi^2} \approx 3.151211,\ S_8 = \frac{3}{7} S_4^2 = \frac{\Gamma^{16}(\frac{1}{4})}{20856176640000\pi^4} \approx 4.2557732, \tag{A.11}$$

$$S_{12} = \frac{\Gamma^{24}(\frac{1}{4})}{7028736000\pi^6} \approx 3.938849,\ S_{16} = \frac{\Gamma^{32}(\frac{1}{4})}{7028736000\pi^8} \approx 4.015695. \tag{A.12}$$

Here, the first equation (A.7) and the following series for the transcendental number $\Gamma\left(\frac{1}{4}\right)$ are used

$$\Gamma^8\left(\frac{1}{4}\right) = 960\pi^6\left(\frac{1}{45} + \frac{16}{3}\sum_{m=1}^{\infty}\frac{m^3 e^{-2\pi m}}{1-e^{-2\pi m}}\right). \tag{A.13}$$

This and other exact formulas follow from the relations on the Weierstrass invariants established in [132] and works cited therein.

A.2 EISENSTEIN'S SERIES AND WEIERSTRASS' ELLIPTIC FUNCTIONS

Introduce the designation $\boldsymbol{\omega} = m_1\boldsymbol{\omega}_1 + m_2\boldsymbol{\omega}_2$ for elements of the considered lattice Q for shortness. The Eisenstein function of order m is introduced by the series

$$E_l(z) = \sum\nolimits_{\omega\in Q}^{(e)} \frac{1}{(z-\boldsymbol{\omega})^l}, \tag{A.14}$$

where the Eisenstein summation is used for the conditionally convergent series (A.14) in the case $l = 1, 2$.

The Weierstrass functions are defined by the absolutely and almost uniformly convergent series [10]

$$\wp(z) = \frac{1}{z^2} + \sum_{\boldsymbol{\omega}\in Q}{}' \left[\frac{1}{(z-\boldsymbol{\omega})^2} - \frac{1}{\boldsymbol{\omega}^2}\right], \tag{A.15}$$

$$\zeta(z) = \frac{1}{z} + \sum_{\boldsymbol{\omega}\in Q}{}' \left[\frac{1}{z-\boldsymbol{\omega}} + \frac{1}{\boldsymbol{\omega}} + \frac{z}{\boldsymbol{\omega}^2}\right], \tag{A.16}$$

where the prime as above means that the term $\boldsymbol{\omega} = 0$ is skipped in the summation. The Weierstrass σ-function is introduced by means of the infinite product

$$\sigma(z) = z\prod_{\boldsymbol{\omega}\in Q}{}' \left[\left(1 - \frac{z}{\boldsymbol{\omega}}\right)\exp\left(\frac{z}{\boldsymbol{\omega}} + \frac{1}{2}\left(\frac{z}{\boldsymbol{\omega}}\right)^2\right)\right]. \tag{A.17}$$

Its logarithmic derivative coincides with the ζ-function, $\zeta(z) = \frac{\sigma'(z)}{\sigma(z)}$. The σ-function belongs to the class of θ-functions [190]. The most favorable method to compute the Weierstrass functions is using fast formulas for θ-functions.

The Weierstrass functions and the Eisenstein functions are related by the formulas [231]

$$E_1(z) = \zeta(z) - S_2 z, \quad E_2(z) = \wp(z) + S_2. \tag{A.18}$$

The functions $\wp(z)$ and $E_2(z)$ are double periodic. The function $\zeta(z)$ has the following jumps per cell [10]

$$\zeta(z+\boldsymbol{\omega}_j) - \zeta(z) = \delta_j \quad (j = 1, 2), \tag{A.19}$$

where $\delta_j = 2\zeta\left(\frac{\omega_j}{2}\right)$ satisfy Legendre's identity $\delta_1\omega_2 - \delta_2\omega_1 = 2\pi\mathrm{i}$. Using (A.4) and (A.1) we express the jumps (A.19) by S_2

$$\delta_1 = S_2\omega_1, \quad \delta_2 = S_2\omega_2 - \frac{2\pi\mathrm{i}}{\omega_1}. \tag{A.20}$$

It follows from (A.19)–(A.20) that the jumps of $E_1(z)$ have the form

$$E_1(z+\omega_1) - E_1(z) = 0, \quad E_1(z+\omega_2) - E_1(z) = -\frac{2\pi\mathrm{i}}{\omega_1}. \tag{A.21}$$

The high-order Eisenstein functions are related to the Weierstrass function $\wp(z)$ by the identities [231]

$$E_2(z) = \wp(z) + S_2, \quad E_l(z) = \frac{(-1)^l}{(l-1)!}\frac{d^{l-2}\wp(z)}{dz^{l-2}}, \quad l = 3, 4, \ldots. \tag{A.22}$$

Every function (A.22) is doubly periodic and has a pole of order l at $z = 0$. The Eisenstein functions of the even-order $E_{2m}(z)$ can be presented in the form of the series [231]

$$E_{2m}(z) = \frac{1}{z^{2m}} + \sum_{k=1}^{\infty} s_k^{(m)} z^{2(k-1)}, \quad m = 1, 2, \ldots, \tag{A.23}$$

where

$$s_k^{(m)} = C_{2m+2k-3}^{2m-1} S_{2(m+k-1)} \equiv \frac{(2m+2k-3)!}{(2m-1)!(2k-2)!} S_{2(m+k-1)}. \tag{A.24}$$

In particular,

$$E_2(z) = \frac{1}{z^2} + \sum_{k=1}^{\infty} (2k-1) S_{2k} z^{2(k-1)}. \tag{A.25}$$

The Eisenstein functions are related by the differentiation formula

$$E_l'(z) = -l E_{l+1}(z). \tag{A.26}$$

The relation (A.26) implies the following form of the nth derivative of E_l

$$E_l^{(n)}(z) = (-1)^n \frac{(l+n-1)!}{(l-1)!} E_{l+n}(z). \tag{A.27}$$

Differentiating (A.23) and using (A.26) with $l = 2m$ we arrive at the Laurent series for the odd-order Eisenstein function

$$E_{2m+1}(z) = \frac{1}{z^{2m+1}} - \sum_{k=2}^{\infty} s_k^{(m)} z^{2k-3}, \quad m = 1, 2, \ldots. \tag{A.28}$$

Two expansions (A.23) and (A.28) can be written in the form

$$E_n(z) = \frac{1}{z^n} + (-1)^n \sum_{l=0}^{\infty} C_{l+n-1}^{n-1} S_{l+n} z^l, \quad n = 2, 3, \ldots. \tag{A.29}$$

Laurent's series for the function $E_1(z)$ has to be separately written

$$E_1(z) = \frac{1}{z} - \sum_{k=1}^{\infty} S_{2k} z^{2k-1}. \tag{A.30}$$

The function $\wp(z)$ satisfies the differential equation

$$[\wp'(z)]^2 = 4\wp^3(z) - g_2\wp(z) - g_3, \tag{A.31}$$

where the constants $g_2 = 60\, S_4$ and $g_3 = 140\, S_6$ are called the invariants of $\wp(z)$.

Special attention is paid to the square array in the bulk of the book. In this case

$$\delta_1 = \pi, \quad \delta_2 = -\pi \mathrm{i}. \tag{A.32}$$

The Eisenstein summation method (A.3) was used by Rayleigh for the macroscopically isotropic square array [206] in order to avoid the arising of the additional induced charge discussed in Chapter 3. The summation method corresponded to the direction of the external field along the x_1-axis. In order to extend the method to general macroscopically anisotropic composites, we have to solve the second boundary value problem for the second external field applied, for instance, along the x_2-axis. The rotation of the square periodicity cell Q around the angle $\frac{\pi}{2}$ changes the method of summation. The corresponding summation has to be introduced as follows

$$\sum_{m_1,m_2}{}^{*} := \lim_{M_1\to+\infty} \lim_{M_2\to+\infty} \sum_{m_1=-M_1}^{M_1} \sum_{m_2=-M_2}^{M_2}. \tag{A.33}$$

The rectangular cell with the periods $\boldsymbol{\omega}_1 > 0$ and $\boldsymbol{\omega}_2 = \mathrm{i}\boldsymbol{\omega}_1^{-1}$ was considered in [208] where it was proved that

$$S_2^* = 2\pi - S_2. \tag{A.34}$$

Here, S_2 and S_2^* were calculated by the summation methods (A.3) and (A.33), respectively.

The Eisenstein function of the second order is transformed according to the first formula (A.22). The rotation of Q around the angle $\frac{\pi}{2}$ yields the transformation $z \mapsto \mathrm{i}z$, hence,

$$E_2^*(\mathrm{i}z) = \wp(\mathrm{i}z) + S_2^*. \tag{A.35}$$

The Weierstrass function is not changed due to its representation by the absolutely convergent series (A.15). The relation $\wp(\mathrm{i}z) = -\wp(z)$ holds for the square array, since in this case

$$\wp(z) = \frac{1}{z^2} + \sum_{p=1}^{\infty} (4p-1) S_{4p} z^{2(2p-1)}. \tag{A.36}$$

Adding (A.34) and (A.35) and using the first formula (A.22) we obtain

$$E_2^*(\mathrm{i}z) = 2\pi - E_2(z). \tag{A.37}$$

Moreover, $S_2^* = \pi$ since $S_2 = \pi$ for the square array.

The higher-order Eisenstein functions are represented by the absolutely convergent series (A.14). Therefore,

$$E_l(\mathrm{i}z) = \mathrm{i}^{-l} E_l(z) \quad \text{for} \quad l \geq 3. \tag{A.38}$$

References

1. Example code from this book:. https://github.com/rkycia/AnisotropyOfMetamaterialsBook. Accessed: 27.09.2024.
2. Opencv library. https://opencv.org/. Accessed: 29.07.2024.
3. Pcam dataset. https://github.com/basveeling/pcam. Accessed: 29.07.2024.
4. Scientific python lectures. https://lectures.scientific-python.org/. Accessed: 29.07.2024.
5. Scikit learn library. https://scikit-learn.org. Accessed: 29.07.2024.
6. Tensorflow library. https://www.tensorflow.org/. Accessed: 29.07.2024.
7. Uci ml hand-written digits datasets. https://archive.ics.uci.edu/dataset/80/optical+recognition+of+handwritten+digits. Accessed: 29.07.2024.
8. P.M. Adler and V. Mityushev. Effective medium approximation and exact formulae for electrokinetic phenomena in porous media. *Journal of Physics A: Mathematical and General*, 36(2):391, 2002.
9. G. Akhalaia, G. Giorgadze, G. Makatsaria, and N. Manjavidze. Deformation of complex structures and boundary value problem with shift. *Analysis as a Life: Dedicated to Heinrich Begehr on the Occasion of his 80th Birthday*, pages 1–17, 2019.
10. N.I. Akhiezer. *Elements of the theory of elliptic functions*. American Mathematical Soc., 1990.
11. M.J. Akhter, W. Kuś, A. Mrozek, and T. Burczyński. Mechanical properties of monolayer mos2 with randomly distributed defects. *Materials*, 13(6):1307, 2020.
12. B. Alali and R. Lipton. Optimal lower bounds on local stress inside random media. *SIAM Journal on Applied Mathematics*, 70(4):1260–1282, 2009.
13. H. Ammari, H. Kang, and K. Touibi. Boundary layer techniques for deriving the effective properties of composite materials. *Asymptotic Analysis*, 41(2):119–140, 2005.
14. I Andrianov, J. Awrejcewicz, G.A. Starushenko, and V.A. Gabrinets. Refinement of the maxwell formula for a composite reinforced by circular cross-section fibres. part ii: using padé approximants. *Acta Mechanica*, 231(12):5145–5157, 2020.
15. I. Andrianov, J. Awrejcewicz, G.A. Starushenko, and V.A. Gabrinets. Refinement of the maxwell formula for composite reinforced by circular cross-section fibers. part i: using the schwarz alternating method. *Acta Mechanica*, 231(12):4971–4990, 2020.
16. I. Andrianov, S. Gluzman, and V. Mityushev. *Mechanics and Physics of Structured Media: Asymptotic and Integral Equations Methods of Leonid Filshtinsky*. Academic Press, 2022.
17. I. Andrianov and V. Mityushev. Exact and "exact" formulae in the theory of composites. In *Modern Problems in Applied Analysis*, pages 15–34. Springer, 2018.
18. I. Andrianov and A. Shatrov. Padé approximants, their properties, and applications to hydrodynamic problems. *Symmetry*, 13(10):1869, 2021.
19. I.V. Andrianov and J. Awrejcewicz. *Asymptotic Methods for Engineers*. CRC Press, 2024.
20. I.V. Andrianov, V.I. Olevskyi, I.V. Shapka, and T.S. Naumenko. Technique of padé-type multidimensional approximations application for solving some problems in mathematical physics. In *AIP Conference Proceedings*, volume 2025. AIP Publishing, 2018.

21. I.V. Andrianov, G.A. Starushenko, and V.A. Gabrinets. Percolation threshold for elastic problems: self-consistent approach and padé approximants. *Advances in Mechanics of Microstructured Media and Structures*, pages 35–42, 2018.
22. N.S. Bakhvalov and G. Panasenko. *Homogenisation: averaging processes in periodic media: mathematical problems in the mechanics of composite materials*, volume 36. Springer Science & Business Media, 2012.
23. D.I. Bardzokas, L.A. Filshtinskii, and L.V. Shramko. Homogeneous solutions of the electroelasticity equations for piezoceramic layers in r 3. *Acta mechanica*, 209(1-2):27–41, 2010.
24. D.I. Bardzokas, M.L. Filshtinsky, and L.A. Filshtinsky. *Mathematical methods in electro-magneto-elasticity*, volume 32. Springer Science & Business Media, 2007.
25. D.M. Barnett and J. Lothe. Dislocations and line charges in anisotropic piezoelectric insulators. *physica status solidi (b)*, 67(1):105–111, 1975.
26. B.E. Bejnordi, M. Veta, Paul J. Van D., B. Van Ginneken, N. Karssemeijer, G. Litjens, J. Van Der Laak, et al. Diagnostic assessment of deep learning algorithms for detection of lymph node metastases in women with breast cancer. *Jama*, 318(22):2199–2210, 2017.
27. P.A. Belov, R. Marques, S.I. Maslovski, I.S. Nefedov, M. Silveirinha, C.R. Simovski, and S.A. Tretyakov. Strong spatial dispersion in wire media in the very large wavelength limit. *Physical Review B*, 67(11):113103, 2003.
28. P.A. Belov, S.A. Tretyakov, and A.J. Viitanen. Nonreciprocal microwave band-gap structures. *Physical Review E*, 66(1):016608, 2002.
29. Y. Benveniste. Correspondence relations among equivalent classes of heterogeneous piezoelectric solids under anti-plane mechanical and in-plane electrical fields. *Journal of the Mechanics and Physics of Solids*, 43(4):553–571, 1995.
30. M.J. Beran. *Statistical Continuum Theories*, volume New York. Interscience Publishers, 1968.
31. D.J. Bergman. Analytical properties of the complex effective dielectric constant of a composite medium with applications to the derivation of rigorous bounds and to percolation problems. In *AIP Conference Proceedings*, volume 40, pages 46–62. American Institute of Physics, 1978.
32. D.J. Bergman. The dielectric constant of a composite material—a problem in classical physics. *Physics Reports*, 43(9):377–407, 1978.
33. D.J. Bergman. Bounds for the complex dielectric constant of a two-component composite material. *Physical Review B*, 23(6):3058, 1981.
34. D.J. Bergman. Rigorous bounds for the complex dielectric constant of a two-component composite. *Annals of Physics*, 138(1):78–114, 1982.
35. D.J. Bergman. The self-consistent effective medium approximation (sema): New tricks from an old dog. *Physica B: Condensed Matter*, 394(2):344–350, 2007.
36. L. Berlyand, A.G. Kolpakov, and A. Novikov. *Introduction to the network approximation method for materials modeling*. Cambridge University Press, 2013.
37. L. Berlyand and V. Mityushev. Generalized clausius–mossotti formula for random composite with circular fibers. *Journal of Statistical Physics*, 102(1):115–145, 2001.
38. L. Berlyand and V. Mityushev. Increase and decrease of the effective conductivity of two phase composites due to polydispersity. *Journal of Statistical Physics*, 118:481–509, 2005.
39. P. Bisegna, R. Luciano, et al. Variational bounds for the overall properties of piezoelectric composites. *Journal of the Mechanics and Physics of Solids*, 44(4):583–602, 1996.

40. B. Bojarski and V. Mityushev. R-linear problem for multiply connected domains and alternating method of schwarz. *Journal of Mathematical Sciences*, 189(1):68–77, 2013.
41. V. Bonnaillie-Noël, M. Dalla Riva, M. Dambrine, and P. Musolino. Global representation and multiscale expansion for the dirichlet problem in a domain with a small hole close to the boundary. *Communications in Partial Differential Equations*, 46(2):282–309, 2021.
42. M. Born and E. Wolf. *Principles of optics: electromagnetic theory of propagation, interference and diffraction of light*. Elsevier, 2013.
43. J. Bravo-Castillero, R. Guinovart-Diaz, F.J. Sabina, and R. Rodriguez-Ramos. Closed-form expressions for the effective coefficients of a fiber-reinforced composite with transversely isotropic constituents–ii. piezoelectric and square symmetry. *Mechanics of Materials*, 33(4):237–248, 2001.
44. J. Bravo-Castillero, R. Rodríguez-Ramos, R. Guinovart-Díaz, F.J. Sabina, A.R. Aguiar, U.P. Silva, and J.L. Gómez-Muñoz. Analytical formulae for electromechanical effective properties of 3–1 longitudinally porous piezoelectric materials. *Acta Materialia*, 57(3):795–803, 2009.
45. D.A.G. Bruggeman. Berechnung verschiedener physikalischer konstanten von heterogenen substanzen. i. dielektrizitätskonstanten und leitfähigkeiten der mischkörper aus isotropen substanzen. *Annalen der physik*, 416(7):636–664, 1935.
46. T. Buchukuri, O. Chkadua, and D. Natroshvili. Mathematical problems of generalized thermoelectro-magneto-elasticity theory. *Mem. Differ. Equ. Math. Phys*, 68:1–166, 2016.
47. T. Buchukuri, O. Chkadua, D. Natroshvili, and A.-M. Sändig. Solvability and regularity results to boundary-transmission problems for metallic and piezoelectric elastic materials. *Mathematische Nachrichten*, 282(8):1079–1110, 2009.
48. R.B. Burckel. Iterating analytic self-maps of discs. *The American Mathematical Monthly*, 88(6):396–407, 1981.
49. F. Capolino. *Theory and phenomena of metamaterials*. CRC press, 2017.
50. A. Cherkaev. Optimal three-material wheel assemblage of conducting and elastic composites. *International Journal of Engineering Science*, 59:27–39, 2012.
51. A. Cherkaev. *Variational methods for structural optimization*, volume 140. Springer Science & Business Media, 2012.
52. A. Cherkaev and L. Gibiansky. Variational principles for complex conductivity, viscoelasticity, and similar problems in media with complex moduli. *Journal of Mathematical Physics*, 35(1):127–145, 1994.
53. A. Cherkaev and A.D. Pruss. Effective conductivity of spiral and other radial symmetric assemblages. *Mechanics of Materials*, 65:103–109, 2013.
54. L.I. Chibrikova. Boundary value problems for a rectangle. *Kazan. Gos. Univ. Uchen. Zap.*, 123(10):15–39, 1963.
55. T.C. Choy. *Effective medium theory: principles and applications*, volume 165. Oxford University Press, 2015.
56. R.M. Christensen. A critical evaluation for a class of micro-mechanics models. *Journal of the Mechanics and Physics of Solids*, 38(3):379–404, 1990.
57. Ch.E. Clayton and P. Muggli. Advanced accelerator concepts: Tenth workshop. In *Advanced Accelerator Concepts: Tenth Workshop*, volume 647, 2002.
58. R.W. Cohen, G.D. Cody, M.D. Coutts, and B. Abeles. Optical properties of granular silver and gold films. *Physical Review B*, 8(8):3689, 1973.

59. R. Connelly and W. Dickinson. Periodic planar disc packings. *Philosophical Transactions of the Royal Society A: Mathematical, Physical and Engineering Sciences*, 372(2008):20120039, 2014.
60. A.I. Curatolo, O. Kimchi, C.P. Goodrich, and M.P. Brenner. The assembly yield of complex, heterogeneous structures: a computational toolbox. *bioRxiv*, pages 2022–06, 2022.
61. G. Cybenko. Approximation by superpositions of a sigmoidal function. *Mathematics of control, signals and systems*, 2(4):303–314, 1989.
62. R. Czapla, W. Nawalaniec, and V. Mityushev. Effective conductivity of random two-dimensional composites with circular non-overlapping inclusions. *Computational Materials Science*, 63:118–126, 2012.
63. S. Czarnecki and T. Lewiński. Shaping the stiffest three-dimensional structures from two given isotropic materials. *Computer Assisted Mechanics and Engineering Sciences*, 13(1):53–83, 2006.
64. M. Dalla Riva, Massimo Lanza de Cristoforis, and P. Musolino. *Singularly Perturbed Boundary Value Problems*. Springer, 2021.
65. M. Dalla Riva, P. Musolino, and R. Pukhtaievych. Series expansion for the effective conductivity of a periodic dilute composite with thermal resistance at the two-phase interface. *Asymptotic Analysis*, 111(3-4):217–250, 2019.
66. L. Demkowicz. *Computing with hp-adaptive finite elements: volume 1 one and two dimensional elliptic and Maxwell problems*. CRC press, 2006.
67. P. Drygaś, S. Gluzman, V. Mityushev, and W. Nawalaniec. *Applied analysis of composite media: analytical and computational results for materials scientists and engineers*. Woodhead Publishing, 2019.
68. R.O. Duda and P.E. Hart. Use of the hough transformation to detect lines and curves in pictures. *Communications of the ACM*, 15(1):11–15, 1972.
69. A.M. Dykhne. Conductivity of a two-dimensional two-phase system. *Sov. Phys. JETP*, 32(1):63–65, 1971.
70. G. Eisenstein. Beiträge zur theorie der elliptischen functionen. *Crelles Journal*, 35:153–247, 1847.
71. Ch. Engström. Inverse bounds and bulk properties of complex-valued two-component composites. *SIAM Journal on Applied Mathematics*, 67(1):194–213, 2006.
72. M. Ferrari. Asymmetry and the high concentration limit of the mori-tanaka effective medium theory. *Mechanics of Materials*, 11(3):251–256, 1991.
73. L.A. Filshtinskii. *Problems of thermal conduction and thermoelasticity for a plane weaken by a double periodic array of equal circular holes*. Thermal Stresses in Elements of Constructions, Naukova Dumka, Institute of Mechanics, Academy of Sciences USSR, Kiev, 103-112, 1964.
74. L.A. Fil'shtinskii and Yu.D. Kovalyov. Concentration of mechanical stresses near a hole in a piezoceramic layer. *Mechanics of composite materials*, 38(2):121–124, 2002.
75. L.A. Fil'shtinskii and Yu.V. Shramko. Averaging the physical properties of fibrous piezocomposites. *Mechanics of composite materials*, 34(1):87–93, 1998.
76. L.A. Fil'shtyns' kyi, Yu.V. Shramko, and D.S. Kovalenko. Averaging of the magnetic properties of fibrous ferromagnetic composites. *Materials Science*, 46:808–818, 2011.
77. J. Flusser. On the independence of rotation moment invariants. *Pattern recognition*, 33(9):1405–1410, 2000.
78. J. Flusser and T. Suk. Rotation moment invariants for recognition of symmetric objects. *IEEE Transactions on Image Processing*, 15(12):3784–3790, 2006.

79. F.D. Gakhov. *Boundary value problems*. Courier Corporation, 1990.
80. F.L. Galeener. Submicroscopic-void resonance: the effect of internal roughness on optical absorption. *Physical Review Letters*, 27(7):421, 1971.
81. A. Galka, J.J. Telega, and R. Wojnar. Homogenization and thermopiezoelectricity. *Mechanics Research Communications*, 19(4):315–324, 1992.
82. A. Galka, J.J. Telega, and R. Wojnar. Some computational aspects of homogenization of thermopiezoelectric composites. *Comput. Assist. Mech. Eng. Sci*, 3(2):133–154, 1996.
83. A. Galka and R. Wojnar. Layered piezoelectric composites: macroscopic behaviour. In *Smart Structures: Requirements and Potential Applications in Mechanical and Civil Engineering, J. Holnicki-Szulc; J. Rodellar, (Eds.), Proceedings of the NATO Advanced Research Workshop, Pultusk, Warsaw, Poland, 16-19 June 1998 Series: Nato Science Partnership Subseries: 3 (closed), Vol. 6*, pages 79–88. Kluwer Academic Publisher, Dordrecht, 1999.
84. B. Gambin. Influence of microstructure on properties of elastic, piezoelectric and thermoelastic composites. In *Warszawa*. IFTR Reports, 2006.
85. B. Gambin and A. Galka. Boundary layer problem in a piezoelectric composite. In *IUTAM Symposium on Anisotropy, Inhomogeneity and Nonlinearity in Solid Mechanics*, pages 327–332. Kluwer, Netherlands, 1995.
86. Aurélien Géron. *Hands-on machine learning with Scikit-Learn, Keras, and TensorFlow*. "O'Reilly Media, Inc.", 2022.
87. S.K. Ghosh and A. Böker. Self-assembly of nanoparticles in 2d and 3d: Recent advances and future trends. *Macromolecular Chemistry and Physics*, 220(17):1900196, 2019.
88. G. Giorgadze and G. Gulagashvili. Riemann-hilbert boundary value problem with piecewise constant transition function. *Journal of Dynamical and Control Systems*, pages 1–11, 2022.
89. G. Giorgadze and N. Manjavidze. On some constructive methods for the matrix riemann–hilbert boundary-value problem. *Journal of Mathematical Sciences*, 195:146–174, 2013.
90. R. Glowinski, S.J. Osher, and Wotao Yin. *Splitting methods in communication, imaging, science, and engineering*. Springer, 2017.
91. S. Gluzman. Asymptotics and summation of the effective properties of suspensions, simple liquids and composites. *Symmetry*, 14(9):1912, 2022.
92. S. Gluzman. Critical index for conductivity, elasticity, superconductivity. results and methods. In *Mechanics and Physics of Structured Media*, pages 141–175. Elsevier, 2022.
93. S. Gluzman, V. Mityushev, and W. Nawalaniec. *Computational analysis of structured media*. Academic Press, 2017.
94. K. Golden and G. Papanicolaou. Bounds for effective parameters of heterogeneous media by analytic continuation. *Communications in Mathematical Physics*, 90(4):473–491, 1983.
95. R.C. Gonzalez. *Digital image processing*. Pearson education india, 2009.
96. T. Gric, S.G. Sokolovski, N. Navolokin, O. Semyachkina-Glushkovskaya, and E.U. Rafailov. Metamaterial formalism approach for advancing the recognition of glioma areas in brain tissue biopsies. *Optical Materials Express*, 10(7):1607–1615, 2020.
97. Eh.I. Grigolyuk and L.A. Fil'shtinskij. Perforated plates and shells. *M.: Nauka*, 1970.
98. Eh.I. Grigolyuk and L.A. Fil'shtinskij. Periodic piecewise homogeneous elastic structures. *Nauka, Moscow*, 1992.

99. B. Grzybowski, K. Fitzner, J. Paczesny, and S. Granick. From dynamic self-assembly to networked chemical systems. *Chemical Society Reviews*, 46(18):5647–5678, 2017.
100. R. Guinovart-Díaz, P. Yan, R. Rodríguez-Ramos, J.C. López-Realpozo, C.P. Jiang, J. Bravo-Castillero, and F.J. Sabina. Effective properties of piezoelectric composites with parallelogram periodic cells. *International Journal of Engineering Science*, 53:58–66, 2012.
101. A.N. Guz and et al. *Mechanics of Coupled Fields in Structural Elements*. Naukova Dumka, Kiev, 1988.
102. Z. Hashin. Analysis of composite materials—a survey. *Journal of Applied Mechanics*, 50(3):481–505, 1983.
103. Z. Hashin and Sh. Shtrikman. A variational approach to the theory of the effective magnetic permeability of multiphase materials. *Journal of applied Physics*, 33(10):3125–3131, 1962.
104. R. Haynes, S. MacLachlan, Xiao-Chuan Cai, L. Halpern, Hyea Hyun Kim, A. Klawonn, and O. Widlund. *Domain decomposition methods in science and engineering XXV*. Springer, 2020.
105. J. Helsing. Integral equation methods for elliptic problems with boundary conditions of mixed type. *Journal of Computational Physics*, 228(23):8892–8907, 2009.
106. R. Hill. A self-consistent mechanics of composite materials. *Journal of the Mechanics and Physics of Solids*, 13(4):213–222, 1965.
107. E. Honein, T. Honein, and G. Herrmann. On two circular inclusions in harmonic problems. *Quarterly of applied mathematics*, pages 479–499, 1992.
108. J. Howse and J. Minichino. *Learning OpenCV 4 Computer Vision with Python 3: Get to grips with tools, techniques, and algorithms for computer vision and machine learning*. Packt Publishing Ltd, 2020.
109. Ming-Kuei Hu. Visual pattern recognition by moment invariants. *IRE transactions on information theory*, 8(2):179–187, 1962.
110. A. Hurwitz and R. Courant. *Vorlesungen über allgemeine funktionentheorie und elliptische funktionen: Geometrische funktionentheorie*. Springer, 1964.
111. Chyanbin Hwu. *Anisotropic elastic plates*. Springer Science & Business Media, 2010.
112. J.D. Jackson. *Classical electrodynamics*. John Wiley & Sons, 2021.
113. V.V. Jikov, S.M. Kozlov, and O.A. Oleinik. *Homogenization of differential operators and integral functionals*. Springer Science & Business Media, 2012.
114. A. Kaehler and G. Bradski. *Learning OpenCV 3: computer vision in C++ with the OpenCV library*. "O'Reilly Media, Inc.", 2016.
115. S.K. Kanaun and V. Levin. *Self-consistent methods for composites: Vol. 1-2*. Springer Science & Business Media, 2007.
116. J.B. Keller. A theorem on the conductivity of a composite medium. *Journal of Mathematical Physics*, 5(4):548–549, 1964.
117. R. Klette. *Concise computer vision*, volume 233. Springer, 2014.
118. W.T. Koiter. Some general theorems on doubly-periodic and quasi-periodic functions. In *Proc. Kon. Ned. Akad. Wet*, pages 120–128, 1959.
119. A.A. Kolpakov and A.G. Kolpakov. *Capacity and transport in contrast composite structures: asymptotic analysis and applications*. CRC Press, 2009.
120. A.S. Kosmodamianskii and A.P. Kravchenko. Distribution of electrostatic fields of direct piezoeffect in crystal half-plane with an elliptic hole. *Teoreticheskaya i Prikladnaya Mekhanika*, 16:66–70, 1985.

121. A.S. Kosmodamianskii, A.P. Kravchenko, and V.N. Lozhkin. The action of a point electric charge on the boundary of a piezoelectric half-plane weakened by an elliptic hole. *Izv. Akad. Nauk Arm. SSR. Mekh.*, 30(1):13–20, 1977.
122. W. Krauth. *Statistical mechanics: algorithms and computations*, volume 13. OUP Oxford, 2006.
123. E. Kröner. *Statistical continuum mechanics*, volume 92. Springer, 1972.
124. V.V. Kruglyak, R.J. Hicken, A.N. Kuchko, and V.Yu. Gorobets. Spin waves in a periodically layered magnetic nanowire. *Journal of applied physics*, 98(1), 2005.
125. B.A. Kudriavtsev, V.Z. Parton, and V.I. Rakitin. Fracture mechanics of piezoelectric materials. rectilinear tunnel crack on the boundary with a conductor. *Journal of Applied Mathematics and Mechanics*, 39(1):136–146, 1975.
126. J. Kurtz, D. Pardo, M. Paszynski, W. Rachowicz, and A. Zdunek. *Computing with Hp-Adaptive Finite Elements: Volume 2, Frontiers: Three Dimensional Elliptic and Maxwell Problems with Applications. Applied Mathematics and Nonlinear Science.* CRC Press, 2008.
127. L.D. Landau and E.M. Lifshits. *Electrodynamics of continuous media*. Pergamon Press Oxford, 1946.
128. R. Landauer. Electrical conductivity in inhomogeneous media. In *AIP conference proceedings*, volume 40, pages 2–45. American Institute of Physics, 1978.
129. M. Lanza de Cristoforis and P. Musolino. Analytic dependence of a periodic analog of a fundamental solution upon the periodicity parameters. *Annali di Matematica Pura ed Applicata (1923-)*, 197:1089–1116, 2018.
130. S.G. Lekhnitskii. *Theory of elasticity of an anisotropic elastic body*. Mir Publishers, Moscow, 1981.
131. T. Ligocki. Application of structural sums to study collective behavior of bacteria. *Communications of the ECMS. Proceedings, ECMS D. Grzonka, N. Rylko, G. Suchacka, V. Mityushev (Editors)*, 38(1):304–309, 2024.
132. Ch.-B. Ling. Extension of summation of series of hyperbolic functions. *Journal of mathematical analysis and applications*, 131(2):451–464, 1988.
133. R. Lipton. Optimal lower bounds on the electric-field concentration in composite media. *Journal of Applied Physics*, 96(5):2821–2827, 2004.
134. R. Lipton and M. Stuebner. Optimal design of composite structures for strength and stiffness: an inverse homogenization approach. *Structural and Multidisciplinary Optimization*, 33(4-5):351–362, 2007.
135. R. Luciano and E.J. Barbero. Formulas for the stiffness of composites with periodic microstructure. *International Journal of Solids and Structures*, 31(21):2933–2944, 1994.
136. P. Luzzini and P. Musolino. Perturbation analysis of the effective conductivity of a periodic composite. *Networks & Heterogeneous Media*, 15(4), 2020.
137. G. Matheron. Eléments de théorie sur les milieux poreux, 1967.
138. J.C. Maxwell. Viii. a dynamical theory of the electromagnetic field. *Philosophical transactions of the Royal Society of London*, (155):459–512, 1865.
139. J.C. Maxwell. *A treatise on electricity and magnetism*, volume 1. Clarendon press, 1873.
140. J.C. Maxwell-Garnett. Xii. colours in metal glasses and in metallic films. *Philosophical Transactions of the Royal Society of London. Series A, Containing Papers of a Mathematical or Physical Character*, 203(359-371):385–420, 1904.
141. M. Maździarz, A. Mrozek, W. Kuś, and T. Burczyński. Anisotropic-cyclicgraphene: A new two-dimensional semiconducting carbon allotrope. *Materials*, 11(3):432, 2018.

142. V. Maz'ya, A. Movchan, M. Nieves, et al. *Green's kernels and meso-scale approximations in perforated domains*, volume 2077. Springer, 2013.
143. L.N. McCartney. Applications of maxwell's methodology to the prediction of the effective properties of composite materials. In *Multi-Scale Continuum Mechanics Modelling of Fibre-Reinforced Polymer Composites*, pages 179–216. Elsevier, 2021.
144. D.R. McKenzie, R.C. McPhedran, and G.H. Derrick. The conductivity of lattices of spheres-ii. the body centred and face centred cubic lattices. *Proceedings of the Royal Society of London. A. Mathematical and Physical Sciences*, 362(1709):211–232, 1978.
145. R.C. McPhedran. Transport properties of cylinder pairs and of the square array of cylinders. *Proceedings of the Royal Society of London. A. Mathematical and Physical Sciences*, 408(1834):31–43, 1986.
146. R.C. McPhedran and D.R. McKenzie. The conductivity of lattices of spheres i. the simple cubic lattice. *Proceedings of the Royal Society of London. A. Mathematical and Physical Sciences*, 359(1696):45–63, 1978.
147. R.C. McPhedran and G.W. Milton. Bounds and exact theories for the transport properties of inhomogeneous media. *Applied Physics A*, 26(4):207–220, 1981.
148. R.C. McPhedran and G.W. Milton. Transport properties of touching cylinder pairs and of the square array of touching cylinders. *Proceedings of the Royal Society of London. A. Mathematical and Physical Sciences*, 411(1841):313–326, 1987.
149. Ch. Menzel, Th. Paul, C. Rockstuhl, Th. Pertsch, S. Tretyakov, and F. Lederer. Validity of effective material parameters for optical fishnet metamaterials. *Physical Review B—Condensed Matter and Materials Physics*, 81(3):035320, 2010.
150. S.G. Mikhlin. *Integral equations: and their applications to certain problems in mechanics, mathematical physics and technology*. Elsevier, 2014.
151. M. Milgrom and Sh. Shtrikman. Linear response of two-phase composites with cross moduli: Exact universal relations. *Physical Review A*, 40(3):1568–1575, 1989.
152. G.W. Milton. Bounds on the complex dielectric constant of a composite material. *Applied Physics Letters*, 37(3):300–302, 1980.
153. G.W. Milton. Bounds on the complex permittivity of a two-component composite material. *Journal of Applied Physics*, 52(8):5286–5293, 1981.
154. G.W. Milton. Bounds on the transport and optical properties of a two-component composite material. *Journal of Applied Physics*, 52(8):5294–5304, 1981.
155. G.W. Milton. The theory of composites (cambridge monographs on applied and computational mathematics) cambridge university press. *Cambridge, UK*, 2002.
156. V. Mityushev. Functional equations and its applications in mechanics of composites. *Demonstratio Math*, 30:64–70, 1997.
157. V. Mityushev. Transport properties of finite and infinite composite materials and rayleigh's sum. *Archives of Mechanics*, 49(2):345–358, 1997.
158. V. Mityushev. Transport properties of regular arrays of cylinders. *ZAMM*, 77:115–120, 1997.
159. V. Mityushev. Convergence of the poincaré series for some classical schottky groups. *Proceedings of the American Mathematical Society*, 126(8):2399–2406, 1998.
160. V. Mityushev. Steady heat conduction of a material with an array of cylindrical holes in the nonlinear case. *IMA journal of applied mathematics*, 61(1):91–102, 1998.
161. V. Mityushev. Transport properties of two–dimensional composite materials with circular inclusions. *Proceedings of the Royal Society of London. Series A: Mathematical, Physical and Engineering Sciences*, 455(1987):2513–2528, 1999.
162. V. Mityushev. Transport properties of doubly periodic arrays of circular cylinders and optimal design problems. *Appl. Math. Optimization*, 44:17–31, 2001.

163. V. Mityushev. R-linear problem on torus and its applications to composites. *Complex Variables, Theory and Application: An International Journal*, 50(7-11):621–630, 2005.
164. V. Mityushev. Representative cell in mechanics of composites and generalized eisenstein–rayleigh sums. *Complex Variables and Elliptic Equations*, 51(8-11):1033–1045, 2006.
165. V. Mityushev. Exact solution of the R-linear problem for a disk in a class of doubly periodic functions. *Journal of Applied Functional Analysis*, 2(2), 2007.
166. V. Mityushev. Conductivity of a two-dimensional composite containing elliptical inclusions. *Proceedings of the Royal Society A: Mathematical, Physical and Engineering Sciences*, 465(2110):2991–3010, 2009.
167. V. Mityushev. Riemann-hilbert problems for multiply connected domains and circular slit maps. *Computational Methods and Function Theory*, 11:575–590, 2012.
168. V. Mityushev. Cluster method in composites and its convergence. *Applied Mathematics Letters*, 77:44–48, 2018.
169. V. Mityushev. Effective properties of two-dimensional dispersed composites. part ii. revision of self-consistent methods. *Computers & Mathematics with Applications*, 121:74–84, 2022.
170. V. Mityushev. High-order contrast bounds for piezoelectric constants of two-phase fibrous composites. *Multiscale Modeling & Simulation*, 21(4):1644–1666, 2023.
171. V. Mityushev and P.M. Adler. Longitudinal permeability of spatially periodic rectangular arrays of circular cylinders ii. an arbitrary distribution of cylinders inside the unit cell. *Zeitschrift für angewandte Mathematik und Physik ZAMP*, 53:486–514, 2002.
172. V. Mityushev, I. Andrianov, and S. Gluzman. L.A. Filshtinsky's contribution to applied mathematics and mechanics of solids. In *Mechanics and Physics of Structured Media*, pages 1–40. Elsevier, 2022.
173. V. Mityushev and P. Drygas. Effective properties of fibrous composites and cluster convergence. *Multiscale Modeling & Simulation*, 17(2):696–715, 2019.
174. V. Mityushev and W. Nawalaniec. Basic sums and their random dynamic changes in description of microstructure of 2d composites. *Computational Materials Science*, 97:64–74, 2015.
175. V. Mityushev, D. Nosov, and R. Wojnar. Two-dimensional equations of magneto-electro-elasticity. In *Mechanics and Physics of Structured Media*, pages 63–98. Elsevier, 2022.
176. V. Mityushev, Kycia R., W. Nawalaniec, and N. Rylko. *Introduction to Mathematical Modeling and Computer Simulations*. Francis & Taylor, 2nd edition, 2025.
177. V. Mityushev and N. Rylko. Optimal distribution of the nonoverlapping conducting disks. *Multiscale Modeling & Simulation*, 10(1):180–190, 2012.
178. V. Mityushev and N. Rylko. A fast algorithm for computing the flux around non-overlapping disks on the plane. *Mathematical and Computer Modelling*, 57(5-6):1350–1359, 2013.
179. V. Mityushev and N. Rylko. Maxwell's approach to effective conductivity and its limitations. *Quarterly Journal of Mechanics and Applied Mathematics*, 66(2):241–251, 2013.
180. V. Mityushev and N. Rylko. Effective properties of two-dimensional dispersed composites. part i. schwarz's alternating method. *Computers & Mathematics with Applications*, 111:50–60, 2022.
181. V. Mityushev, N. Rylko, and M. Bryla. Conductivity of two-dimensional composites with randomly distributed elliptical inclusions. *ZAMM-Journal of Applied Mathematics and Mechanics/Zeitschrift für Angewandte Mathematik und Mechanik*, 98(4):554–568, 2018.

182. V. Mityushev and Zh. Zhunussova. Optimal random packing of spheres and extremal effective conductivity. *Symmetry*, 13(6):1063, 2021.
183. V.V. Mityushev. Plane problem for the steady heat conduction of material with circular inclusions. *Archives of Mechanics*, 45(2):211–215, 1993.
184. V.V. Mityushev. Generalized method of schwarz and addition theorems in mechanics of materials containing cavities. *Archives of Mechanics*, 47:1169–1182, 1995.
185. V.V. Mityushev and S.V. Rogosin. *Constructive Methods for Linear and Nonlinear Boundary Value Problems for Analytic Functions*, volume 108. CRC Press, 1999.
186. O.F. Mosotti. Discussione analitica sull'influenza che l'azione di un mezzo dielettrico ha sulla distribuzione dell'elettricità alla superficie di più corpi elettrici disseminato in esso. *Mem. Soc. Ital. Sc.(Modena)*, 14:49, 1850.
187. H. Moulinec, P. Suquet, and G.W. Milton. Convergence of iterative methods based on neumann series for composite materials: Theory and practice. *International Journal for Numerical Methods in Engineering*, 114(10):1103–1130, 2018.
188. A.B. Movchan, N.V. Movchan, and Ch.G. Poulton. *Asymptotic models of fields in dilute and densely packed composites*. World Scientific, 2002.
189. M. Mrevlishvili and D. Natroshvili. Investigation of nonclassical transmission problems of the thermo-electro-magneto elasticity theory for composed bodies by the integral equation method. *Transactions of A. Razmadze Mathematical Institute*, 173:103–130, 2019.
190. D. Mumford and C. Musili. *Tata lectures on theta. I (Modern Birkhäuser classics)*. Birkhäuser Boston Incorporated, 2007.
191. D.C. Munson. A note on lena. *IEEE Transactions on Image Processing*, 5(1):3–3, 1996.
192. N.I. Muskhelishvili. *Some basic problems of the mathematical theory of elasticity*, volume 15. Noordhoff Groningen, 1953.
193. N.I. Muskhelishvili. *Singular integral equations: boundary problems of function theory and their application to mathematical physics*. Courier Corporation, 2008.
194. V.Ya. Natanzon. On stresses in a tensioned plate with holes located in the chess order. *Matematicheskii sbornik*, 42:617–636, 1935.
195. W. Nawalaniec. Classifying and analysis of random composites using structural sums feature vector. *Proceedings of the Royal Society A*, 475(2225):20180698, 2019.
196. S. Nemat-Nasser and M. Hori. *Micromechanics: overall properties of heterogeneous materials*. Elsevier, 2013.
197. S. Nemat-Nasser, T. Iwakuma, and M. Hejazi. On composites with periodic structure. *Mechanics of materials*, 1(3):239–267, 1982.
198. A.N. Norris. A differential scheme for the effective moduli of composites. *Mechanics of materials*, 4(1):1–16, 1985.
199. A.N. Norris, P. Sheng, and A.J. Callegari. Effective-medium theories for two-phase dielectric media. *Journal of Applied Physics*, 57(6):1990–1996, 1985.
200. D. Nosov, L. Filshtinsky, and V. Mityushev. Cracks in two-dimensional magneto-electro-elastic medium. In *Mechanics and Physics of Structured Media*, pages 41–61. Elsevier, 2022.
201. N. Otsu et al. A threshold selection method from gray-level histograms. *Automatica*, 11(285-296):23–27, 1975.
202. V.Z. Parton and B.A. Kudryavtsev. *Electro-Magneto-Elasticity of Piezoelectric and Conducting Materials*. Nauka, Moscow, 1988.

203. M. Paszkuta, A. Krezel, and N. Rylko. Application of shape moments for cloudiness assessment in marine environmental research. *Remote Sensing*, 14(4):883, 2022.
204. W.T. Perrins, D.R. McKenzie, and R.C. McPhedran. Transport properties of regular arrays of cylinders. *Proceedings of the Royal Society of London. A. Mathematical and Physical Sciences*, 369(1737):207–225, 1979.
205. S. Raschka, Yuxi Hayden Liu, and V. Mirjalili. *Machine Learning with PyTorch and Scikit-Learn: Develop machine learning and deep learning models with Python.* Packt Publishing Ltd, 2022.
206. Lord Rayleigh. Lvi. on the influence of obstacles arranged in rectangular order upon the properties of a medium. *The London, Edinburgh, and Dublin Philosophical Magazine and Journal of Science*, 34(211):481–502, 1892.
207. R. Rodriguez-Ramos, F.J. Sabina, R. Guinovart-Diaz, and J. Bravo-Castillero. Closed-form expressions for the effective coefficients of a fiber-reinforced composite with transversely isotropic constituents–i. elastic and square symmetry. *Mechanics of Materials*, 33(4):223–235, 2001.
208. N. Rylko. Transport properties of a rectangular array of highly conducting cylinders. *Journal of Engineering Mathematics*, 38(1):1–12, 2000.
209. N. Rylko. Dipole matrix for the 2d inclusions close to circular. *ZAMM-Journal of Applied Mathematics and Mechanics/Zeitschrift für Angewandte Mathematik und Mechanik: Applied Mathematics and Mechanics*, 88(12):993–999, 2008.
210. N. Rylko. Structure of the scalar field around unidirectional circular cylinders. *Proceedings of the Royal Society A: Mathematical, Physical and Engineering Sciences*, 464(2090):391–407, 2008.
211. N. Rylko. Effective anti-plane properties of piezoelectric fibrous composites. *Acta Mechanica*, 224(11):2719–2734, 2013.
212. N. Rylko. Representative volume element in 2d for disks and in 3d for balls. *Journal of Mechanics of Materials and Structures*, 9(4):427–439, 2014.
213. N. Rylko. A pair of perfectly conducting disks in an external field. *Mathematical Modelling and Analysis*, 20(2):273–288, 2015.
214. N. Rylko, P. Kurtyka, O. Afanasieva, S. Gluzman, E. Olejnik, A. Wojcik, and W. Maziarz. Windows washing method of multiscale analysis of the in-situ nano-composites. *International Journal of Engineering Science*, 176:103699, 2022.
215. N. Rylko, M. Stawiarz, P. Kurtyka, and V. Mityushev. Study of anisotropy in polydispersed 2d micro and nano-composites by elbow and k-means clustering methods. *Acta Materialia*, page 120116, 2024.
216. K. Schulgasser. Relationships between the effective properties of transversely isotropic piezoelectric composites. *Journal of the Mechanics and Physics of Solids*, 40(2):473–479, 1992.
217. A.H. Sihvola and Jin Au Kong. Effective permittivity of dielectric mixtures. *IEEE Transactions on Geoscience and Remote Sensing*, 26(4):420–429, 1988.
218. B. Smith, P.E. Björstad, and W. Gropp. Domain decomposition: parallel multilevel methods for elliptic partial differential equations. *(No Title)*, 1998.
219. J.P. Straley. Thermoelectric properties of inhomogeneous materials. *Journal of Physics D: Applied Physics*, 14(11):2101, 1981.
220. S.P. Suetin. Padé approximants and efficient analytic continuation of a power series. *Russian Mathematical Surveys*, 57(1):43, 2002.
221. K. Szymanska and N. Rylko. Structural investigation of clouds. *Communications of the ECMS. Proceedings, ECMS D. Grzonka, N. Rylko, G. Suchacka, V. Mityushev (Editors)*, 38(1):316–320, 2024.

222. C.-H. Teh and R.T. Chin. On the detection of dominant points on digital curves. *IEEE Transactions on pattern analysis and machine intelligence*, 11(8):859–872, 1989.
223. J.J. Telega. Stochastic homogenization: convexity and nonconvexity. In *Nonlinear Homogenization and Its Applications to Composites, Polycrystals and Smart Materials*, pages 305–347. Springer, 2004.
224. S. Tokarzewski and J.J. Telega. S-continued fraction method for the investigation of a complex dielectric constant of two-components composite. *Acta Applicandae Mathematica*, 49(1):55–83, 1997.
225. V.Yu. Topolov, Ch.R. Bowen, and P. Bisegna. *Piezo-Active Composites: Microgeometry-Sensitivity Relations*, volume 271. Springer, 2018.
226. S. Torquato. *Random Heterogeneous Materials: Microstructure and Macroscopic Properties*, volume New York. Springer-Verlag, 2002.
227. M. Trenti and P. Hut. N-body simulations (gravitational). *Scholarpedia*, 3(5):3930, 2008.
228. B.S. Veeling, J. Linmans, J. Winkens, T. Cohen, and M. Welling. Rotation equivariant cnns for digital pathology. In *Medical Image Computing and Computer Assisted Intervention–MICCAI 2018: 21st International Conference, Granada, Spain, September 16-20, 2018, Proceedings, Part II 11*, pages 210–218. Springer, 2018.
229. I.N. Vekua. *Generalized analytic functions*. Elsevier, 2014.
230. A.F. Villán. *Mastering OpenCV 4 with Python: a practical guide covering topics from image processing, augmented reality to deep learning with OpenCV 4 and Python 3.7*. Packt Publishing Ltd, 2019.
231. A. Weil. *Elliptic functions according to Eisenstein and Kronecker*, volume 88. Springer Science & Business Media, 1999.
232. G.M. Whitesides and B. Grzybowski. Self-assembly at all scales. *Science*, 295(5564):2418–2421, 2002.
233. O. Wiener. Die theorie des mischkörpers für das feld der stationären strömung. *Abhandlungen der Sachsischen Gesellschaft der Akademischen Wissenschaften in Mathematik und Physik*, 32:507–604, 1912.
234. R. Wojnar. Homogenization of piezoelectric solid and thermodynamics. *Reports on Mathematical Physics*, 40(3):585–598, 1997.
235. R. Wojnar, S. Bytner, and A. Galka. Effective properties of elastic composites subject to thermal fields. *Thermal Stresses V*, pages 257–265, 1999.
236. S. Yakubovich, P. Drygas, and V. Mityushev. Closed-form evaluation of two-dimensional static lattice sums. *Proceedings of the Royal Society A: Mathematical, Physical and Engineering Sciences*, 472(2195):20160510, 2016.
237. Jiashi Yang. *Special topics in the theory of piezoelectricity*. Springer Science & Business Media, 2010.
238. R.W. Zimmerman. Effective conductivity of a two-dimensional medium containing elliptical inhomogeneities. *Proceedings of the Royal Society of London. Series A: Mathematical, Physical and Engineering Sciences*, 452(1950):1713–1727, 1996.

Index